BIOLOGICAL CONTROL

BIOLOGICAL CONTROL

Proceedings of an AAAS Symposium
on Biological Control, held at Boston, Massachusetts
December 30-31, 1969

Edited by
C. B. HUFFAKER

Professor of Entomology and
Entomologist in the Experiment Station
Division of Biological Control
Department of Entomology and Parasitology
The University of California, Berkeley

and

Director, International Center for Biological Control
The University of California
Berkeley and Riverside

ℚℙ PLENUM PRESS • NEW YORK–LONDON • 1971

First Printing—July 1971
Second Printing—August 1972
Third Printing—January 1973
Fourth Printing—March 1974
Fifth Printing—September 1976

Library of Congress Catalog Card Number 79-157149
SBN 306-30532-1

© 1971 Plenum Press, New York
A Division of Plenum Publishing Corporation
227 West 17th Street, New York, N. Y. 10011

United Kingdom edition published by Plenum Press, London
A Division of Plenum Publishing Company, Ltd.
Davis House (4th Floor), 8 Scrubs Lane, Harlesden, NW 10 6 SE, England

Printed in the United States of America

PREFACE

The explosive increase in the world's human population, with consequent need to feed an ever-increasing number of hungry mouths, and the largely resultant disturbances and pollution of the environment in which man must live and produce the things he needs, are forcing him to search for means of solving the first problem without intensifying the latter. Food production requires adequate assurance against the ravages of insects. In the last three decades short-sighted, unilateral and almost exclusive employment of synthesized chemicals for insect pest control has posed an enormous and as yet unfathomed contribution to the degradation of our environment, while our insect pest problems seem greater than ever. Properly viewed, pest control is basically a question of applied ecology, yet its practice has long been conducted with little regard to real necessity for control, and in some cases, with little regard to various detrimental side-effects or long-term advantage with respect, even, to the specific crop itself. This book deals fundamentally with these questions.

The development of pesticide resistance in many of the target species, against which the pesticides are directed, has occasioned an ever-increasing load of applications and complexes of different kinds of highly toxic materials. This has been made even more "necessary" as the destruction of natural enemies has resulted, as a side effect, in the rise to pest status of many species that were formerly innocuous. The application of broad-spectrum pesticides thus has many serious and self-defeating features. Yet the greatest fault is the environmental pollution by chemical residues or the immediate harmful action on non-target species (including the natural enemies serving to control the insect pests), fish, birds, and other wildlife, and even man himself.

The need for greater use of non-chemical methods of pest control has been a recurrent theme from many quarters. The President's Science Advisory Committee report "Restoring the Quality of Our Environment," The National Academy of Sciences' volume, "Scientific Aspects of Pest Control," the formal worldwide program of the International Biological Program (IBP), and many other sources emphasize this need.

Some novel and highly successful results such as the sterile male technique developed by Dr. E. F. Knipling have recently been obtained, and promising avenues of pest control include the use of biochemical controls (e.g., hormones and pheromones) highly specific to the insects. However, a feasible non-chemical method of control is already available—i.e., biological control (the action of parasites, predators and pathogens). These agents, together with intrinsic resistance factors, constitute Nature's own method. This method is often grossly underrated, under-researched, and only minimally applied. This statement applies significantly to much of the United States and Western Europe where pesticides have been most widely used. In these regions, many entomologists highly competent in their own disciplines commonly dismiss biological control as a significant option even though little or no real effort has been made and little insight offered upon which such a negative attitude could be based. Such men appear to assume, erroneously, that the method has been adequately tried but has failed.

If we are to reverse the trend toward an ever-intensified overloading of the environment with polluting and highly toxic pesticides, we must show that biological control, combined with restricted usage of selective chemicals, use of resistant varieties and other integrative measures can, in fact, solve many of our pest problems without resort to such disturbing and polluting chemicals. Biological control, where effective, is cheap, usually persistent, without need for recurrent expense, entails no significant genetic counter-attack in the pests in nature (with reference to insect pests and insectan parasites or predators), does not occasion the rise to pest status of forms normally innocuous, does not add to the ever-growing problem of man's pollution of the environment, and is not attendant with the serious toxic hazards to the workers using the methods, to consumers of the products, or to our cherished and declining wildlife. Moreover, because of the expense in the use of other methods, it is often the only method available in underdeveloped countries—a contributing reason why the IBP is sponsoring this method. Biological control, moreover, is compatible with enlightened integrated control programs wherein restricted use of chemicals combined with cultural and other ecological methods are employed. In fact, biological control is usually a key aspect of integrated control programs, for this technique is manipulatable and augmentable, whereas other major aspects of natural control, e.g., the weather, are not.

The objective of this book is to present this approach as a significant and realistic option in our pest control programs.

The science and theoretical basis of biological control has made rapid strides in recent years, and there have been many recent successful applications of the method. At the same time, concepts have been advanced that challenge the premises and indict the practices in this field. This book examines these concepts, premises and practices and lays before the scientific world and the lay public an expose of the potentialities for a much wider employment of biological control.

The book is essentially a *Proceedings,* being a collection of papers presented as a symposium of the Ecological Society of America and the American Association for the Advancement of Science held at Boston, Massachusetts, December 30 and 31, 1969.

Each paper is presented as a chapter, for convenient cross referencing. The papers have, however, been arranged, coordinated and edited for cohesiveness and unity of philosophy, for the most part. A unified document is presented, representing recent outstanding developments in both theory and application. The unity centers around the concept that biological control could be far more successful and find far greater use in pest control if adequate support were available and intensified effort could be made along sound ecological lines.

The emphasis is on documented cases of biological control and the use of biological control in developing integrated control programs around the world. The scope of the examples, geographically, systematically, and ecologically is sufficient to suggest that with persistence and imagination, biological control can be utilized anywhere. General procedures and the facilities used in this work, and even the important and fundamental areas embracing the systematics and biologies of the major groups of natural enemies available, are necessarily omitted. For an authoritative account of these areas, the reader is referred to the books "Entomophagous Insects" by C. P. Clausen, published in 1940 by McGraw Hill Book Company and "Biological Control of Insect Pests and Weeds," edited by Paul DeBach and published in 1964 by Reinhold Publishing Corporation, N.Y. and Chapman and Hall Ltd., London.

This book is divided into four logical sections. In Section I, the background to the problems posed by past practices in the use of pesticides (and a preview of the possibilities to be elaborated in later sections) are treated in the opening paper. Other papers deal with the theory, ecological basis, and technical means of assessing the action of biological control agents.

Section II presents outstanding recent examples wherein *classical* biological control has borne successful results, some of them involving a unique or novel approach. They deal with biological control of weeds in both terrestrial and aquatic environments, with scale insects, pests in glasshouse crops, a major threatening forest and orchard pest in eastern Canada, and a unique means of mass production and distribution of a parasite lacking good powers of dispersion in the control of a most severe pest of range grasses in hot, dry land regions of the U.S.A., Brazil, and other countries.

Section III is included because so many ecologists and economic entomologists around the world have long looked at biological control as something that just pertains to instances where *exotics* are involved. This section emphasizes that biological control exists all around us—that indigenous species, no less than exotics, may be very capable natural control agents. The cases considered in this section are restricted to the U.S.A. and Canada.

Section IV represents the culmination of all that is presented in Sections I, II, and III. It highlights the sorts of informational in-put required in a systems approach to ascertaining the strategies and tactics of an ecological pest control and points to the "pitfalls" along the way. Integrated control workers have been working at this for many years and in some outstanding cases, much headway has been made. These are reported as exemplary of what might be accomplished on a grander scale. The examples include control of cotton pests in California, of apples and peaches in Washington and California, use of a pathogen in the control of that real headache, the codling moth, and striking results obtained in recent years in Malaysia and Israel.

Lastly, I will make no effort to acknowledge individually the genuine and enthusiastic assistance and response from a great many people to whom I am grateful and who have contributed in many ways not obvious here to give the book its value.

I do wish, however, to express my deep appreciation to all the authors, to Dr. William S. Osburn, Jr., whose invitation led to development of the symposium, to Drs. F. R. Lawson, Philip S. Corbet and Maurice Tauber for their excellent chairing of sessions of the symposium, to Drs. P. S. Messenger and Robert van den Bosch who greatly assisted me in the planning, to Dr. J. E. Laing for much of the indexing, and to Mrs. Nettie Mackey and Barton Matsumoto for much technical assistance, and finally to Miss Shirley Tiangsing for excellent typing and computerized composition.

C. B. Huffaker, Editor
Division of Biological Control
Department of Entomology and
    Parasitology
University of California
Berkeley, California

THE AUTHORS

L. A. Andres: *Leader, Biological Control of Weeds Investigations, Agricultural Research Service, U.S. Department of Agriculture, Albany, California.*

J. C. Boling: *Currently, Department of Entomology, Kansas State University, Manhattan.*

L. Bravenboer: *Entomologist, Proefstation voor de Groenteen Fruitteelt onder Glas, Naaldwijk, The Netherlands.*

L. E. Caltagirone: *Associate Entomologist, Division of Biological Control, Department of Entomology and Parasitology, University of California, Berkeley.*

D. L. Dahlsten: *Associate Professor of Entomology, Division of Biological Control, Department of Entomology and Parasitology, University of California, Berkeley.*

Paul DeBach: *Professor of Biological Control, Division of Biological Control, University of California, Riverside.*

R. L. Doutt: *Professor of Biological Control, and Acting Dean, College of Agriculture, University of California, Berkeley.*

D. G. Embree: *Research Officer, Canadian Forestry Service, Fredericton, New Brunswick.*

L. A. Falcon: *Assistant Insect Pathologist, Division of Entomology, Department of Entomology and Parasitology, University of California, Berkeley.*

R. D. Goeden: *Assistant Professor of Biological Control, Division of Biological Control, University of California, Riverside.*

D. Gonzalez: *Assistant Entomologist, Department of Entomology, University of California, Riverside.*

G. R. Gradwell: *Lecturer in Entomology, Hope Department of Entomology, Oxford University, Oxford, England.*

K. S. Hagen: *Professor of Entomology, Division of Biological Control, Department of Entomology and Parasitology, University of California, Berkeley.*

Isaac Harpaz: *Head, Department of Entomology, Hebrew University of Jerusalem, Rehovot, Israel.*

ix

S. C. Hoyt: *Entomologist, Tree Fruit Research Center, Washington State University, Wenatchee.*

C. B. Huffaker: *Professor of Entomology, Division of Biological Control, Department of Entomology and Parasitology, University of California, Berkeley.*

N. W. Hussey: *Head, Department of Entomology, Glasshouse Crops Research Institute, Littlehampton, England.*

C. E. Kennett: *Associate Specialist, Division of Biological Control, University of California, Berkeley.*

T. F. Leigh: *Entomologist, Department of Entomology, University of California, Davis.*

C. R. MacLellan: *Research Officer, Department of Entomology, Canada Department of Agriculture Research Station, Kentville, Nova Scotia.*

A. W. MacPhee: *Head, Department of Entomology, Canada Department of Agriculture Research Station, Kentville, Nova Scotia.*

J. J. Marony, Jr.: *Currently, Museum of Zoology, Louisiana State University, Baton Rouge.*

P. S. Messenger: *Professor of Entomology, Division of Biological Control, Department of Entomology and Parasitology, University of California, Berkeley.*

F. D. Parker: *Research Entomologist, Entomology Research Division, Agricultural Research Service, U.S. Department of Agriculture, Columbia, Missouri.*

R. L. Rabb: *Professor of Entomology, Department of Entomology, North Carolina State University, Raleigh.*

David Rosen: *Lecturer, Hebrew University of Jerusalem, Rehovot, Israel.*

M. F. Schuster: *Texas A&M University Agricultural Research and Extension Center at Weslaco.*

R. F. Smith: *Chairman, Department of Entomology and Parasitology, University of California, Berkeley.*

R. W. Stark: *Professor of Entomology, Division of Entomology, Department of Entomology and Parasitology, University of California, Berkeley.*

V. M. Stern: *Professor of Entomology, Department of Entomology, University of California, Riverside.*

R. van den Bosch: *Professor of Entomology and Chairman, Division of Biological Control, Department of Entomology and Parasitology, University of California, Berkeley.*

G. C. Varley: *Professor of Entomology and Head, Hope Department of Entomology, Oxford University, Oxford, England.*

Brian Wood: *Entomologist, Chemara Research Station, Oil Palm Division, Layang Layang, Johore, Malaysia.*

CONTENTS

SECTION II: OUTSTANDING RECENT EXAMPLES

OF CLASSICAL BIOLOGICAL CONTROL

Chapter 9. THE BIOLOGICAL CONTROL OF THE WINTER
           MOTH IN EASTERN CANADA BY INTRODUCED
           PARASITES
           D. G. Embree

Chapter 10. BIOLOGICAL CONTROL OF RHODESGRASS
            SCALE BY AIRPLANE RELEASES OF AN
            INTRODUCED PARASITE OF LIMITED
            DISPERSING ABILITY
            Michael F. Schuster, J. C. Boling, and J. J. Marony, Jr.

SECTION III: THE UNHERALDED NATURALLY-OCCURRING BIOLOGICAL CONTROL

SECTION IV: BIOLOGICAL CONTROL AS A KEY ELEMENT

IN THE SYSTEMS APPROACH TO PEST CONTROL

Chapter 19. DEVELOPMENT OF INTEGRATED CONTROL
                PROGRAMS FOR PESTS OF TROPICAL
                PERENNIAL CROPS IN MALAYSIA
                Brian J. Wood

SECTION I

THE THEORY, ECOLOGICAL BASIS AND ASSESSMENT OF BIOLOGICAL CONTROL

Chapter 1

THE PESTICIDE SYNDROME–DIAGNOSIS AND SUGGESTED PROPHYLAXIS

R. L. Doutt and Ray F. Smith

Department of Entomology and Parasitology
University of California, Berkeley

THE PESTICIDE SYNDROME

The word "pesticide" is a fabricated term of recent invention. Like the vast array of chemical poisons which it appropriately designates, it is both synthetic and modern. Pesticides emerged rather slowly and in a crude form out of the nineteenth century. Then with surprising suddenness they evolved into compounds having spectacular toxicity. They proliferated enormously. They spread over the entire globe. They are characteristic of our time and have confronted us with another environmental crisis of this decade, outstanding in terms of massiveness, extensiveness and rate of change.

An engineer (Rosenstein, 1968) has written that "Today we live in constant threat of man-created, irreversible phenomena. Man can literally change the face of the earth and the composition of his environment before the public and its protective agencies are aware of pending danger." This statement is particularly applicable to pesticides, for in the case of DDT the lag time between its first use and the public's general awareness of the danger it presents to the environment has been in the order of a quarter of a century. People are now greatly alarmed, and this is over a pesticide which a few years ago won a Nobel Prize for its developer! It is particularly ironic that Sweden, the country that presented the Nobel Prize for DDT has now severely restricted its use!

Today there is probably not a single square centimeter of the earth's surface that has not felt the impact of man. We have recently become

3

accustomed to thinking that even though this miniscule patch of earth may be remote it could have easily been caressed by our polluted air or could have been the landing pad for radioactive fallout from our diverse follies. We are not, however, quite so accustomed to realizing that this same square centimeter has very likely also received a molecular quantum of pesticide. A decade ago we would have considered this entirely improbable, but now our ignorance is being rapidly dispelled for we are told that a substantial amount of pesticides enter the atmosphere either vaporized or absorbed on dust which may fall out in rain at a great distance from the area of application (Frost, 1969).

The development of these synthetic pesticides is truly a success story. The astounding advances in modern science and technology have put these powerful chemical weapons into the bristling arsenal of all pest control practitioners. Their global use reflects the marketing abilities of corporate entities on an international scale. This is a remarkable achievement and a tribute to industrial and commercial enterprise. In the United States it is a billion dollar industry.

It is not surprising that the new pesticides had widespread and immediate acceptance. The pesticide salesman has no natural enemy in the agribusiness jungle. Uncontrolled pests can be an economic catastrophe, so the grower is constantly uneasy. His fear of loss from pests is a very real and almost a tangible thing. He is not likely to risk, through inaction, his source of income for long if he believes that his cash crop is exposed and vulnerable to attack. With very little persuasion he runs scared. In seeking protection from real or imagined dangers the grower finds that commercial pesticides are made immediately and easily available to him. An efficient sales organization is at hand to assist his purchase, even to extending credit, and in all ways encouraging the use of pesticides and furthering the myth that the only good bug is a dead bug.

Our society finds it proper that there should be reasonable profit in manufacturing, distributing, selling and applying pesticides. Nevertheless, it is unsettling in professional meetings to hear that "profit is the name of the game." It sounds vaguely unethical to an audience of research scientists, particularly those with any ecological orientation. On the other hand, the customer for many years seemed satisfied. On the occasions when a compound proved disappointing the farmer was conditioned to look for a newer, advertised as better, chemical. There was no encouragement to try an entirely different control strategy.

This led to a closed, circular, self-perpetuating system with a completely unilateral method of pest control. Alternative measures were not explored or, if known, they were ignored. The chemical cart ran away with the biological horse. Even worse was the fact that no one bothered to ask the sensible question of whether a treatment was really necessary or could be justified economically. *This we term the pesticide syndrome.* Spray schedules were based on the calendar and not on the pest populations. This incredible nonsense was masked by the prudent sounding expression "preventive treatments," and the grower was encouraged to think of this large annual expenditure as "insurance." The sales of pesticides were predictable by the calendar. This greatly simplified the logistics of supplying an agricultural district with chemical poisons and keeping these inventories in balance with the demand.

This procedure has been a very splendid, smoothly functioning commercial structure in the agribusiness community. It is unfortunate that it has suddenly become unglued. At first there were only minor hints of trouble, but these rapidly grew more alarming. Commercial chemical products, once touted as panaceas, became ineffectual on the newly resistant pest populations. This was followed by an increasingly accelerated effort by the chemical industry to develop substitutes. Cross-resistance appeared. Chemically-induced pests arose as another monstrous new phenomenon. Non-target organisms reacted in many spectacularly disturbing ways. Pest control expenses increased. This pesticide structure has continued to crumble in spite of the frantic efforts to keep it patched.

Under the pesticide syndrome none of this was at first openly admitted. Instead, it was business as usual. The gentle, though insistent, voice of Rachel Carson was met with arrogance. The stupid and simplistic question "Bugs or people?" was symptomatic of the limited thinking of those afflicted by the pesticide syndrome. Alternative methods and biological control remained as a tiny enclave in the hostile chemical world of pest control.

A change of attitude is often the hallmark of progress, and slowly attitudes began to change. The dissident murmurs from the group disdainfully scorned by industry spokesmen as "bee, bird and bunny lovers" grew into a public clamor. During the past year it has reverberated in legislative halls throughout the nation. As more and more startling facts about persistent pesticides were discovered, the formerly sacred chemical cow began to transform into an odorous old fish. This is a sad spectacle to behold, even by professionals in biological control who were so long ridiculed by the dominating chemical control proponents as a lunatic fringe of economic entomologists. It is therefore tempting to react by joining the highly vocal anti-chemical band, but biological control, even with its newly acquired aura of shabby respectability, has never lost respect for its powerful old adversary. Neither have we in biological control ever changed our professional opinion that pesticides when properly used are a great boon to humanity. Our philosophy, which the pesticide people never felt any necessity to adopt, has always urged entomologists to take a basic ecological approach to solving pest problems and to evaluate *all* control measures that could possibly be applied in a particular situation.

It is important to emphasize that professional practitioners of biological control today are concerned about the pendulum of public opinion swinging too far in the current anti-pesticide climate. There are indications of disturbingly negative, restrictive and extremist positions being taken against pesticides. There are anti-pesticide crusaders and hordes of "instant ecologists" who have appeared to bury the chemical Caesar, not to honor him. There are some strange political undercurrents and an uncommon amount of emotion involved. All of this simply increases the difficulty of finding sensible solutions to the pesticide dilemma.

THE SUGGESTED PROPHYLAXIS

Disregarding the sound and fury that has been generated over the pesticide issue, the fact remains that problems with their use in agriculture, public health, and forestry are of such a magnitude that almost everywhere a reappraisal is necessary. Actually this has quietly been under way for a number of years in many parts of the world, so there is now a substantial amount of experience with alternative paths and from this a very clearly defined concept for future research programs has arisen.

This change in concept and the outlines of a sensible alternative system are clearly seen in analyses of crop protection associated with cotton production throughout the world and with the program being developed for grape production in California (*vide infra*).

The strategy of pest population management, as opposed to that of preventive treatments, offers the most promising solution to many of the present difficulties. This strategy stems from the notion that pest control is adequate if populations are held at tolerable levels. The unrealistic idea that pest control must be equated with total elimination is rejected. It is recognized that the most powerful control forces are natural ones, some of which can be very effectively manipulated to increase the mortality of a pest species. The pesticide is considered as the ultimate weapon to be held in reserve until absolute necessity dictates its use. This is a sophisticated use of ecological principles, and is both intellectually satisfying on theoretical grounds and extremely effective in application. This in general terms is the concept known and practiced as integrated control.[1]

[1] A Food and Agricultural Organization of the United Nations (FAO) panel of experts

There is still considerable misunderstanding of what is meant by integrated control. It is not merely a workable marriage of chemical and biological controls. Its use goes far beyond this. The best features of all control measures are utilized and applied within the framework of practical farm operations. There are several unique aspects to the research upon which an integrated control program is built. One is philosophical since the researchers must believe that chemical treatments ought not to be applied until they are clearly needed. Another aspect is organizational in that the program is best developed by a team of specialists representing diverse disciplines but all of whom are channeling their research toward a common goal. The third and fundamental aspect is that the program is based on a sophisticated understanding of the ecology of the ecosystem involved.

Most pesticides are applied to the control of arthropod pests of agricultural crops so the following analyses are restricted to farm situations. However, the basic concepts easily extrapolate to pest problems of forests and range lands. The agricultural entomologist who becomes involved in the development of an integrated control program must first work from the fundamental concept of the ecosystem. However, it is important to recognize that an apple orchard, potato patch, wheat field or vineyard is a very special type of ecosystem which we may call an agro-ecosystem. It is different from a naturally occurring, undisturbed situation in that its sustained existence depends upon manipulation of energy flows directed by man. The biotic environment is drastically reduced in complexity, and through the conventional farm operations a few selected energy paths are maximized to produce the greatest output from a single species of cultigen.

All ecosystems are extremely complex and even the simplified agro-ecosystem is composed of many sub-systems. Examples of these have been well analyzed by Smith and Reynolds (1971). It is usually extremely difficult to separate one component and evaluate its role and its influence in the system but this can be done, and now is being done in integrated control programs. Experimental methods for evaluating the role of natural enemies are detailed by DeBach and Huffaker (Chapter 5).

Examples of the development of integrated control programs and the

---

defined integrated control as a pest management system that, in the context of the associated environment and the population dynamics of the pest species, utilizes all suitable techniques and methods in as compatible a manner as possible and maintains the pest populations at levels below those causing economic injury (Anon., 1968). It emphasizes the fullest practical utilization of the existing mortality and suppressive factors in the environment; it is not dependent upon any specific control factor but rather coordinates the several applied techniques appropriate for the particular situation, with the natural regulating and limiting elements of the environment.

problems involved are found in the integrated control of cotton pests in Peru (Smith, 1969) and the research effort on integrated control of grape pests in California (e.g., Doutt and Nakata, 1965; Flaherty and Huffaker, 1970).

The Cotton Picture

For cotton, the pattern of crop protection may go through a series of five phases, or it may stop at a particular phase, or cotton production may cease before all phases are established. The sequence can be described as follows:

1) *Subsistence Phase.* The cotton, usually grown under non-irrigated conditions, is part of a subsistence agriculture. Normally the cotton does not enter the world market and is usually processed by native hand weavers. The yields are below 200 pounds of lint per acre, often much lower (e.g., Afghanistan, Chad, Mozambique, Uganda and parts of Ecuador). There is no organized program of crop protection. Whatever crop protection is available results from natural control, inherent plant resistance in the cotton, hand picking, cultural practices, rare pest treatments, and luck. In some cases, the subsistence phase may exist side by side with later phases (e.g., in Ecuador, where Indians grow Criollo cotton in the subsistence phase on hills adjacent to valleys containing later development phases).

2) *Exploitation Phase.* In many newly irrigated areas in developing countries, cotton is often one of the crops first planted to exploit the new land resource. Crop protection measures are introduced to protect these large and more valuable crops. These measures are justified to take full advantage of the irrigation scheme and to maximize yields. Crop protection methods are sometimes needed to protect a new variety having qualities of increased yield and better fiber but poorer insect resistance. Unfortunately, in most cases, in the exploitation phase the crop protection measures are dependent solely on chemical pesticides. They are used intensively and often on fixed schedules. At first these schemes are successful and the desired high yields of seed and lint are obtained.

3) *Crisis Phase.* After a variable number of years in the exploitation phase and heavy use of pesticides, a series of events occurs. More frequent applications of pesticides are needed to get effective control. The treatments are started earlier in the cotton growing season and extended later into the harvest period. It is notable that the pest populations now resurge rapidly after treatment to new higher levels. The pest populations gradually become so tolerant of a particular pesticide that it becomes useless. Another pesticide is substituted and the pest populations become tolerant to it too, but this

happens more rapidly than with the first chemical. At the same time, pests that never caused damage in the area previously, or only occasionally, become serious and regular ravagers of the cotton fields. This combination of pesticide resistance, pest resurgence, and the unleashing of secondary pests, or the induced rise to pest status of species previously innocuous, occasions a greatly increased application of pesticides and causes greatly increased production costs.

4) *Disaster Phase.* The heavy pesticide usage increases production costs to the point where cotton can no longer be grown profitably. At first only marginal land and marginal farmers are removed from production. Eventually, cotton is no longer profitable to produce in any parts of the area. A number of Central American countries are now in this disaster phase. In the United States, this disaster phase has been postponed a bit by cotton subsidies.

5) *Integrated Control or Recovery Phase.* In only a few places has the integrated control phase been developed. The most noteworthy examples are in several valleys in Peru. In the integrated control phase, a crop protection system is employed comprising a variety of control procedures rather than pesticides alone. Attempts are made to modify the environmental factors that permit the insects to achieve pest status, and the fullest use is made of biological control and other natural mortality.

The pattern outlined above can be seen in the case history of cotton pests in the Canete Valley of Peru (Smith, 1969). The Canete Valley is one of some forty coastal valleys in Peru. These valleys are self-contained micro-agro-ecosystems isolated from each other by severe stretches of desert. There is almost no rainfall in coastal Peru and the extent and timing of agricultural crops is determined largely by available water in the intermittent rivers. The valleys differ in size, climate, soils, water supply and quality, crops, and pests. In the 1920's the Canete Valley shifted from an emphasis on cultivation of sugar cane to cotton. Approximately two-thirds of the valley is devoted to cotton. Maize, potatoes, beans and other crops are also important. The Canete story is a classic example of the pest control problems that can beset the grower if he ignores ecology and relies unilaterally on chemical pesticides.

During the period from 1943 to 1948, the chemical control of cotton pests was based mainly on arsenicals and nicotine sulphate, although some organic chlorides were used toward the end of the period. The main insect pests, a "gusano de la hoja" (*Anomis texana* Riley), the "picudo peruano" (*Anthonomus vestitus* Bohm), and the "perforador pequeno de la bellota" (*Heliothis virescens* F.), gradually increased until there was a severe outbreak in 1949. It was presumed that this outbreak, and the associated heavy aphid infestations, were brought on by the widespread use of ratoon cotton (second and third year cotton) and the introduction of organic insecticides. During the years 1943 to 1948, the average annual yield in the Canete Valley ranged from 466 to 591 kilograms per hectare (415 to 526 lbs/acre). In 1949, with the heavy *Heliothis* and aphid outbreak, the average yield dropped to 366 kilograms per hectare (326 lbs/acre).

The next period lasted roughly from 1949 to 1956. Ecologically oriented entomologists such as Juan E. Wille recommended a variety of cultural controls and a return to inorganic and botanical insecticides. This advice was not followed in full. One-year ratoon cotton was permitted and a heavy reliance was placed on the new exciting organic insecticides (mainly DDT, BHC and Toxaphene). Some cultural practices were modified to increase yields. New strains of *Tanguis* cotton were introduced and more efficient irrigation practices initiated.

These procedures were very successful initially. Yields nearly doubled. The average yield in the Valley went from 494 kilograms per hectare (440 lbs/acre) in 1950 to 728 kilograms per hectare (648 lbs/acre) in 1954. Farmers were enthusiastic and developed the idea that there was a direct relationship between the amount of pesticides used and the yield, i.e., the more pesticides the better. The insecticides were applied like a blanket over the entire Valley. Trees were cut down so the airplanes could easily treat the fields. Birds that nested in these trees disappeared. Beneficial insect parasites and predators disappeared. As the years went by, the number of treatments was increased; also, each year the treatments were started earlier because of the earlier attacks of the pests. In late 1952, BHC was no longer effective against aphids. In the summer of 1954, toxaphene failed to control the leafworm, *Anomis*. In the 1955-1956 season, *Anthonomus* reached high levels early in the growing season; then *Argyrotaenia sphaleropa* Meyrich appeared as a new pest. *Heliothis virescens* then developed a very heavy infestation and showed a high degree of resistance to DDT. Organophosphorous compounds were substituted for the chlorinated hydrocarbons. The interval between treatments was progressively shortened from a range of 8-15 days down to 3 days. Meanwhile a whole complex of previously innocuous insects rose to serious pest status. Included were *Argyrotaenia sphaleropa* Meyrich, *Platynota* sp., *Pseudoplusia rogationis* Guen, *Pococera atramenalis* Led., *Planococcus citri* (Risso) and *Bucculatrix thurberiella* Busck. In nearby similar valleys, where the new organic insecticides were not used, or only in small amounts, these insects did not become pests. Finally, with most of the pests resistant to the available pesticides and the insects rampant in the cotton fields, the 1955-1956 season was an economic disaster for the growers of the Canete Valley. Millions of bales were lost. In spite of the large amounts of insecticides used in attempts to control the pests, yields dropped to 332 kilograms per hectare (296 lbs/acre).

The development of integrated control followed this severe crisis. When the Canete growers realized that they could not solve their pest problems with insecticides alone, but had, instead, produced a catastrophe, they appealed to their experiment station for help. A number of changes in pest control practices were made, including certain cultural practices. Cotton production was forbidden on marginal land. Ratoon cotton was prohibited. The Canete Valley was repopulated with natural enemies from neighboring valleys and beneficial forms were fostered in other ways. Uniform planting dates, timing and methods of plowing, and a cotton-free fallow period were established. Use of synthetic organic insecticides in specific fields was prohibited except by approval of a special commission. There was a return to arsenicals and nicotine sulphate. These changes were, with the approval of the growers, codified as regulations and enforced by the Ministry of Agriculture. As a result of this new integrated control program, there was a rapid and striking reduction in the severity of the cotton pest problem. The whole complex of formerly innocuous species, which cropped up in the organic insecticide period, reverted to their innocuous status. Furthermore, the intensity of the key pest problems also diminished and so there was an over-all reduction in direct pest control costs. By the next year, the yield was back to 526 kilograms per hectare (468 lb/acre) and since has varied from 724 to 1,036 kilograms per hectare (644 to 922 lb/acre). These are the highest yields in the history of the Valley.

In the ecologically similar valleys of Chincha and Pisco to the south of the Canete Valley, the insect situation was very similar except that the drop in yield occurred one year later. The economic crash in Chincha Valley was even more dramatic than in the Canete Valley. The warmer, drier Ica Valley did not experience the crash of these foggy coastal valleys. The Ica growers used fewer organic insecticides in the early fifties, and in 1957-1958, they started following regulations based on the integrated control program of the Canete Valley. Other valleys, especially those away from the Central Coast, have still different agro-ecosystems with different pest complexes and histories.

The Grape Picture in California

The grape industry in California is based on selected varieties of the Old World species, *Vitis vinifera*. Because of the topography of the State and the sharp gradients in climate from the temperate coast into the more extreme conditions of the interior valleys, these commercial grapes grow in remarkably diverse environments and this is reflected in the different patterns of pest problems. In the more arid parts of the State, such as the southern San Joaquin Valley, the grapes survive solely because of irrigation and the continual care by man. There are two species of native wild grapes in California, both of which are found occurring naturally along streams and riparian areas where there is available water and generally an overstory of trees. One species, *Vitis californica,* ranges through central and northern California and the study of its associated fauna has been an important adjunct to the research program on grape pests in central California and demonstrates the need to look at the pest's ecology in developing a balanced pest control program.

In 1961 the grape growers considered the grape leafhopper, *Erythroneura elegantula* Osborn, to be the key pest in the San Joaquin Valley where the bulk of the California grapes are grown. At first DDT had been a very successful pesticide against this species but the leafhoppers then developed resistance. The growers then switched to other newer pesticides, and the leafhoppers in turn began again to exhibit resistance. At the same time, it was suspected that the leafhopper treatments were causing increases in tetranychid mites (a suspicion that was supported by later studies, e.g., Flaherty and Huffaker, in press), and it was abundantly clear that the leafhopper control in California was an expensive operation for the total cost in 1961 attributed to the grape leafhopper was over eight million dollars. This was the existing situation when the research was initiated on integrated control. From the outset, this research has been a team effort between pesticide entomologists and biological control specialists, with an active supporting group including farmers, professional viticulturists and a plant pathologist.

The initial research included an effort to determine the population level at which leafhoppers require treatment. This is complicated by the fact that some grape varieties are very versatile and may be used for wine, raisins, or table grapes as the individual grower may decide. Theoretically, the growers could tolerate a higher level of leafhoppers on grapes destined for wine or raisins than on a crop grown for the fresh fruit market where appearance is an important factor.

The actual level at which a treatment is required to prevent loss has not yet been accurately ascertained, but the grapes can tolerate far higher populations of leafhoppers than anyone previously thought possible. To communicate this to the persons who make field decisions on the treatment of grape pests we have simply said that low populations do not require treatment and then we define that as an average of 10 nymphs/leaf for the first brood and 5 nymphs/leaf for the second brood. This is a very conservative figure because populations twice this size have not resulted in any measurable reduction in either quality or yield of the grapes. Its use has enabled growers to avoid insecticide applications for an entire growing season in many vineyards, and this has permitted the maximum effectiveness of a parasitic wasp which attacks the eggs of the grape leafhoppers.

The story of this wasp is ecologically fascinating and illustrates the use of existing mortality factors in the environment in the development of an integrated control program. This parasite is a tiny mymarid, *Anagrus epos* Girault, which is scarcely 0.3 mm in length. The female wasp searches the grape leaves for an egg of the host leafhopper in which she deposits an egg from which the young parasite develops. During the growing season the parasite will produce nine or ten generations to three of the host leafhopper. This rapid development, coupled with the ability to locate and to attack most host eggs, make this parasite very effective in suppressing its host.

At the beginning of this grape research program, it was learned that the parasite occurred early in the season in some vineyards but not in others. Also, it soon became evident that the parasite must have an alternative host during the winter period. Eventually this was found to be a non-economic leafhopper, *Dikrella cruentata* Gillette, that occurs on native and cultivated species of blackberry, *Rubus* spp. When this was discovered, the riddle of the early season occurrence in certain vineyards and not in others was solved, because these vineyards were found invariably to be associated with some nearby *Rubus*.

It became equally clear that when the commercial vineyards were established in California many were planted miles away from any blackberry patch. In this situation the leafhopper could spend the entire year in the vineyards, but the egg parasite could only exist through the winter in a blackberry refuge so that the two species were effectively separated by a barrier of time and space. This permitted the leafhopper to increase without its natural control agent. To overcome this, the practice of planting a blackberry refuge adjacent to a vineyard has been started, and these overwintering refuges have a measurable effect over a two-mile radius.

An additional favorable aspect is that the parasite seems to be unaffected by the sulfur treatments which must be applied in practically all vineyards for mildew control. It is suspected that this is a case of a parasite having developed resistance to a pesticide. The basis for this thinking is that while the parasite is undoubtedly a native species which evolved with its hosts on the wild blackberries and grapes that grow together along the California rivers, it was never seen in any commercial vineyards until 1931. The only plausible barrier would have been the sulfur treatments which were everywhere applied. Each season the parasite would have had the opportunity to move into a vineyard and breed in enormous numbers as it does today, except that the sulfur treatments possibly were very disturbing. In this situation, repeated every year, it is possible that finally through natural selection an *Anagrus* strain tolerant to sulfur was able to break through the barrier. Once this was achieved, the opportunities for producing enormous populations of the grape leafhoppers in vineyards were immediate and extensive. Theoretically, this would in a very short time result in a population consisting entirely of sulfur tolerant wasps. We think this probably occurred but we have no proof. At any rate, this is only of academic interest for the important fact is that the parasite is present and effective and can be maintained throughout the year by providing nearby an overwintering refuge of *Rubus*.

In summary, the integrated control program being developed for grape pests in California has already eliminated much of the insecticide applications through the simple but sensible procedure of basing decisions for treatment on existing pest population levels instead of a calendar date. This first step has substantially reduced costs and has improved the quality of the vineyard environments. The natural suppressive factors which cause mortality in the grape pests are utilized to their fullest extent, and through the cultural practices the vineyard environment is manipulated to maintain natural enemies throughout the year. Emerging from this pest management effort is an alleviation of serious spider mite problems which were mostly chemically induced.

CONCLUSION

These experiences in Peru and California strikingly illustrate that pest control cannot be analyzed or developed as an exercise in control of a given pest alone; rather, it must be considered and applied in the context of the ecosystem in which a number of pest, or potential pest, populations exist and in which control actions are taken. Whatever the new technologies which may be developed for control or management of pest insects in the future, they must be applied with a knowledge of the operation of the pertinent ecosystem. This means that in addition to the fundamental studies on the new methodologies, a series of parallel investigations should be carried out to provide a solid base of information on the significant factors in the ecosystem influencing the pest populations and possible potential pests. Both of these case histories show how quickly an agricultural industry can recover from the pesticide syndrome, and they offer hope to others afflicted by this malady so peculiar to the 20th century.

LITERATURE CITED

Anonymous. 1968. FAO report of the second session of the FAO panel of experts on integrated pest control. Rome, Italy, 19th-24th September, 1968, 48 pp.

Doutt, R. L., and J. Nakata. 1965. Parasites for control of grape leafhopper. Calif. Agr. 19: 3.

Flaherty, D. L. and C. B. Huffaker. 1970. Biological control of Pacific mites and Willamette mites in San Joaquin Valley vineyards. Hilgardia 40 (10): 267-330.

Frost, J. 1969. Earth, air, water. Environment 11 (6): 14-33.

Rosenstein, A. B. 1968. A study of a profession and professional education. Reports group. School of Engineering and Applied Science. Univ. Calif., Los Angeles. 169 pp.

Smith, R. F. 1969. The new and the old in pest control. Proc. Accad. Nazion. dei Lincei, Rome (1968), 366(128): 21-30.

Smith, R. F., and H. T. Reynolds. 1971. Effects of manipulation of cotton agro-ecosystems on insect populations. *In* The Careless Technology—Ecology and International Development, M. T. Farvar and J. P. Milton (eds.). Natural History Press, New York.

Chapter 2

THE NATURAL ENEMY COMPONENT IN NATURAL CONTROL AND THE THEORY OF BIOLOGICAL CONTROL

C. B. Huffaker, P. S. Messenger, and Paul DeBach

Divisions of Biological Control
University of California, Berkeley and Riverside

INTRODUCTION

The theory of natural control of animal abundance has had an extended and sometimes heated history. The role played by natural enemies has likewise been subject to debate. We hope here to cast light on both questions, and to place the action of enemies, both endemic and introduced ones, as discussed at length in other chapters of this book, in perspective relative to natural control as a general phenomenon. We first deal with the concept of natural control itself, with the classification of the components of the process, with the two principal theories or viewpoints, with the various expressions of the density-related mechanism, and lastly, with the role of natural enemies, (a) in the process of regulation, (b) in control as distinct from regulation, and (c) with the damping processes either inherent or extrinsic to enemy action but required as supplemental feature(s) in maintaining stability. The role of natural enemies is also discussed relative to recent challenges and indictments of the supposed theory and practices of biological control practitioners.

NATURAL CONTROL—THE BALANCE OF NATURE

Doutt and Debach (1964) aptly describe what we mean by *natural control* or the *balance of nature:*
"Every student of biology is aware that in any given locality certain

16

species are more or less consistently abundant, others are less common, and finally, some species are so rarely encountered as to become collectors' items. This condition among the resident species tends to exist year after year, albeit relative and absolute numbers vary, but on the average really substantial changes rarely occur in the numerical relationship between the several species inhabiting a given, more or less stable, environment."

A similar description by MacFadyen (1957) is given subsequently.

This quotation from Doutt and DeBach describes a status of contained variation or *characteristic* abundance. The process by which it is accomplished, i.e., natural control, is defined as the maintenance of population numbers (or biomass) within certain upper and lower limits by the action of the *whole* environment, necessarily including an element that is density-induced, i.e., *regulating,* in relation to the *conditions* of the environment and the *properties* of the species.

The reference level of density about which such populations vary was referred to by Nicholson (1933, 1954) as the "steady state" and by Smith (1929, 1935) as the "equilibrium density" (see further Huffaker and Messenger, 1964.) While the mean densities of populations that persist at such characteristic levels are not technically equivalent to these terms of Nicholson and Smith, we can, however, utilize the concept of equilibrium density for the purpose of better understanding a population's dynamics and its natural control, through the use of models and experimentation.

Odum (1959) clearly illustrates what we mean by characteristic abundance in his statement that a forest area may have 10 birds per hectare or 2000 arthropods per sq. meter, but *never* the reverse!

Obviously, different populations exhibit very different degrees of contained variation in numbers over a given span of time. Some populations may vary only slightly from year to year, as with Carrick's (1963) magpies, *Gymnorhina tibicen* near Canberra, Australia. This bird's numbers are controlled in a density-dependent way related to their territorial behavior and complex social structure, by which surplus individuals are denied the privilege of breeding. Natality and mortality are thus neatly balanced at a density which imposes no noticeable shortage of food, and both predation and disease have little impact. If, however, the population were deliberately and greatly depleted artificially or by a natural catastrophe, we can assume that the density-imposed stresses accounting for the balance in births and deaths would be lessened and the population would then grow until it attained its characteristic abundance in relation to its behavioral and social characteristics and the availability and nature of breeding habitats.

Populations in widely varying environments are expected to vary widely unless they are under control by factors other than competition for food or suitable habitats or shelters—factors which vary greatly with the varying meteorological conditions. Thus, the beet leafhopper, *Circulifer tenellus* (Baker) in western United States and in arid areas of Eurasia and Africa, or the plague locust, *Schistocerca gregaria* (Forskål) in various regions of the world, may vary many thousand-fold between their lowest and greatest abundance. Far from refuting our concept of natural control, such examples simply illustrate the fact that regulation is *in relation to* the conditions of the environment (*vide infra*). Huffaker and Messenger (1964) discussed the interrelationships between the degree of stability in the physical conditions and the two quite different aspects of a species' population dynamics, i.e., *changes* in density and the *magnitude* of mean density.

THE PROCESS OF NATURAL CONTROL

There are two principal views of control of population size. One view may be summarized as control by the resistive forces of the environment such that, by chance, environmental favorability and unfavorability balance out, and any animal population existing there waxes and wanes accordingly. Such a state of affairs has been held to occur in some populations by Uvarov (1931), Thompson (1939), and Andrewartha and Birch (1954).

The second view concerns control by density-dependent processes. In this view, a population increases in density in a favorable environment, but as it does so it brings about or induces repressive forces, from the environment or from the population itself, which tend to prevent further increase, and commonly, in fact, eventually cause through excessive reaction a decline in numbers; as population density decreases, such repressive forces, external or internal to the population, relax in their actions and tend to allow the population to recover and to increase again.

Hence, relative to these two views, one sees a state of "balance" occurring in the environment, the other sees "balance" occurring in the population. One sees no necessary connection between mortality or natality and density, whereas the other holds this connection to be wholly necessary. Only in the second, or density-dependent view, does the concept of "regulation" apply.

The Concept of Control by Density-unrelated Factors

Control in the sense of temporary suppression or limitation can of course occur from any adverse action. The view of real long-term or natural control unrelated to density (Uvarov, 1931; Thompson, 1939, 1956; and Andrewartha and Birch, 1954) implies that populations may be held within the long-term ranges of densities observed by the ceaseless change in the favorableness or magnitude of conditions and events unrelated to the population's own density, the latter exerting no necessary or special influence. Density-independent resistance factors, the heterogeneity of the environment in time and place, and the play of chance in the action of such factors, is considered sufficient.

We reject this hypothesis because: (1) it concerns itself mainly with *changes* in density, largely leaving out the causes of the magnitude of mean density; (2) being concerned mainly with changes in density, the view ignores or denies the fact of characteristic abundance, and provides no logical explanation of why some species are always rare, others common and still others abundant, even though each may respond similarly to changes in the weather, for example; (3) to accomplish such long-term natural "control," it is presumed that the ceaseless change from favorability to unfavorability of the environment, in terms of a species' tolerances, can be so delicately balanced (a knife-edged balance) over long periods of time as to keep a population in being without density-related stresses coming into play to stop population increase at high densities, and without any lessening of these density-related stresses coming into play at low densities to reduce the likelihood of extinction; (4) such knife-edged balance relative to a species' adaptations, i.e., a balanced favorability and unfavorability, is essentially control by chance, and is incompatible with the view of adaptive improvement resulting through evolution. If, as Birch (1960), claims, there is a tendency for adaptive increase in "r" or any adaptive tendency to overcome environmental stresses (increased fecundity, better protective behavior, resistance to natural enemies, all assumed unrelated to density), this would immediately destroy this precise balance unless a corresponding repressive change in environment were at the same time automatically elicited. Otherwise, the improved population would increase without limit; (5) last, and most important, the special significance of density-dependent processes (which the view denies as essential) can be easily demonstrated except for the most unmanageable and violently fluctuating examples by applying a heavy mortality or adding enormous numbers to the populations, or by removing mortality factors, or adding additional resources and then following the population trends afterwards.

The Concept of Density-Dependent Regulation

The concept of density-dependent regulation owes its origin to people like Malthus (1803), and much later Woodworth (1908), Howard and Fiske (1911), Nicholson (1933), and Smith (1935).

Regarding the concept of a knife-edged balance hypothesis (*vide supra*), as Huffaker and Messenger (1964) stated, "Hence it appears to us mandatory that a suitable theory of natural control of animal abundance should include some provision whereby the tendency of the animals to adapt to, and hence in numbers to overcome, stresses from the environment, in particular the physical factors in the environment, be met by reaction of some sort or other. To us, again, the mechanism of reactive change in the intensity of action of a given controlling factor elicited by change in abundance, i.e., density-dependent control factors, provides the simplest and most elegant explanation of how such adaptational possibilities are counterbalanced." The objections to the view of environmental balance by chance (*vide supra*) do not apply to the concept of density-dependent regulation of numbers. This view offers an explanation of the fact of characteristic abundance and, as stated above, its reality can be demonstrated rather easily (*vide infra*). It is consistent with the observation that extreme densities are inevitably met by starvation or lack of space, if not by predation, disease or other catastrophe associated with such abundance, and with many studies showing that such extreme densities are often prevented by density-related processes operating at much lower densities.

In this view, natural control embraces all factors of the environment directly or indirectly influencing natality and mortality or movements into and out of the population. However, as stated previously, the collective natural control process must include at least one factor that is density-dependent. There are three broad general components in the relationship involving a population and its environment; they are *inseparable* parallel parameters: (1) The properties of the population (subject to some short-term change genetically, but basically set historically by evolutionary progress; (2) the conditioning framework concerning environmental capacity, and interim density-unrelated impact on population size; (3) the regulation of population size by density-induced mechanisms *in relation to* the conditions and, of course, the properties of the species.

Demonstrations of Characteristic Abundance and Density-Dependence. Four examples clearly explain this concept.

a) Quite comparable to Odum's example cited above, Varley and Gradwell (1958) cited the case of species of insects on European oaks that are always rare, common, or abundant, respectively. Each species has its own characteristic density even though two or more species of contrasting mean densities might follow one another very closely in their *changes* in density, in response perhaps to some meteorlogical factor.

b) The application of pesticides or other severe depressive action is nearly always followed by a resurgence of the population to its former density. If density-unrelated forces alone were involved in the natural "control" of such species, they would continue to vary as before but forever after at the new, greatly reduced mean density.

c) Conversely, a large addition of stock to a population that is at its natural characteristic density, results in rather rapid decline to the former level. And again, the same logic applies against the view of "control" by chance according to which view the population should never decline at all, but continue ever-after to vary as before but about the new, greatly expanded mean density!

d) A change in any significant environmental factor (as by artificial irrigation, use of a pesticide affecting an efficient enemy, the introduction of a new efficient enemy, or a change in the properties of the species) will occasion a change in the characteristic density. This has been demonstrated many times, and certain examples will be given in subsequent chapters.

These four examples prove the existence of characteristic abundance, and are inconsistent with the hypothesis of "control" by chance, for characteristic abundance itself is inconsistent with this hypothesis. In b) and c) it is shown that characteristic abundance is density-related for density-geared compensations are exhibited, while d) shows that the regulating process accounting for characteristic abundance *determines* population size only *in relation to* the conditions of the environment (and the properties of the species). Thus, regulation is control in the relative sense; control in the absolute sense embraces the whole process of natural control (*vide supra*). To make this point clear, we take a population that is regulated by competition for its food, and hypothesize that we can very gradually increase the required food supply over a period of many years (or the species' properties in efficiency of use of a given amount of food may be hypothesized as gradually and steadily increasing). If there were no limit to the extent to which we could increase the food supply, the population would have no limit. There would be no control in the absolute sense although the population is under *relative* control at all times, i.e., density-induced *regulation,* in terms of amounts of food present at any one time and the properties of the species for utilizing this food (Fig. 1).

The Regulatory Mechanism and the Instruments. Not only is the more logical concept of density-dependence easily demonstrated but it rests upon the axiomatic premise that as organisms increase in numbers they automatically use up or defile the things they need or bring about or favor reaction in the environment inimical to them (e.g., food and residences are depleted, parasites, predators or pathogens are generated or attracted, waste products or

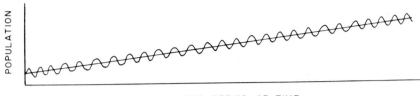

Fig. 1. The relationship of "control" in the relative sense, as through competition for a resource (see oscillations), and "control" in the absolute sense (here no control at all but, rather a steady increase in mean density), as the resource is here hypothesized as being indefinitely increased.

harmful heat accumulate.)

As a definition, density-dependent or regulating action includes the actions of repressive environmental factors, collectively or singly, which so intensify as the population increases beyond a characteristic high level and so relax as density falls that population increase beyond characteristic maxima is prevented and decrease to extinction is made less likely. The process could not *guarantee* against extinction.

Thus, a negative feed-back process between density and rate of increase is involved. The mechanism comprises a *collective* process. The emphasis is on the process, not on a particular instrument brought into play. Moreover, a regulating process is not necessarily to be identified for any short interval of time; when population cycling is involved, the process must be viewed as embracing a whole cycle at least, not intervals of time within a cycle. White and Huffaker (1969a,b) and Hassell and Huffaker (1969b) identified several regulatory elements, each of which formed parts of the process in the regulation and establishment of cyclic patterns in the dynamics of the Mediterranean flour moth, *Anagasta kuhniella* (Zeller), in laboratory ecosystems (*vide infra*).

Wynne-Edwards (1962) postulated that all populations possess intrinsic properties which regulate their numbers at levels so low that stress from shortages of food or action of predators or disease, for example, do not have significant roles in their natural control. Such examples fall naturally into our concept of density-dependent regulation. However, if he were right in this contention that *all* populations are controlled this way this book would never have been written. Certainly, those more highly evolved organisms that exhibit complex social structures and strong territoriality may, by virtue of these traits, have acquired considerable independence of these other forms of stress. But they are likely not always so immune as Wynne-Edward's presumes. Lower forms of animals like the insects for the most part appear to lack such highly refined mechanisms of density regulation. Moreover, they lack the ability of homoiotherms to regulate their own body temperatures and are thus more subject to stress from extreme cold, for example. (Other devices, e.g., diapause, are utilized). Their reproductive abilities are much more dependent upon prevailing conditions; hence, changes in their abundance exhibit more striking correlations with changes in meteorological conditions than is true for homoiotherms. It is to be expected, therefore, that the insects are greatly subject to the primitive stresses of starvation, disease and predation. But this by no means preempts the role of density-dependent regulation.

THE NATURE OF REGULATION BY NATURAL ENEMIES

The more reliable, efficient regulating predator or parasite is one that has a reciprocal density-dependent relation with its host or prey (Huffaker and Messenger, 1964). That is, the host is regulated by its enemy, and its enemy in turn is limited by the number of hosts. This means that such a phytophagous pest is predator- or parasite-limited, rather than food-limited, for the higher level reciprocal relation, when it exists, precludes the one below it in the food chain. If starvation or lack of hosts is perhaps an uncommon regulatory avenue in the phytophagous insects, certainly it is the rule in the predatory and parasitic forms, which perhaps largely accounts for the fact that the great majority of phytophagous insects are relatively innocuous (see later chapters). We cannot, however, assume that phytophagous insects are seldom limited by their food supply. Forms that serve as efficient regulators of their plant hosts are certainly so limited. There may be many more of these than commonly thought, for example, in complex highly diverse tropical rain forests where such biological control agents may even be responsible for the great diversity (Janzen, 1970). Examples of biological control of weeds illustrate this potentiality (Chapter 6).

Here we are more concerned with insect pests of crops. If they are to be *reliable* and *efficient* regulatory agents, the natural enemies of crop pests must necessarily be food-limited to a considerable degree.

Biological control rests on the premise that the densities of many noxious species of plants and animals are amenable to control, probably regulation, largely by their natural enemies (parasites, predators and pathogens), or may be so controlled or regulated by the enemies of their relatives in other lands if they are brought together. Thus, both naturally-occurring and manipulated biological control are embraced.

The Kinds of Natural Enemies

Both adults and young of many predatory species feed on the same prey species. In general, a number of prey individuals are required for a young predator to complete its development. However, the adults are not always carnivorous; some species require, instead, certain liquid nourishment high in protein (e.g., Hagen, 1950; Doutt, 1964). Both immature and adult predators must search for their prey and commonly there is not the close synchronization in physiological and life-history relationships found to be true relative to the parasites (parasitoids) and their hosts.

In general, entomophagous parasites search for hosts only as adults and the adult does not itself kill the victim (it may paralyze it), but instead deposits an egg or eggs in or on it. The developing larvae devour the host and a single individual suffices for complete development of one or more parasites.

These relationships and others pose interesting questions regarding the merits of predators versus parasites as biological control agents. Doutt and DeBach (1964) discussed this question and the related questions of degree of host specificity and powers of prey consumption. Huffaker *et al.* (1969) considered the latter points relative to control of tetranychid mites by predators. Empirical evidence and theoretical considerations seem to favor parasites over predators for a number of reasons (Doutt and DeBach, 1964), e.g., in general, parasites are more host-specific, more closely adapted and synchronized in interrelationships, their lower food requirements permit them to maintain a balance with their prey species at a lower prey density, and the young (the less active stages) do not have to search for their food. Yet, as those authors noted, there are many cases of biological control by predators and many more could perhaps be established by close study. It would be a great mistake to discount the value of predators. Each situation merits independent evaluation. The extensive naturally-occurring biological control of spider mites throughout the world, where nutritional factors are favorable, has been largely due to *predators;* no parasites are known and pathogens have a secondary role. Again, however, those predatory species of lower powers of consumption appear to be the best (Huffaker *et al.,* 1969).

Four main characteristics are pertinent to the efficiency of a parasite or predator, and excellent control implies superior properties relative to the situation. These are, (1) its adaptability to the varying physical conditions of the environment, (2) its searching capacity, including its general mobility, (3) its power of increase relative to that of its prey, and relative to its power of prey consumption, and (4) other intrinsic properties, such as synchronization with host, its host specificity, degree of discrimination, ability to survive host-free periods, and special behavioral traits that alter its performance as related to density or dispersion of its host and its own population.

Searching capacity is dominant, but it would not possess high searching capacity unless it also possessed certain of these other inherent properties. (See further "Models of Host-Parasite (Parasitoid) or Predator-Prey Interactions," this chapter; Chapters 3 and 4; and Nicholson, 1933, 1954; Hagen, 1964; Doutt, 1964; Huffaker and Kennett, 1969).

Lag effects in the effective numerical response of a parasite to change in host abundance are of great importance and are derivative of the above characteristics. Again, reliable control is itself evidence that the lag period involved is not too long or that it occurs at such low pest densities that increase to economic levels does not occur during the lag period.

Pathogens are often conspicuous as a force in the decimation of animal populations (e.g., Steinhaus, 1954; Tanada, 1964). We know a great deal about the kinds of pathogens, their virulence, their infectivity, host specificity, general pathology and ways of studying them in the laboratory, but their role in epizootics in nature is far from clear. Since, in general, epizootics appear to be associated with high concentrations of host populations, and consequent conditions of lowered resistance, they seem to belong in the category of density-dependent agencies. For human epidemics this is an inescapable conclusion. There are, however, so many circumstances other than high host density that must be met concomitantly for some epizootics to occur that we cannot view pathogens generally as a *very reliable* regulator of population size. Good correlations with host density in many cases would be elusive. This, however, does not refute an inherency of density-dependent action. The test is whether or not, given the same conditions in all other respects, mortality from disease would be greater in the average population at high density than in one at low density.

In discussing this question, Tanada (1964) referred to work of Ullyett and Schonkon (1940) on the diamond moth, *Plutella maculipennis* (Curtis), in South Africa. Ullyett and Schonkon concluded, "Although the *spread* of the fungus among the population of hosts is largely dependent upon the density of the latter the *appearance* of the disease is wholly dependent upon the extraneous factor of weather conditions." Tanada correctly noted that weather factors in such cases should not, however, be considered as causing pathogens to act as a density-independent factor but only in permitting or not permitting pathogens to exert an inherently density-dependent role. He cited cases suggesting that the lag effects of such density-dependent action carry over as nearly annihilative mortality on subsequent low density populations, producing a prolonged control.

Environmental factors may act not only on active infections but may also activate latent and chronic infections. There appears to be no relationship of weather factors to certain diseases (viruses) while humidity, free water and temperature are important in triggering epizootics in other diseases (especially fungi). However, Tanada noted that there is "very little quantitative data on the specific factors and mechanisms of action of environmental factors in epizootics of insect populations." He added that in many cases the occurrence of disease is the result of a combination of factors and ". . . it is difficult to separate the importance of the individual factors." These and other related questions are considered further in Chapter 15.

The virulence and capacity to spread of a pathogen may vary from time to time (mainly the bacteria and fungi). Moreover, the host population may vary in its susceptibility to disease and this may be of more fundamental importance than the changes in the virulence of the pathogen (Webster, 1946).

Most studies on survival capacities of pathogens relate to laboratory conditions; studies on survival in natural habitats have customarily been limited to short-term studies—not long enough to ascertain if the period of survival or persistence in the virtual absence of hosts is related to the periodism often observed in periodic outbreaks! Until we know a great deal more from field studies on the ecology of pathogens, we are not likely to soon learn how we may more reliably use pathogens in our pest control programs, one of the important goals of insect pathology (Tanada, 1964) (see further Chapter 15).

The remainder of this chapter deals largely with parasites and predators, as the corresponding relationships for pathogens have not received research attention.

Models of Host-Parasite (Parasitoid) or Predator-Prey Interactions

The basic host-parasite (or predator-prey) interaction is first treated in terms of the functional, numerical and over-all response, and then in relation to the damping processes that appear to account for the degree of population stability that is often observed in nature.

The Functional, Numerical and Over-all Response. A first step in assaying the performance of a predator or parasite is to learn how it performs as an individual, i.e., the way in which it searches for prey, perceives prey, and accepts or refuses given examples for attack. Here we consider the quantitative and theoretical approaches to the net result from the performance of such individuals. Such functional responses of individuals are essential for clear understanding and proper approach to modeling of host-parasite or predator-prey interactions. Modeling may in turn lead to greater insight and to development of more realistic strategies and tactics of pest control, as the guidelines to a systems analysis approach are developed. (See Section IV, especially Chapter 14.)

In Fig. 2a (from Huffaker *et al.*, 1968) we illustrate four hypotheses of behavioral or functional response, in terms of number of attacks by the individual as related to prey density. In Fig. 2b we plot the same data in terms of the more meaningful proportionate kill.

In Thompson's (1924, 1939) conception, predators (including parasitoids) have no difficulty in finding their prey; the number attacked is limited only by the consumption capacity (appetite) of the predator and the supply of eggs available in the parasite. The number attacked is therefore a constant unrelated to the density of prey, as shown in Fig. 2a.

In the Nicholson conception (Nicholson, 1933; Nicholson and Bailey, 1935), the number of attacks an individual can make for any given density of

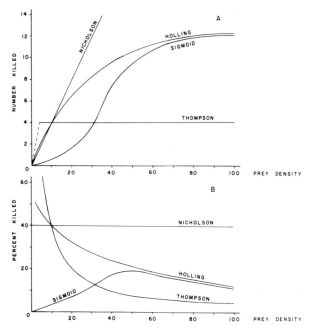

Fig. 2. (A)—The four postulated types of functional response of predators (including entomophagous parasites) to prey density. (B)—The same types of functional response expressed in terms of proportionate kill. (After Huffaker *et al.*, 1968.)

hosts is dependent only upon its searching capacity, which is considered a constant for the species; the host or prey population is considered as held at such low densities that satiation of appetite (predator) or exhaustion of the egg supply (parasite) would not be a factor. Thus, numbers killed are directly proportional to ease in finding hosts or prey—i.e., to their density (Fig. 2a). The percentage killed is then constant (Fig. 2b).

The empirically established performance of such parasites (e.g., Burnett, 1954, 1958; Messenger, 1969) or predators (e.g., Morris, 1963) is best represented by the disc equation of Holling (1959); numbers killed increase, but at a progressively reduced rate, as prey density increases, while the percentage killed declines, but less precipitously than under the Thompson conception.

A sigmoid response is exhibited by some vertebrate predators when offered opportunity of shifting from other prey, or as a result of improved skill, as a given prey increases in density. Such a sigmoid relation presents the only response implying regulating possibilities *as far as the functional response alone* is concerned (but see below, Haynes and Sisojevic). The increasing steepness of this curve in the low prey-density range of Fig. 2a indicates what Fig. 2b clearly shows, i.e., an increasing percentage kill over that range of prey density. It should be noted, however, that tests with invertebrate predators have not presented them with the opportunity of shifting from other prey as has been true relative to vertebrate performance. Recently, however, Haynes and Sisojevic (1966) demonstrated an intriguing attack curve for the spider, *Philodromus rufus* Walckenaer, preying on *Drosophila* adults. The curve increased in slope with prey density, within a given range (Fig. 3). Shifts from alternate prey were not involved. Different patterns of such increasing slope were obtained for hungry versus satiated spiders. For either state, the time required to capture diminished with density of prey, but so also did time for feeding, for uncaptured flies disturbed the feeding spider, reducing the time spent feeding (less than full utilization resulted) and allowing more time for capture. Gradual reduction in capture time, with prey density, dominated the over-all response up to a fly density approaching eight flies per gallon, but capture time then remained constant at 18 minutes at higher fly densities. The reduction in time used in feeding on one fly then became the dominant feature and the time released, allowing more captures, resulted in the second rise (increasing slope) to a new plateau. When the measure of density was "corrected" to one of "activity density," the attack response was the typical one for invertebrates for hungry spiders, but not for satiated ones. Thus, this example illustrates an inherently regulating type of functional response in an invertebrate predator, although of some subtlety and complexity. Sandness and McMurtry (1970) present similar results for two predatory phytoseiid

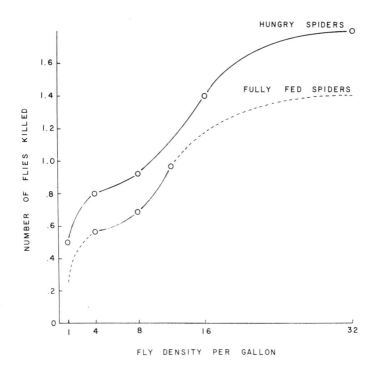

Fig. 3. Influence of fly density on flies killed in a three-hour period by male spiders. (After Haynes and Sisojevic, 1966)

responses. Moreover, Embree (1966) interpreted data on the performance of a tachinid parasite (*Cyzenis albicans* (Fall.)) of the winter moth in eastern Canada as representing a sigmoid performance.

Since entomophagous parasites and predators seldom if ever achieve their major control effect by virtue of the functional response alone, we must turn to some other parameter or derivative of the functional response for insight into the control potential. This is a basic reason, aside from statistical ones, why Morris' (1959, 1963) technique for appraising the action of an entomophagous parasite is faulty; the technique will not assess the results from the numerical response which is a generation to generation phenomenon, not a within-generation affair (see further, Southwood, 1967; Varley and

Gradwell, 1970; Hassell and Huffaker, 1969*a*; and Chapter 4).

The numerical response of entomophagous parasites and predators is the key aspect of their performance, though it is a derivative, in large part, of the functional response. Successful biological control following release of natural enemies customarily requires time for three or more generations of reproductive increase.

The numerical response may come about in three ways: (1) a rather immediate response by concentration which is not clearly distinct from the sigmoid type of functional response involving a shift from other prey and *positions* in the environment, (2) a rather immediate numerical response through improved survival, and (3) a delayed reproductive response.

Even in cases where proportionate kill, brought about by a fixed functional response, is inversely density-dependent, as with the Thompson model, annihilation may result through the numerical response. The Nicholson model, where the functional response is linear, generates both control and a *tendency* to regulation through the numerical response alone. It is arguable that since this model generates oscillations of increasing amplitude leading to annihilation, only a temporary tendency to regulation is inherent in the model itself.

The biological control caused by a parasite or predator may be so effective that *searching* remains an overwhelmingly dominant feature and this, plus the reproductive power of the host, may largely determine the equilibrium densities about which the two coupled populations oscillate. It is conceivable that within the low range of densities representative of excellent biological control, the responsible parasites or predators do not commonly approach the limits of their egg supply or appetites, nor spend more than an inconsequential amount of handling time. However, for parasites that require time to develop subsequent eggs following oviposition, the time involved is equivalent to an extension of handling time.

Huffaker and Kennett (1966) found that in the spring season, when *Aphytis maculicornis* (Masi) is mainly active against olive scale, *Parlatoria oleae* (Colvee), it is numerous enough, searches thoroughly enough, and possesses an excess of eggs, such that it could parasitize a great many more scales than it does, were it not for the fact that previous parasitism of hosts by *Coccophagoides utilis* Doutt has rendered them unacceptable. In fact, *A. maculicornis* are so numerous relative to the host population that less than one immature host is available per female parasite (only one egg is laid per host). The "handling time" (Holling, 1959) would be inconsequential and no interval comparable to it, such as time for development of a subsequent egg, seems of any importance in this case. *In stable situations, therefore, a parasite's searching capacity far outweighs its power of increase or its fecundity, for at the hypothetical steady state, only one female egg/female parent is required to maintain control.* However, limited egg supply may be a factor following summer decimation in *A. maculicornis* numbers and increase in host numbers.

On the other hand, in situations where good control is not maintained, and there is a substantial lag period, particularly if there is a host "escape" due to extraneous factors, and the host species builds up to relatively high densities, the question of a limitation in egg supply (or true handling time, or satiation in the predator, or time to develop subsequent eggs) could well be important. This introduces, again, some unreality in the Nicholson model. Furthermore, at times of escape, rapidity of increase in the enemy species can be very important in quickly regaining control of the pest species. If relatively stable conditions pertain, a Nicholson-type model may be quite useful as a means of assessing the nature of the interactions (e.g., Hassell and Varley, 1969; Hassell and Huffaker, 1969b; Chapter 4).

Damping Processes and Regulation by Natural Enemies. Based on the simple Nicholson model, the consequence that the host and parasite populations oscillate with increasing amplitude, with annihilation resulting eventually, means that the parasite is not alone a self-sufficient *regulating* factor. However, the delayed density-dependent action of the parasite is to us clearly *of the nature* of a regulating factor, even in this simple hypothesis, for successive densities oscillate about the steady state even as the amplitude goes ever more severe. If this change in amplitude were quite gradual, the parasite could be said to be regulating over a rather long period of time, i.e., before annihilation occurs. Moreover, density-induced parasite action is the regulating feature relative to the steady state situation. It is therefore *controlling,* causing *changes* in density and tending to *regulate* in terms of the conditions, so long as the system stays in being.

Furthermore, the importance of the parasite as a regulating instrument is seen by the fact that the addition of a relatively small amount of extrinsic, direct density-dependent (damping) action or of some intrinsic damping feature can be sufficient to stabilize the system, and the level at which this occurs primarily determined by the action of the enemy rather than the additional feature. For example, in Fig. 4a (unpublished data of C. B. Huffaker and R. Stinner), an equilibrium log density of a hypothetical Nicholsonian host-parasite system is shown. The arithmetic equilibrium host density is 48.4. We may now hypothesize a case wherein all characteristics are the same except that a low intensity of action of a direct density-dependent factor is added, acting after the parasite. This could represent competition for food of a given intensity that could come into play at a given threshold density. As seen in Fig. 4b, with appropriate values the system may even be stabilized at the *same* equilibrium densities for both host and parasite of Fig. 4a, where only the parasite is involved. In contrast, Fig. 4c presents the results when the same values for direct density-dependent action (competition) operate in the absence of parasitism. The equilibrum host density here is 276.8, vastly greater than that in the systems containing the parasite. It would appear then that such a parasite is a powerful control factor, and that its action is also regulative in nature, even if alone insufficient to stabilize.

In real situations it is obvious that densities would not increase in the manner described by the Nicholson model without various factors entering the picture. Such factors, of a density-dependent type, may include other natural enemies, intraspecific competition in the host, or certain performance of the given natural enemy. Thus, the density-dependent regulating process in this case, and in many cases, is a *collective* process and involves the collective action of all agencies that increase in intensity as the density of the population acted upon increases (Huffaker, 1958; Huffaker and Messenger, 1964; White and Huffaker, 1969a,b). However, one particular agency may have a quite dominant role (e.g., competition for food, or action of an efficient enemy).

In the case of entomophagous parasites and predators, we now consider the possibilities by which the theoretical, inherently annihilative, predominantly delayed density-dependent action may be damped. An explanation of these features seems essential in bringing theory in line with what appears overwhelmingly implied in the empirical results of biological control.

In the Nicholson-Bailey model, or some other model incorporating lag effects inherently associated with such numerical response phenomena, the increasing amplitude that is a direct consequence of these lag effects[1] must be damped if stability is to be achieved. Thus, if theory and the status of a rather persistent, stable control of many pest species by their natural enemies are to be brought into accord, damping processes tending to stabilize must be built into theory and used in modeling. Nicholson (1933, 1954) introduced the idea of population fragmentation, a commonly observed phenomenon in biological control, as an explanation of how predator/prey systems are saved from extinction by processes inherent to his model. Just how this may be accomplished, and how it can be modeled, was not made clear.

Recently, several studies have been made in which the over-all generation to generation interaction, including both the functional (behavioral) and numerical responses, have been considered. In these studies, both intrinsic and extrinsic damping features have been important components. (Tinbergen and

---

[1]Introducing a lag effect into the Lotka-Volterra model also produces this result (Wangersky and Cunningham, 1957).

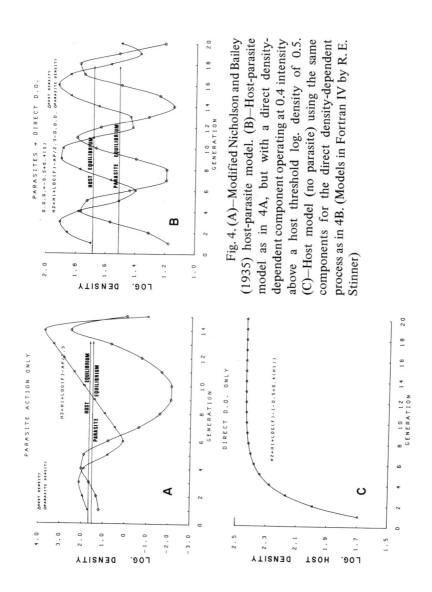

Fig. 4. (A)—Modified Nicholson and Bailey (1935) host-parasite model. (B)—Host-parasite model as in 4A, but with a direct density-dependent component operating at 0.4 intensity above a host threshold log. density of 0.5. (C)—Host model (no parasite) using the same components for the direct density-dependent process as in 4B. (Models in Fortran IV by R. E. Stinner)

Klomp, 1960; Huffaker *et al.,* 1963; Hassell and Huffaker, 1969b; Griffiths and Holling, 1969; Holling and Ewing, 1970).

Extrinsic damping features include such factors as (1) competition for food or shelter in any degree on the part of the prey (or host) species, and (2) the super-imposition upon the specific predator-prey system of significant action by a general predator whose actions are geared in a density-dependent way such as to destroy more of the primary predator than of the prey, relative to their respective equilibrium densities, during the crash phase of the interaction. Tinbergen and Klomp (1960) illustrated the latter and the stabilizing influence of competition for food was illustrated by Huffaker *et al.* (1963) and Hassell and Huffaker (1969b).

A damping feature inherent to the action of the natural enemy itself is not altogether necessary in order that the enemy be a good control or regulating agent in the field. As already discussed (*vide supra*), incorporation of a degree of direct density-dependent action can furnish the necessary damping action. Moreover, recent work alluded to above points to a number of features intrinsic to the interaction itself that serve as damping features.

In these studies two rather distinct approaches were used. In the approach of Hassell and Huffaker (1969b) and Hassell and Varley (1969), the components of the attack process (as used in a Holling approach) were not utilized; rather, the modeling was based on the Nicholsonian approach. In the former paper, it was found that direct and/or indirect interference between parasites (*Venturia canescens* (Gravenhorst)) served as a damping feature; this resulted in a variation in the "area of discovery" (assumed constant under the Nicholson conception) and the quantitative description of this variation furnished a significant feature of the predictive equation for both host and parasite populations. From this finding, Hassell and Varley (1969) developed a general model incorporating the idea of a varying "area of discovery," the upper limit of which is a constant (the quest constant), i.e., at a parasite density of 1 per unit area.

Until it is tested against a considerable number of real population interactions the utility of the Hassell-Varley method will not be known. It certainly offers an easier route for some purposes. If the few factors used do in fact dominate the situation, insight would be much enhanced. The method does not require the precise determination of the thirty or so parameters required by the model of Holling and Ewing (1970). This makes it easier to use and insight is more condensed. It may still lack reality and generality.

The model of Holling and associates is based on close, component by component, simulation modeling, the basic biological parameters having been ascertained by experimentation or from field studies. No attempt has yet been made to compare such model-generated populations with real interacting populations. The method, however, offers the potential for study in great depth. It can only be tested against real population interactions as the necessary input of information becomes available. It remains a question whether or not the attainment of the necessary data from field populations will be practicable for many situations, and if so, whether the dimensions will be so many as to becloud insight. Holling and Ewing (1970) further say of their most simplified version, "The model is appallingly inelegant because it is realistic." So long as it is not intractable, however, this should not deter its use.

The method of Holling has been used as a test of the role of a number of factors in producing stability in the predator/prey system, i.e., damping processes. Let us note first, however, that intrinsic behavior leading to a sigmoid functional response (Haynes and Sisojevic, 1966; Sandness and McMurtry, 1970), this being a direct density-dependent feature, could serve to damp the over-all interaction even if alone it were not sufficient to produce either control or regulation. Griffiths (1969) studied the indirect effects of interference (in the form of superparasitism) and of imperfect coincidence in a parasite of the European pine sawfly, *Neodiprion sertifer* (Geoffroy), which does not exhibit discrimination. Under the basic Holling model not incorporating the effects of superparasitism and with perfect temporal coincidence, the parasite reduced host numbers to zero in the 10th generation. When the superparasitism effect was added the, system was stabilized. However, when the temporal coincidence between host and parasite populations was less than 0.5, the parasite had little influence on host density. In another system, a marked stabilizing effect (at higher densities) resulted when perfect temporal coincidence was combined with 0.9 or 0.75 spatial coincidence. These latter features can be visualized as the refuge aspect of damping mechanisms. Griffiths thus illustrated through simulation modeling that superparasitism (indirect interference) and lack of perfect temporal or spatial coincidence (synchronization) *may* serve as intrinsic damping features.

Griffiths and Holling (1969) concluded from a consideration of 44 sets of data from the literature that the distribution of parasite attacks relative to the host population (as measured by the negative binomial function) is a constant for the species for a set of conditions—i.e., independent of host density or parasite density. However, when discrimination occurs they noted that the value of k, the negative binomial exponent, which describes the distribution of attacks among hosts, increases as more and more attacks are generated. They noted that discrimination is a form of interference. Perfect discrimination produces the same effect as plain predation (devouring of prey), for either parasitization by a perfectly discriminating parasite or predation would effectively "remove" hosts, reducing the number of attacks possible by a given individual. This operates through the exploitation component. Thus, the initial assumption that k is a constant can be relaxed.

Additional studies on the role of k in interaction with other factors are under study by Holling and associates. Holling and Ewing (1970) explored the potential damping power of different constant values for k. With all other factors held constant, values for k of 1.5 and above gave annihilative interactions. However, for lower values of k, representing increasing contagion of attacks, the annihilative tendency was delayed until eventually (in their examples, k near 0.8) stability resulted. Starting densities, in these cases, are important, however, and the outcome for any given case will vary according to the initial densities of host and parasite. When contagion of attacks was very high, k equal 0.2, a larger combination of initial densities led to stability. If the contagion of attacks becomes too great, i.e., attacks are too concentrated, ability to control the host is lost (they escape). (Fig. 5).

Summarizing, it would appear that any action that results in a sufficiently greater adverse effect on the predator than on the prey during the crash phases of the interactions when there are "too many" predators would add a damping tendency to the interaction, and the degree of this effect would determine its sufficiency or insufficiency for stabilization. Extrinsic damping tendencies include, (1) competition for food or shelter on the part of the host (or prey), (2) physical refuges affording some protection for the host, and (3) actions of other natural enemies superimposed on the given host/parasite system where the above stricture is satisfied. Intrinsic damping tendencies include, (1) a functional response that is directly density-dependent in effect—i.e., sigmoid (even if insufficient alone to regulate), (2) intrinsic behavior or limitations leading to partial host protection through imperfect temporal or spatial synchrony, (3) mutually interfering contacts, cannibalism, superparasitism, wound-killing or host-feeding (by parasite adults) that may satisfy the above strictures, (4) an appropriate contagion of attacks, and (5) general density-related movements into and away from sub-population units of the habitat. The latter would seem to apply where high densities of predators, generated in the crash phase of the prey (or host) at a given locale, and emigration away (and losses), are disproportionately heavy (a mutual interference effect). Contrariwise, immigrations of predators into such sub-population areas would seem to be disproportionately meaningful as a damping feature when occurring during the nadir phase of the interaction.

The Role of Natural Enemies in Control, Regulation, and Change in Density of Hosts

Having illustrated the contention that there are many ways by which the

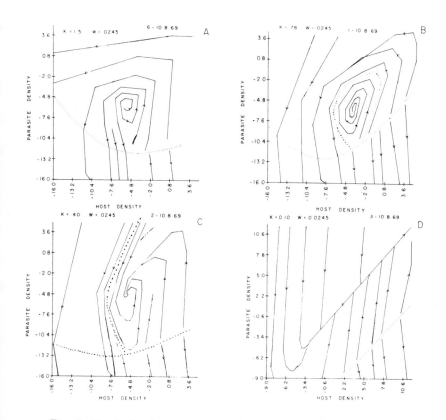

Fig. 5. Simulation host-parasite models exploring the potential damping power of different constant values of the negative binomial exponent, k, representing increasing contagion of attacks. (After Holling and Ewing, 1970)

theory of population regulation and of host-parasite and predator-prey interactions can conceptually be brought into accord with the general experience of regulation of pest species by their natural enemies, we wish now to consider in more detail the role of natural enemies relative to other factors. We do not here consider the horizons opened up by the possibilities of expanded use of natural enemies. This was done in Chapter 1. The two main avenues, introduction of new enemies and improved use of ones already resident, are illustrated in Sections II, III and IV.

Important evidence of the role of natural enemies in control of endemic pests has resulted from the recent era of synthetic pesticides. An extensive but scattered literature exists, some examples proving and others strongly suggesting, that the destruction of formerly effective natural enemies by use of such chemicals has resulted in the rise to pest status of many forms formerly innocuous. This area was introduced in Chapter 1 and will be dealt with particularly in Sections III and IV.

Biological control is recognized as a substantial means of pest *control* even by those who deny that natural enemy action is truly *regulating* in nature. Indeed, it is far easier to establish that they can *control* their host's densities than that they *regulate* those densities (i.e., as proved, density-related factors). Economically, the question is only academic, but we contend that good natural enemies not only act as a substantial mortality factor and thereby limit or reduce (control) the prey's population size, but they do so in a manner that tends to stabilize both their prey's density and their own density.

We use the laboratory examples of Hassell and Huffaker (1969*b*) and White and Huffaker (1969*a,b*) to illustrate that the parasite *Venturia canescens* acted to *control,* to *regulate* and to cause marked *changes* in density of its host, the Mediterranean flour moth over a period of 23 host generations (46 parasite generations) in two room *ecosystems* and several small sleeve-cage ecosystems. None of these effects can be attributed to weather factors as these conditions were held nearly constant and were the same in both control and parasite-present ecosystems.

In the absence of the parasite (control systems), the moth fully utilized the grain medium serving as food whereas, depending upon the nature of the dispersion arrangements, amounts of grain per unit, et cetera, the grain was variably conserved in the presence of the parasite. The *control* effect is thus established and it varied from about 15 per cent to about 50 per cent. Under these conditions this parasite is a rather inefficient natural enemy, but since the grain in the controls was fully utilized in much less time than it was in the systems, the effectiveness is greater than the above comparison suggests. In related but different ecosystems of Flanders (1968), considerable grain was conserved by this parasite when available to the moths for two years.

While biotic interactions involving uneven usage of the food (supplied regularly) in the control systems produced marked *changes* in density even after the host populations attained their capacity (based on rate of adding grain), these changes were much less pronounced and quite distinct from those observed in the parasite-present systems. The action of the parasites initiated and maintained a sharp distinctness in the generations while in the control systems there was a strong overlapping of generations. Moreover, a marked cycling in the size of the generations was produced while no such cycling occurred in the controls. These changes were induced largely through the inherent, delayed density-dependent action of the parasite, a fact that was clearly evident from the patterns of parasitization intensity and host population survival, generation by generation, and furthermore by statistical and graphic analyses of the data (Huffaker *et al.*, 1968; Hassell and Huffaker, 1969*b*).

Fig. 6 (from White and Huffaker, 1969b) presents a sample interval of a sleeve-cage system showing the degree of conservation of food, the discreteness of the generations, and the definite four-generation cycling of this system.

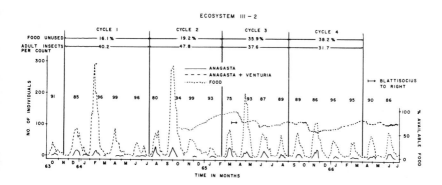

Fig. 6. Four-generation cycling in populations of flour moths *Anagasta kuhniella* (Zeller) and the parasite *Venturia canescens* (Grav.) in a sleeve-cage ecosystem in which the discreteness of generations was initiated and maintained by parasite action. (After White and Huffaker, 1969*b*)

Fig. 7a (from Hassell and Huffaker, 1969*b*) presents a model of the room A ecosystem showing a definite damping of the initial intensity of the oscillations in that system, wherein competition for food was apparently the main damping mechanism, but an interference cómponent was also present (*vide supra*). In Fig. 7b, the delayed density-dependent nature of the parasite's action in the room ecosystem is shown by the counter-clockwise spiralling during the first eight generations, i.e., before other factors became sufficiently strong to mask the relation. The same relationship is shown in Fig. 7c (from Huffaker *et al.*, 1968) for the sleeve-cage system.

These results show that the parasite did in fact, (1) produce a control effect, (2) cause very marked changes in density, and (3) acted as a delayed density-dependent factor. The question of whether such a delayed density-dependent factor should be considered truly regulating was discussed previously. Density-dependent action on the parasite (mutual interference) together with the fact that parasite and host numbers are coupled, in any event, can *result* in regulation of the host (Chapter 4).

Coming now to the role of natural enemies relative to other natural control factors, space will permit consideration of only a few aspects. Natural enemies are the main factor of natural control that we have much power to manipulate. We can and do alter cultural practices, breed for pest resistance, and use various other forms of artificial controls, i.e., pesticides, irrigation, burning, et cetera. One of the main natural control factors, the weather, however, is not subject to manipulation. We can only manipulate the microenvironmental "weather" that is itself subject to the pervading weather.

Moreover, for our pest species that are consistently troublesome year after year, it is obvious that the weather is not a sufficient natural control factor. By the same token, the existing natural enemies under the existing conditions are not either. However, (1) we may change the conditions, perhaps involving cessation of especially detrimental pesticide treatments or improving the habitat or continuity of food for the enemies, or (2) introduce additional enemies. These aspects are dealt with in subsequent chapters.

It is quite essential to be aware that a natural enemy may be the main regulating agent even though it causes only a minor part of the host mortality. This would be true, however, only if other regulatory factors have even less effect and the density-independent mortality is high and consistent (Chamberlin, 1941; Huffaker, 1966). Thus, if the other mortality were high and consistent, only a small amount of density-geared mortality would be required to balance births and deaths, and this could occur at a lower host density (contrasted to the situation where the other mortality were low and consistent). In general, however, a fairly quick numerical response and a good capacity to find hosts is required to prevent escape to higher, non-economic

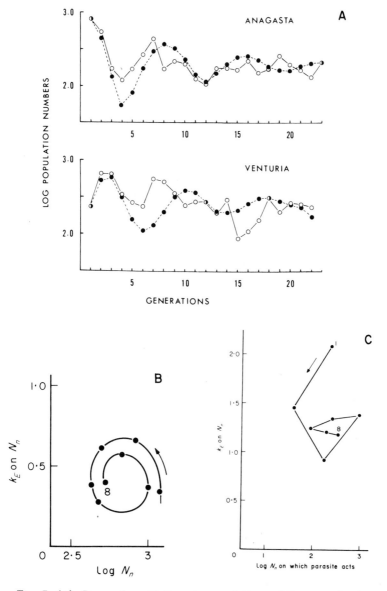

Fig. 7. (A)–Damped oscillations in populations of flour moths *Anagasta kuhniella* (Zeller) and the parasite *Venturia canescens* (Grav.) in a Room Eco-system (see text). (After Hassell and Huffaker, 1969*b*). (B) and (C)–Counter-clockwise spiralling in intensity of *V. canescens* attacks on *A. kuhniella* in small sleeve-cage ecosystems, suggesting delayed density-dependent action. (After Huffaker *et al.*, 1968)

densities in relatively *changing* situations.

In stable situations, as remarked earlier, searching capacity is far more important than power of increase in the enemy, whereas in unstable situations, the enemy would require a high power of increase to quickly overtake and subdue an otherwise escaping prey population, the escape tendency resulting from some temporary short-circuiting of the enemy's effectiveness. This is illustrated by work on olive scale in California. The parasite *Aphytis maculicornis* is rendered quite ineffective by the weather each summer, but due to its high power of increase it can usually overtake the increasing host population when conditions are again favorable in the fall and spring. This description applies to the situation prior to the introduction of a second, supplementing parasite, concerning which see Chapter 7, and also Huffaker and Kennett (1966).

This example illustrates a curious paradox. A parasite may serve as the key-factor for change while at the same time being the main regulatory factor, although the latter would seem to preclude the former (Varley and Gradwell, 1963). In this case *A. maculicornis* is the key-factor for change only because it is so strongly affected by the weather (*vide supra*), whereas it regulated its host's density at a very low level due mainly to its ability to find hosts when they are extremely scarce (interrelated with its power of increase and some damping feature). A parasite may also be the key regulating agent even though this action is masked statistically by other mortality, and another mortality factor is the key-factor for change. Experimental methods of analysis are required to appraise complex situations. These are dealt with in Chapter 5.

CHALLENGING CONCEPTS RELATIVE TO THE THEORY AND PRACTICE OF BIOLOGICAL CONTROL

Several controversial ideas have recently been vigorously advanced that directly, or by implication, challenge certain premises and indict some practices in biological control. These include, (1) the idea that the premise of biological control rests primarily on the concept of community stability, with the latter being dependent upon species diversity, (2) that multiple introductions of enemy species are likely to lead inherently to a decreased degree of control as contrasted to introduction of only the "best" species, (3) that polyphagous species are better enemies than highly specific ones, (4) that only pests that damage the marketable item indirectly (so-called "indirect pests") are amenable to biological control, (5) that there have been few successful cases of biological control, and the endeavor is therefore unpromising, and (6) that where natural enemies have long been in association with their host or prey species, a "harmonious" relationship will have evolved in which the degree of control is not substantial.

Because these criticisms are either faulty in conception or are in disagreement with the evidence, we here examine them. Certain aspects of these questions are also dealt with in later chapters.

Species Stability and Community Stability

The premise of biological control rests only in the fact that in nature there exist, for many of our noxious and potentially noxious species, natural enemies that are capable of controlling them. This applies to endemics all around us, as well as to introduced species. Virtually all of the most outstanding cases of biological control have involved rather host-specific enemies (Huffaker, 1964, p. 632; Huffaker and Messenger, 1964, pp. 100,109; Doutt and DeBach, 1964, p. 122; DeBach, 1971). Huffaker and Messenger (1964, p. 109) emphasized that in crop situations, in particular, control usually rests on the actions of rather specialized enemies.

It is strange, then, that Turnbull (1967) and Watt (1965) appear to base their critiques concerning biological control and the practice of making multiple introductions, in part at least, on community diversity relationships. Turnbull stated, "The philosophy behind this practice [of biological control importations] is that community stability is a product of diversity of the environment." But community stability achieved through diversity is not necessarily an objective of biological control. To be sure, community stability is often a secondary feature and the existence of a diversity of enemy species would serve as a balance-wheel tending to maintain some degree of control of the given pest when a more important enemy is rendered ineffective in time or place, for whatever reason (Smith, 1929; Solomon, 1957; Huffaker *et al.,* 1969, p. 153). The latter authors stated, "Theoretically, general predators tend to serve as regulators of community stability while the specialists tend to regulate single species stability." They added, "There is no sharp line; the two are interrelated. . ." General predators may also damp a more specific host-parasite interaction (*vide supra*). Thus, Turnbull's view of this philosophy misleadingly presents a half-truth only marginally related to the root premise Itself.

DeBach (in press) elaborated on this question: "The factors responsible for community stability are not necessarily the same or even similar to those responsible for single species population regulation . . . ", and again, "One really effective enemy may be . . . and often is . . . enough to regulate a host species at low population levels even though its colonization may eliminate other natural enemy species and actually reduce the diversity of natural enemies." Huffaker and Kennett (1966, p. 332) suggested that the evolutionary result may well be that a species upon which has evolved a highly specific (stenophagous) and effective enemy will, because of that very effectiveness, not be found to support a wide complex of enemies. Thus, the efficient enemy would tend to control the host at a density too low to support such a complex of less efficient, more euryphagous enemies, a condition that has been observed in a number of actual cases of biological control (Chapter 7; DeBach and Sundby, 1963; Huffaker and Kennett, 1969). Huffaker and Kennett (1966) considered wrong Zwolfer's (1963) interpretation, and Watt's (1965) support, by implication, that host species are abundant because they have too many enemies competing with each other to the detriment of the over-all control effect. In fact, this interpretation of the associations found may well have *cause* and *effect* reversed.

Thus, we do not introduce natural enemies to achieve diversity alone. We do so in an effort to establish a single good species or to find a combination of species that will prove to be better than what we already have. The history of a few thousand examples (sequences of introductions at *local places*) attests to the soundness of this objective, and, by implication, refutes the claim that due to competition between species of enemies, a reduced degree of control will result from multiple introductions, as contrasted to the introduction of the "best" species alone (see next topic).

In a later paper, Turnbull (1969) forcefully develops an important idea which we have also long held, as to its basic core.

In brief, he correctly holds that only the primitive, persistent or climax community stage is truly stable and the over-all effect of natural regulating forces will tend to return the system to that state when for any reason there is a departure. He holds that since man seeks to permanently maintain his crop areas in developmental or artificial states, his efforts run counter to these regulating forces. He contends, perhaps correctly, that the great northern boreal forest biome as a whole is stable and that the episodes represented by build-up of uniformity in age-class and over-concentration of a single tree species, followed by corrective destruction by outbreaking forest insect pests or disease, form a part of the over-all stability. We have ourselves long contended that biological control of plant species may well be a root cause of the great diversity of species in tropical rainforests, for example. The destructive action of insects or disease on a group of canopy trees of a given species may be seen by some as disturbing the stability; to us, it could simply be a means by which the characteristic and high degree of over-all stability is maintained.

However, we feel that Turnbull minimizes the part that can be played by single regulating elements, e.g., a host-specific natural enemy of a phytophagous insect. Turnbull stated, "Natural regulation, if it applies to anything, applies to the whole community, not to any particular part of it. Thus if we restore natural regulation to cultured communities we must expect it to apply to all organisms in that community, including the crop organism" Here Turnbull seems to be saying that "natural regulation" is an improper term to use unless the whole community is returned to the climax state. With this we emphatically disagree. In biological control, or in the system of integrated control which he espouses, we *can* use such enemies—ones that are regulating forces for sub-systems or links of their historic natural communities —to do similar jobs for us in crop communities. (See further Southwood and Way, 1970)

Multiple Introductions versus Single Species Introductions

Turnbull and Chant (1961) renewed an old attack on the practice of introducing a sequence of natural enemies. This and later papers by Zwolfer (1963), Watt (1965), and Turnbull (1967) have been taken by administrators and entomologists in positions of decision as a warning that introductions should be made only after elaborate study, and even then, only the single "best" species should be introduced. This implies an elaborate and unnecessary endeavor for which funds availabe will seldom if ever suffice.

Watt's work deals mainly with other questions, but he suggested that his hypothesis supports the ideas of Turnbull and Chant (1961) and Zwolfer (1963). Watt ends, however, by asking some questions and posing a few statements partly as follows (other aspects were considered previously) (from DeBach, 1971):

"Hence, . . . there is no point in releasing a number of different species of biological control agents, as opposed to one species carefully pre-screened."

" . . . why is it worthwhile to add any biological control agents at all to a complex of parasites and predators already unable to control a pest . . . ? "

" . . . it does not make sense to add one more entomophagous species to a complex, unless that species has one of two characteristics:

1. It fills some functional niche not already filled . . .
2. It fills some niche already filled more effectively than species already filling it."

These two points and others throughout his paper indicate that he is not rigidly sold on the ideas he supports. In particular, the two characteristics he stipulates are just a way of describing, (1) a species that would have a supplementing role, and (2) a clearly better species. Thus, boiled down, this fits well with accepted goals in biological control. The main difference is whether or not the introduction of species not conforming to the idea would cause a worsening of the pest situation.

The premise of biological control relative to this is that the introduction of highly specific species other than the "best" one will do no harm. We contend that it is entirely unrealistic to attempt to find and to pre-rank every possible candidate in order to ascertain the "best" one to introduce. Moreover, no one has established any very definite set of criteria by which we can pre-rank given species. Even painstaking bioclimatic studies on parasites of the spotted alfalfa aphid in California, extending over a number of years (Messenger, 1964; Messenger and Force, 1963; Force and Messenger, 1964a, 1964b, 1965), suggest that no single enemy species can be found possessing in higher degree than other enemies those characters considered important (searching rate, searching effectiveness, fecundity, attack rate, descriminatory ability, synchrony with host life cycle, et cetera). The time (many years) and money spent to make such exhaustive studies before making *any* introductions would be imprudent and unnecessary.

We have in the empirical record of biological control ample proof that in general the introduction of a sequence of rather highly specific species is a desirable practice. Moreover, Hassell and Varley (1969) (see also Chapter 4) have recently explored this question theoretically and their findings refute earlier theoretical implications (Nicholson, 1933, 1954; Varley, 1959) that such introductions may well be detrimental. However, Nicholson (pers. comm.) has stated that as former Chief of the Entomology Division of the Australian Council for Scientific and Industrial Research Organization, he did not use his own theoretical implication of a *possible* adverse effect, to deny any otherwise proper introductions.

There are of course two rather distinct aspects of the question. One has to do with the geographic coverage of the hosts' environment by imported natural enemies. Rarely is a specific natural enemy co-extensive with its host. Even if it were, it is unlikly that it would be superior to all other competing enemies over the whole range. The value to be gained in obtaining a complex of enemies, such that a greater degree of control would obtain over wide geographic areas (different enemies superior in different situations) is largely ignored in the criticisms leveled against current practice. The California red scale (DeBach et al., 1962; DeBach and Sundby, 1963; and Chapter 7), the spotted alfalfa aphid (van den Bosch et al., 1959, 1964) and Klamath weed (Huffaker, 1967; Harris et al., 1969) represent examples wherein two or more species have produced a superior over-all control than could be accomplished by the "best" one alone. Many other examples could be cited. This aspect is obvious (see Chapter 3).

The other aspect of the question concerns the possible supplemental or competitive, detrimental effects at a *given place*. As Huffaker and Kennett (1966) commented, "While the gross record of deliberate introductions of enemies to control given pests is largely undocumented as to the degrees of control achieved by the first introduced enemy versus the first two, versus the first three or four, et cetera, two facts seem abundantly clear from experience alone: (1) introduction of an additional enemy has invariably improved upon the degree of over-all control or else it has made little difference, and (2) unless we can make the highly unlikely [ridiculous] assumption that in all cases the first enemy was an inferior one and that each later one added could have done a better job if introduced alone, the suggestion is disproved that the more enemies a species has, the poorer and more unstable will be the degree of control." One would expect, on chance, that the second introduced species would be inferior to the first one-half the time, and hence there should, according to this hypothesis, be as many cases wherein the second species introduced led to a decreased level of control as to an increased level.

Turnbull (1967) discussed competitive displacement at length and used this phenomenon to explain the low ratio of establishment of natural enemies. He did not, however, note that competitive displacement, which he then used as a bludgeon in his continuing attacks against biological control practices, is in fact a boomerang serving to demolish his earlier basic concept (Turnbull and Chant, 1961), i.e., that competitive interference from two or more enemies reduces the total effectiveness below that which the "best" one would achieve alone (the host density is increased).

Before treating the evidence, we concur also with van den Bosch (1968) that in order to emphasize the supposedly neglected competitive displacement (as an explanation of "failures"), Turnbull assumed that exotic enemies are typically introduced into areas already populated by highly competitive, resident enemy species. Van den Bosch noted that this is basically untrue. We introduce species usually against exotic pests that are relatively abundant and there is low competition for them; they have few, or relatively ineffective, enemies attacking them in their area of invasion. Actually, in order to increase the level of control, we often do introduce additional species after the host has been reduced considerably, but not sufficiently, and the degree of competition between the enemies would then be greater. It is after host densities have declined substantially that the intense (displacement) phase of competition is reached. In these situations, this phenomenon may account for some failures of establishment, but not the failure of establishment of a *better* species, or one capable of supplementing a better one already present. As van den Bosch (1968) surmised, it does not serve as a basis for explaining failures of biological control *efforts,* as contrasted to failure of establishment of *certain species* (see further Chapter 7).

The case of olive scale in California illustrates, (1) the supplementing action of two natural enemies working together, (2) the fact that host densities are not raised when two enemies do significantly work together (co-exist), but to the contrary are reduced, and (3) the fact that as very efficient enemy species become established, less efficient competitors may be unable to persist in the rarified host density produced—they may be displaced. This is one of the world's best documented examples of excellent biological control (Doutt, 1954; Huffaker *et al.*, 1962; Huffaker and Doutt, 1965; Huffaker and Kennett, 1966, 1969; Kennett *et al.*, 1966; Broodryk and Doutt, 1966; Kennett, 1967). This control and various ecological aspects are treated in Chapter 7. Here, we emphasize several points. The best species (*A. maculicornis*) was, however, not quite good enough in all situations, although it had been good enough to reduce scale densities in most situations to less than one per cent of their former densities and this had led to the displacement of several competitor species. The introduction of *Coccophagoides utilis* effected an additional marked degree of control, especially in those situation where *A. maculicornis* acting alone gave only marginally reliable control. Three different means of analysis established that *A. maculicornis* is the best species alone, that *C. utilis* is essential as a supplemental factor acting on spring generation scales against which *A. maculicornis* is ineffective (a weather effect), and that both are required contemporaneously to achieve the markedly successful control exhibited (Chapter 7).

Moreover, the way in which these two efficient species displaced all the former exploiters of olive scales is quite imcompatible with the concepts of Turnbull and Chant (1961), Zwolfer (1963), Watt (1965) and Turnbull (1967) as discussed here. Huffaker and Kennett (1969) related the essentials of this process:

> "Here again displacement of competitors has been a characteristic feature and a measure of the superior performance of both *A. maculicornis* and *C. utilis*. (Huffaker and Kennett (1966) discuss the reasons why neither of these two species displaces the other.) Among the several species of imported parasites and predators released in California, four or more appeared definitely to become established, at least in special local situations, i.e., one coccinellid, *Chilocorus bipustulatus* (L.) (Huffaker and Doutt, 1965), and three aphelinid wasps, two internal ones, *Aspidiotiphagus* sp. near *citrinus* (Craw.) and *Coccophagoides utilis,* and one external species or complex of sibling species, *Aphytis maculicornis* (see Doutt, 1954, 1966; Hafez and Doutt, 1954). In addition to these

introduced species that seemed formerly to locally establish and one of them (*Chilocorus*) to thrive when scales were abundant and competition with other species therefore low, a number of indigenous enemies were also known commonly to attack the scale, sometimes very heavily. These included another coccinellid, *Chilocorus orbus* Casey, a nitidulid, *Cybocephalus* sp., another internal aphelinid, *Aspidiotiphagus citrinus* Craw., and a predatory mite, *Hemisarcoptes malus* (Shimer). Host densities were then very high but not in our view because these four enemies and others were interfering with one another or with the best of them (Turnbull and Chant, 1961; Watt, 1965; Zwolfer, 1963), but because not one of them possessed the necessary characteristics to control the scale even if present alone. Prospective "interference" by all these previously present and introduced species as well did not deter the efficient and competitor-displacing action of the only two efficient species among the whole complex, i.e., Persian *A. maculicornis* and *Coccophagiodes utilis*. To all intents and purposes, densities of the scale are now held so low by these latter two parasites [see Chapter 7] that the others appear no longer able to survive; certainly none of them occurs in olive groves except very rarely and in such cases only because of some chemical disruption of the efficiency of *A. maculicornis* and *C. utilis*, when densities are thus allowed to reach higher levels again. Even though hundreds of thousands of olive leaves have been sampled and examined closely and regularly in recent years these other species have not been taken at all. *Chilocorus bipustulatus* was last seen in a disturbed grove in 1963."

A second and quite different case involves the biological control of a weed (Klamath weed or St. Johnswort) in western North America. In this case two (or more) species of enemies have not in any way supplemented one another in the control of the weed in open areas in California at a given locale, whereas this is a likely effect in shady situations in California and even in open areas in parts of Australia (Huffaker, 1967). The control in most open areas in California is achieved exclusively by *Chrysolina quadrigemina* (Suffrian). In shady areas the root borer, *Agrilus hyperici* (Creutzer), exerts some supplemental if not a major role at times. Arguing in the manner of Turnbull and Chant (1961), we would conclude that the weed is more abundant in the shady areas because these two species are able to co-exist there and the interference competition prevents the better species from exerting its potential effect. The fact is, however, that neither species (alone or both together), is as efficient in killing Klamath weeds in the shade as is *C. quadrigemina* alone in the open.

The displacement feature in this situation is a repetition of that pictured for olive scale. The power to displace competing enemy species (if the competition is for hosts—the most likely requisite) is a clear measure of an enemy's control potential. It can only displace a competitor if it can reduce the host's density below the minimum required for survival by the host-specific competitor. While the comparative production of progeny is crucial in this process, the *inherent* reproductive ability is only ancillary. The actual progeny production under the intense competitive situation is the crucial feature. This is more likely to be related to inherent ability to *find* the requisites (hosts at very low densities) than to intrinsic power of increase. Very effective control is itself evidence of excellent searching capacity. Again, Huffaker and Kennett (1969) summarized the case:

> "In this case [in California], three other species of host-specific insects introduced for control of this weed at first did thrive and spread considerably, that is, before they were subjected to competition with *C. quadrigemina*. The gall midge *Zeuxidiplosis giardi* Kieffer also did well in many local situations as did the root borer *Agrilus hyperici* (Creutzer). The other leaf beetle, *Chrysolina hyperici* (Forster), not only did well in certain places; it multiplied, spread, and exhibited control of the weed, but less quickly, everywhere it was colonized; it coexisted with *C. quadrigemina* in great numbers in areas where both were colonized during the time when the weed was still abundant. *C. hyperici*, however, was shown by Huffaker and Kennett (1953) to be a poor competitor of *C. quadrigemina* under the California climatic regime because of its slowness in breaking diapause, and it was observed to decline in relative numbers year by year as the weed became more scarce. For the past eight years or more we have seen no specimen of *C. hyperici*; it appears to have been virtually displaced, and this is a measure of the searching efficiency and control efficiency of *C. quadrigemina*. In Australia, however, where the control effect is not so perfect, and the larval activity of *C. quardigemina* is limited more to the fall period, the spring period being left more to *C. hyperici*, and competition is thus less intense, *C. hyperici* remains common although rarely if ever dominant. Likewise, the gall midge is now essentially extinct nearly everywhere in California. The root borer has been able to hang on rather well in places where the weed holds out in shady situations at a density above that characteristic of action by *C. quadrigemina* in the open. This is an overlap or ecotone situation where both species persist. In deeper shade the root borer has advantage if other conditions are favorable to it."

The case of competitive displacement of California red scale on *Citrus* by congeners of the genus *Aphytis* in Southern California was used by Turnbull (1967) in his critique, but he failed to point out the *results* of the instances of competitive displacement to which he referred. DeBach *et al.* (1962) and DeBach and Sundby (1963) stated, however, that with each displacement of a congener an improved level of biological control was experienced. These instances are reviewed in Chapter 7. They represent still additional examples that are incompatible with the view of the critics here considered.

Polyphagous versus Monophagous or Stenophagous Enemies

Watt (1965) made the odd statement that "...the most unstable biological control agents, and hence those capable of controlling an unstable pest, will be polyphagous." This seems to imply that a pest that is unstable (outbreaks frequently and subsides to low levels) would be unstable regardless of the kinds of enemies that might attact it. The fact that it is so unstable means to us that it is not under reliable and effective biological control. There seems to be a contradiction of terms in the statement itself. In any event, we reject the implications on both empirical and theoretical grounds.

On the record, virtually all of the outstanding results from introductions have been achieved by rather highly prey- (or host-) specific natural enemies. We note here that some stenophagous species may be strictly monophagous relative to a given crop or habitat. As stated earlier, euryphagous species appear to serve in community balance whereas the highly prey-specific forms serve in single species control and balance.

Evolution has served both to promote monophagy and to promote and maintain polyphagy in relation to the organisms and the environment. The value of a broad diet is obvious; if one food (prey) is scarce another can be substituted. By their very nature, specialists are better adapted to utilize a specific prey at low prey densities in maintaining their own populations. They are more closely synchronized in their habits, haunts and seasonal life phases, and are normally better attuned in nutritional needs, reproductive potential and searching behavior to effectively utilize their prey at a minimal prey density. Thus, they are more effective and reliable biological control agents (Huffaker *et al.*, 1969).

Since polyphagous predators (or parasites) can hardly be equally well adapted to utilization of each of its many acceptable prey species, they could hardly be as efficient or as reliable as control or regulating agents, relative to a prey species also attacked by a monophagous predator. One point, however, is worth noting. If their numbers are *consistently* maintained at a good level (they are not *unstable* as in Watt's statement) through utilization of a complex of prey, they could conceivably exert a degree of *control* both consistent and more complete than that achieved by the specialists. Such a form could be said, technically, to be a better *control* factor but a poorer *regulator* (Huffaker *et al.*, 1969).[2] Little evidence, however, exists in support of this conception, but predation by the predatory mite *Typhlodromus caudiglans* Schuster on European red mite, *Panonychus ulmi* (Koch), may be of such a nature (Putnam and Herne, 1966).

Moreover, conditions in a given situation may be such that strict monophagy is not conducive to good control. Alternate prey species or non-prey foods may be required at times when the given prey is unavailable. If a predator maintains a marked predilection for the given prey but utilizes suitable alternate foods in emergency, it may be said to have *maximized its efficiency as a specialist consistent with the situation,* for it is thus maintained in the environment in greater numbers, and better distributed, and thus is at hand to prevent resurgence of the specific prey (Huffaker *et al.*, 1969). (See further Doutt and DeBach, 1964)

Direct and Indirect Pests in Biological Control Work.

Turnbull and Chant (1961) made an overly emphatic distinction between *direct* pests and *indirect* pests. They stated: "Indirect pests are suitable subjects for biological control; direct pests are not." They mean by "direct pests" ones that attack and damage the marketable item. The work on olive scale (*vide supra*) illustrates a clear exception to such an arbitrary statement. While this insect attacks leaves, twigs and woody parts and reduces yield *indirectly* when very abundant (Chapter 7), it selectively settles 10:1 on the fruits in the fall of the year, and a single scale may cause a cull. Nevertheless, a remarkably high degree of commercial control is maintained. What a pity if no effort had been made to control this severe pest of some 200 host plants on the basis that it is a direct pest and therefore not a "suitable subject".

---

[2]K. S. Hagen has suggested that natural selection tends to develop resistance to attack, e.g., through mimicry or camouflage, by such general predators that do not automatically lessen their pressure where the prey reaches dangerously low levels, as do highly-specific species. The latter automatically relax their pressure prior to the point of most intense selection pressure for development of such resistance (see further Chapter 11).

The Claim that Biological Control
is an Unprofitable Endeavor

Perhaps the most damaging claim against biological control, and the one that seems to us to be clearly erroneous, is that which holds that there have been so few successes, and these have been so limited in geographic scope that it is an unprofitable business. This claim has long been used by advocates of reliance on pesticides (e.g., Decker, 1962; McLean, 1967; Hansberry, 1968) and by ecologists and entomologists who have been interested in evaluating biological control results (Taylor, 1955; Turnbull and Chant, 1961; Turnbull, 1967). Taylor wrote: " . . . I know it (biological control) to be the best of all methods of controlling pests when it works, but it seldom works and there is little future in it in continental areas."

Clausen (1958), DeBach (1964), and Simmonds (1959, 1967). rebut Taylor's pessimistic view and that of other critics as well. Both groups of critics skim over or distort the record of complete successes (some 70 or so) and largely ignore the record of partial successes. Decker (1962) stated that biological control has been extensively tried and found wanting; that the case of Vedalia control of cottony cushion scale is about all we have to offer. At the meeting where Decker spoke, one of us (Huffaker, 1962) reminded the gathering that in California alone the following species, where given a chance, have been brought under considerable to good biological control by introduced enemies: cottony cushion scale (*Icerya purchasi* Mask.), California red scale (*Aonidiella aurantii* (Mask.)), yellow scale (*Aonidiella citrina* (Coq.)), San Jose scale (*Quadraspidiotus perniciosus* (Comst.)), purple scale (*Lepidosaphes beckii* (Newm.)), nigra scale (*Saissetia nigra* (Nietn.)), black scale (*Saissetia oleae* (Olivier)), olive scale (*Parlatoria oleae* (Colvee)), citrophilus mealybug (*Pseudococcus gahani* Green), pea aphid (*Acyrthosiphon pisum* (Harris)), and spotted alfalfa aphid (*Therioaphis trifolii* (Monell)). A few additional ones have been added since: grape leaf skeletonizer (*Harrisina brillians* B.&McD.), walnut aphid (*Chromaphis juglandicola* (Kalt.)), and (in the original site) tansy ragwort (*Senecio jacobaea* L.).

Hansberry's (1968) analysis will appeal to the insecticide industry. He noted that the number of papers published in the Journal of Economic Entomology during the year 1967-1968 in which *successful* control of pests was reported totaled 34, and all dealt with the use of pesticides. It might be noted that workers in biological control have, regretably, almost ceased sending manuscripts to this journal and many have dropped their subscriptions, such is its emphasis on pesticides. Hansberry referred to some 1,425 species of pest insects in the United States and Canada, and he considered that the 61 cases of biological control for Canada and the U.S.A. is a paltry achievement. We contend that in view of the long-term solutions, very limited financial support and extreme paucity of biological control specialists in the country, as contrasted to the very large monetary support and veritable army of workers in the pesticide area, this record is commendable. Hansberry made no claim that use of chemicals adequately takes care of these pest species not under biological control. He blandly remarked also, regarding the thousands of innocuous species: "This [their innocuousness] is because most animals and plants are resistant to most insects and all insects are subject to a natural struggle for existence that limits their numbers." This naive description hardly requires comment.

Regarding the utility of biological control, Simmonds (1967) poignantly stated: "There are critics ... who claim that the results achieved are not really commensurate with the effort and money expended, that successes have been few, failures many, that most of those problems where biological control might possibly be effective have already been investigated, and hence that the method has had a poor showing in the past and no future." Simmonds added, "This is *nonsense,* and indicates that these critics have not really taken the trouble to obtain information on the subject they are condemning." While noting that very little attention has been given to the question of the economics of their efforts by biological control workers, Simmonds was able to present a thorough refutation of the claims of these critics. For one case for which a very clear balance sheet is available, on the Island of Principe off the West Coast of Africa, biological control of the coconut scale, *Aspidiotus destructor* Sign., which threatened the entire economy of the Island, was accomplished at an expenditure of about $10,000—a cost borne only once. Conservatively estimated, the savings for the first ten years following the introduction amounted to over $2,000,000 and these benefits continue to accrue. Biological control of the pest weed *Cordia macrostachya* (Jacquin) Roemer and Schultes in Mauritius was accomplished at an expenditure of $28,000. The accrued benefits over the first fifteen years were about $4,200,000. Simmonds presented case after case, the gain on investment being enormous in many instances, e.g., 25,000 per cent for the biological control of sugarcane moth borers in Antigua (see also Chapter 6).

The record presented by DeBach (1964) and in subsequent chapters of this book is adequate refutation of such general criticisms. We note here that more than 250 cases of complete or partial biological control have resulted from introductions, and this omits the roles of a great many indigenous biological control agents (for this evidence, see Section III).

Turnbull (1967) considered that the rating of "partial" successes is based only on the inference that since given natural enemies exist in the environment and are known to attack particular pests, they are therefore exerting "partial" control. This is an irresponsible statement. Some overzealous ratings have undoubtedly been made. Yet, the literature is filled with clear recognition by biological control workers that the great majority of species introduced for control of many pests, although becoming established, have not contributed a worthwhile result (e.g., Clausen, 1956; Clausen (editor, in press)). Using Turnbull's inference, every pest species against which any enemy has been introduced and established would be logged in as a partial success.

Turnbull (1967) also claimed that the *proportion* of species introduced that actually become established is low and biological control is therefore unpromising. In fact, however, the *proportion* of attempted introductions that become established is largely irrelevant. If we obtain a success by finally establishing *one species* that does the job, even though a dozen failed, the number that failed is a minor fact, for the amount of effort going into their importation is usually small relative to the financial gain achieved. For the amount of effort expended, we do not consider the number of successes to have been few—quite the contrary. Chant (1966) stated that the savings to U. S. agriculture for each research dollar invested in biological control has been on the order of 30:1. The effort required is often minimal (Simmonds, 1967). The addition of the necessary supplemental agent in the control of olive scale in California resulted quite coincidental to searches in Pakistan (by DeBach) for parasites of red scale on citrus. Moreover, transfers to other regions of enemy species known to work well in one area of the world have often led to remarkably inexpensive and profitable results. Van den Bosch (1968) admirably dealt with this aspect of Turnbull's criticisms.

Turnbull, however, obviously rejected what he erroneously considered the premise of making importations to be. For he stated "We are not concerned with population diversity but with maximizing a single species—our crop; not with stabilizing pests, but with suppressing them. For these purposes we use specific agents to carry out specific tasks." Here we heartily agree. Turnbull then added: "The chances of finding such agents without knowing what characteristics they should possess are not impossible, but they are certainly remote. In fact, we have little reason to suspect that such organisms exist." This last astonishing sentence alone would seem to be a measure of the bias in Turnbull's attacks on biological control.

It is interesting that Turnbull (1969), however, seemed to roundly contradict this dour prognostication. He stated: "Successes achieved by biological control are among the outstanding achievements of applied biology. Therefore we cannot question the validity of the method. It works! It has worked gloriously in the past, it is working for us in the present, and it will work again in the future."

THE GENETIC FEED-BACK PRINCIPLE AND BIOLOGICAL CONTROL

That communities of plants and animals exhibit a high degree of homeostasis in their space-time features is a widely accepted ecological principle. MacFadyen (1957) stated, "It is generally agreed that the same species are found in the same habitats at the same seasons for many years in succession and that they occur in numbers which are of the same order of magnitude." This illustrates the balance between species in communities as wholes. It has also been rather widely recognized that if two-species or several-species food chain relationships were substantially unstable this would disrupt or preclude the observed over-all community hoemostasis. Further-more, it has been rather generally accepted that the greater the complex of species in such communities the greater is the degree of over-all homeostasis. It is also an accepted principle that true parasitism tends to evolve from the more violent, purely exploitative phase (lethal to the host) through one of tolerance on the part of the host (only the parasite benefits) to a state of mutualism.

In a long series of papers, Professor David Pimentel and associates at Cornell University (e.g., Pimentel, 1961a,b, 1963a,b; Pimentel, Nagel and Madden, 1963; Pimentel and Al-Hafidh, 1965; Pimentel and Stone, 1968; Mad-den and Pimentel, 1965; Nagel and Pimentel, 1963) have utilized these basic principles, together with their own laboratory studies and a much too selective emphasis of examples in the literature of host-parasite relations in nature, to develop a number of concepts, some acceptable and some not. A main thrust is the idea of genetic feed-back as a mechanism operating to regulate interspecies relationships at the *individual, population,* and *community* levels. They stressed and illustrated a much-neglected aspect in the dynamics and natural control of populations—i.e., that populations possess the capacity for significant genetic changes that result in better adaptations to selective factors, and these changes may cause changes in the density or stability at which the populations are regulated.

In a host-parasite or predator-prey interaction, the hypothesis is that the selective pressure exerted by the attacking species on the attacked species in time causes genetic changes in the attacked population, such that it becomes better adapted to resist or to tolerate attacks. Pimentel also states that "co-evolution" may also result in a diminution of the inherent power of the attacking species to harm the attacked species.

We do not challenge this conception as an evolutionary process. We do challenge two main ideas developed by Pimentel and associates that constitute erroneous concepts and inferences concerning the consequences of this process, (1) the idea that genetic feed-back constitutes a distinct mechanism of *regulation* of population size, and (2) the idea that it results automatically in interspecies homeostasis in which the enemy that has evolved with, or been long associated with, its host or prey species will not be an effective control agent—i.e., it does not greatly restrict its host or prey population. It is with these two general conclusions that we are concerned.

Regulation of numbers was discussed earlier in this chapter. Briefly, population size is regulated by actions of repressive environmental factors, such that as density increases these actions intensify and thereby prevent numbers from exceeding some characteristic high level, and as density falls these actions relax. There exists a negative feed-back process involving population density and the actions of these repressive environmental factors. Natural enemies can serve as such regulatory factors, as can intraspecific competition for food, space or other requisites.

Regulation of numbers is not independent of the kinds of animals involved nor of the environment within which the population exists. All three aspects, the intrinsic properties of the species, the capacity of the environment, and the regulatory process(es) occur together, inseparable, and interactive. Hence, regulation does not alone *determine* the characteristic level of abundance or any dynamic displacements therefrom, but rather does so in relation to the survival power and population growth rate capabilities of the species, and the relevant environmental factors which alter these capabilities (*vide supra*).

Pimentel (1961a, 1964, 1968) acknowledges these regulatory processes and factors, but contends that the genetic feed-back mechanism also constitutes such a regulatory factor. So far as a natural enemy and its host are concerned, this concept of the genetic feed-back mechanism holds that the host, through selection, gradually becomes resistant or tolerant to the enemy, such that it becomes more abundant than earlier, while the enemy becomes less abundant, relatively. The process may become cyclic; i.e., the enemy, because of its diminished numbers, exerts less selective pressure, and may then re-attain its earlier level of effectiveness, and so on.

Such a mechanism serves to account for evolutionary changes in the intrinsic capabilities and limitations of the species, as well as the way in which the species responds to the conditioning forces of the environment. What is objectionable about this is Pimentel's contention that such an evolutionary process serves as a regulatory process for populations (see also Milne, 1962). In no case does Pimentel show that as population density increases, say from one generation to the next, the genetic feed-back mechanism brings about a corresponding increase in mortality, and vice versa. In fact, all that the genetic feed-back mechanism does is change the quantitative rules by which the regulative game or process is played. It modifies the properties of the host, for example, and therefore the way the host population responds to mortality factors of the environment, but it does not, as claimed by Pimentel and Al-Hafidh (1965), thereby substitute for or displace regulatory agents, and become a regulatory process itself.

The second general conclusion of Pimentel, that natural enemies long associated with their hosts will exert no very restrictive control over their host (or prey) populations, is quite unwarranted. This is a main thrust of their work, although Pimentel (1963a) commented, contrarily, that none of the changes ensuing tending to co-adaptation of the new host and parasite relation " . . . need result in any important increase in numbers of the pest."

In the laboratory population studies which they used as a basis for support of this consequence, selection was limited to only one factor, parasite attack. In nature, a host population undergoes selective pressure from many factors acting simultaneously or in succession. Two or more enemy species may also be competing for a host species. Additionally, host populations in nature rarely undergo as intensive and continuous a selection from a given parasitic species as took place in their laboratory studies. The host of an effective host-specific enemy is not likely to have its adaptive potential concentrated so exclusively towards resistance against such an enemy, because such an enemy's pressure is automatically relaxed when densities are driven dangerously low; instead, selection pressure at this critical time is concentrated towards developing adaptations to counter the severity of density-independent factors, general predators included (Hagen *et al.*, Chapter 11).

Finally, attack by enemies of the sort we are dealing with is an all-or-none proposition; the host individual either lives or dies. Pimentel's use of such examples as the myxomatosis disease of rabbits in Australia and of an unidentified disease of oyster spat in New England was, to this extent, unfortunate. These are striking examples where genetic adjustments alter substantially the initial regulatory balance between enemy and host. But these enemies are true parasites. Entomophagous parasitoids are predators and they seem, on the contrary, not to have evolved towards a benign toleration situation in respect to their hosts; and insect hosts seem not to have developed tolerance towards these parasitoids. The association between insect hosts and insect parasitoids remains primitive and lethal to the one or the other.[3]

---

[3]A possible selective advantage at the population level against development of such benign interactions—such that a harmonious relationship results even though individual encounters are always lethal—is suggested. A satisfactory genetic explanation of how this might come about is at present lacking.

This argument is not to deny that insect hosts do develop immunity perhaps through evolutionary selection, in respect to their parasitoid enemies. Instances of encapsulation (e.g., Muldrew, 1953; van den Bosch, 1964) wherein a host individual becomes resistant to the enemy does not refute this argument. Such resistance still presents an all-or-none relation; the host or parasitoid either lives or dies. The evolution of such a resistance could represent a stage in the progression from being a host to becoming a non-host, but this is not the question considered by Pimentel in the genetic feed-back mechanism.

Highly specific parasitoids appear to have kept the upper hand, countering trends towards such resistance rather than failing to meet this evolutionary challenge.

It is important to note that there is no clear indication from the experiments of Pimentel and associates that the ability of the parasite *Nasonia vitripennis* Walker to control the host (housefly) populations was much reduced in *freely-interacting* systems, although development of some unknown form of "resistance" (determined by side tests) was clearly evident. This is true, even though Pimentel and Al-Hafidh (1965), (1) obtained a decline of about 50 per cent in the mean parasite density over a period of 1004 days of interaction, (2) the fluctuations became considerably damped, and (3) there was a decline in mean parasitization from about 90 per cent to 83 per cent; for in this experiment host density was *held constant*—only the parasite density was allowed to *freely interact*. The size of the experimental cage (2.5"x3"x4.5") was also inordinately small, considering the few thousands of insects contained.

The experiments of Pimentel and Stone (1968) carried forward the effects of the severe crowding to which the parasite population they used was subjected during over two years of interaction in the experiments of Pimentel and Al-Hafidh. It is true that the parasite could no longer control the host, but the properties concerned (in the host mainly) had been developed under extreme and unnatural selection pressure.

The experiments of Pimentel, Nagel and Madden (1963) present the only example where population size for both host and parasite populations was the result of free interactions. Even though, as with the other experiments, marked changes in "resistance" of the host developed through genetic feed-back, these changes were not sufficient to prevent the parasite from having a real control effect on host population size. For 30-cell systems, the biological control effect was about 90 per cent when host pupae per cell (21) for the parasite-present system is contrasted to host pupae per cell (207) for the parasite-absent system.

Of more direct concern to biological control practice is the conclusion of Pimentel that, through long association, the degree of biological control initially experienced by an effective introduced enemy subsequently diminishes substantially. To our knowledge, no such loss of biological control efficiency has taken place. Many examples of long-standing and effective control by introduced enemies can be cited: the vedalia beetle (and *Cryptochaetum*) for cottony cushion scale in California (80 years), *Cactoblastis* for prickly pear in Australia (40 years), *Rhizobius* for blue gum scale in New Zealand (65 years), *Cyrtorrhinus* for sugarcane leafhopper in Hawaii (50 years), *Patasson* for eucalyptus weevil in South Africa (40 years), *Eretmocerus* for citrus blackfly in Cuba (40 years), *Cryptognatha* for coconut scale in Fiji (40 years), and *Chrysolina* for Klamath weed in California (22 years). (For specific names and further information, see DeBach, 1964.) Episodes of natural enemy disturbances by pesticides after long periods of "co-adjustment" indicate that the phytophagous pests still retain their capabilities to ravage their crop hosts, and results after the restoration of the natural enemies shows that the enemies still retain their capabilities to control their hosts.

Contrary to the view of Pimentel, it is rather generally recognized that enemies become better adapted to control their host, with time, after being introduced into new situations or against new strains or species of an acceptable host (Chapter 3).

Another point related to the same idea is Pimentel's contention that biological control workers would be better advised to deemphasize introductions of natural enemies that have had a long association with the exotic pest species (i.e., in their native home areas) but to emphasize instead the introduction of enemies of related host genera and species to control a particular pest. The idea here is that such natural enemies, not having evolved in association with the target species, would be more effective control agents. In developing this argument, Pimentel (1963a) suggested that insectan parasites introduced to control the Gypsy moth and Japanese beetle in eastern United States failed, and the outbreaks continued, because species that had co-evolved with their hosts had been imported. For this line of reasoning to be admissable in any event, the hosts would have to be outbreaking species in their native regions, which they are not.

Pimentel (1963a) also reviewed 66 instances of successful biological control results from natural enemy introductions, and he considered that 39 per cent of the appropriate instances involved use of enemies of allied species or genera. Some of the examples he cited involve rather general predators that attack various species rather equally, and these provide no support for his general thesis. The cases in general do represent good illustrations of the use of such enemies for control of native pest species. Certainly, these avenues should be explored (Huffaker, 1957; DeBach, 1964). We note, however, that Clausen (1958) stated, "In the history of biological control, instances of successful utilization of parasites from another host genus are exceedingly rare." It seems most unlikely that rather highly host-specific parasitoids would in general be more efficient against other species than against the specific host with which they evolved. The ideal characteristics of an efficient parasitoid species (*vide supra*) would be most nearly met relative to its association with its natural host, and the degree to which this ideal is approached relative to some other related host species would determine whether or not it could also serve as an effective control agent for such other species.

CONCLUSIONS

The employment of indigenous natural enemies is given attention in Sections III and IV (see also DeBach, 1964). The practice of using indigenous species should be greatly expanded through various means of conservation and augmentation of their effectiveness.

As a final conclusion, we suggest that the following guidelines, somewhat abridged from DeBach (in press), are consistent with reasonable theory and have been empirically established on a worldwide basis as sound policy in the practice of making biological control introductions:

1) The discovery and importation of new natural enemies needs much greater emphasis. Many opportunities are being neglected.

2) Native as well as exotic pests are suitable subjects for biological control importations.

3) Direct pests are suitable subjects for biological control, but the probabilities for success may be somewhat lower than with indirect pests.

4) Importation of a diverse complex of natural enemies is the only practical manner of obtaining the best species for a given habitat, or the best combination for such habitat, or the best combination for the entire host range.

5) The competitive displacement principle implies that competition between natural enemies normally is not detrimental to host population regulation; in fact, the displacement of an effective enemy by another means that the second is more effective and will exert a better degree of control.

6) There usually is one *best* enemy for each pest in a given habitat, but a second or third enemy may improve control and may in fact be necessary. The *best* enemy may differ in different habitats, hence there is generally no one best enemy throughout the range. Therefore, all enemies of promise should be imported and tried.

7) From a *practical* standpoint, long-term basic ecological research on a pest need not precede importation of enemies. Such studies are not likely to help a really effective enemy; they add insights and should be conducted after introductions are made.

8) The ultimate success of a given enemy candidate cannot be predicted in advance, but we can select promising ones. An effective natural enemy has the following characteristics: (a) high searching ability, (b) high degree of host specificity or preference, (c) good reproductive capacity relative to the host, and (d) good adaptation to a wide range of environmental conditions. The most essential characteristic is (a), preceding.

9) No geographic area, crop, or pest insect should be prejudged as being unsatisfactory for biological control. The wide variety of successful results now obtained from importation of natural enemies suggests that nearly anything is possible.

LITERATURE CITED

Andrewartha, H. G., and L. C. Birch. 1954. The Distribution and Abundance of Animals. Univ. Chicago Press, Chicago. 782 pp.

Birch, L. C. 1960. Stability and instability in natural populations. New Zealand Sci. Rev. 20: 9-14.

Broodryk, S., and R. L. Doutt. 1966. Studies of two parasites of olive scale, *Parlatoria oleae* (Colvee). II. The biology of *Coccophagoides utilis* Doutt (Hymenoptera: Aphelinidae). Hilgardia 37: 233-254.

Burnett, T. 1954. Influences of natural temperatures and controlled host densities on oviposition of an insect parasite. Physiol. Zool. 27: 239-248.

Burnett, T. 1958. Effect of host distribution on the reproduction of *Encarsia formosa* Gahan (Hymenoptera: Chalcidoidea). Can. Entomol. 90: 179-191.

Chamberlin, T. R. 1941. The wheat jointworm in Oregon, with special reference to its dispersion, injury, and parasitization. U.S. Dept. Agr. Tech. Bull. 784, 48 pp.

Carrick, R. 1963. Ecological significance of territory in the Australian magpie *Gymnorhina tibicens.* Proc. XIII Intern. Ornithological Congr., pp. 740-753.

Chant, D. A. 1966. Integrated control systems. Pp. 193-218. *In* Scientific Aspects of Pest Control. Nat'l Acad. Sci., Nat'l. Res. Council, Publ. 1402.

Clausen, C. P. 1956. Biological control of insect pests in the continental United States. U.S. Dept. Agr. Tech. Bull. 1139, 151 pp.

Clausen, C. P. 1958. Biological control of insect pests. Ann. Rev. Entomol. 3: 291-310.

Clausen, C. P. (Editor and contributor). In press. A World Review of Parasites, Predators and Pathogens Introduced to New Habitats. U.S. Dept. Agr.

DeBach, P. 1964. Successes, trends and future possibilities. Chap. 24. *In* Biological Control of Insect Pests and Weeds, P. DeBach (ed.). Reinhold Publ. Co., N. Y. 844 pp.

DeBach, P. 1964. (Editor and contributor). Biological Control of Insect Pests and Weeds. Reinhold Publ. Co., N. Y. 844 pp.

DeBach, P. 1971. The theoretical basis of importation of natural enemies. Proc. XIII Intern. Congr. Entomol., Moscow (1968) 2: 140-142.

DeBach P., P. J. Landi, and E. B. White. 1962. Parasites are controlling red scale in southern California citrus. Calif. Agr. 16: 2-3.

DeBach, P., and R. A. Sundby. 1963. Competitive displacement between ecological homologues. Hilgardia 34: 105-166.

Decker, G. C. 1962. Pesticide-wildlife relationships. Proc. Congr. Coordinated Program on Wildlife Management and Mosquito Suppression, Yosemite Nat'l. Park, Oct. 15-18, 1962: 1-17. (Mimeo)

Doutt, R. L. 1954. An evaluation of some natural enemies of the olive scale. J. Econ. Entomol. 47: 39-43.

Doutt, R. L. 1964. Biological characteristics of entomophagous adults. Chap 6. In Biological Control of Insect Pests and Weeds, P. DeBach (ed.), Reinhold Publ. Co., N. Y. 844 pp.

Doutt, R. L. 1966. Studies of two parasites of olive scale, Parlatoria oleae (Colvee). I. A taxonomic analysis of parasitic hymenoptera reared from Parlatoria oleae (Colvee). Hilgardia 37: 219-231.

Doutt, R. L., and P. DeBach. 1964. Some biological control concepts and questions. Chap. 5. In Biological Control of Insect Pests and Weeds, P. DeBach (ed.). Reinhold Publ. Co., N. Y. 844 pp.

Embree, D. G. 1966. The role of introduced parasites in the control of the winter moth in Nova Scotia. Can. Entomol. 98: 1159-1168.

Flanders, S. E. 1968. Mechanisms of population homeostasis in Anagasta ecosystems. Hilgardia 39: 367-404.

Force, D. C., and P. S. Messenger, 1964a. Developmental period, generation time and longevity of three parasites of Therioaphis maculata (Buckton) reared at various constant temperatures. Ann. Entomol. Soc. Amer. 57: 405-413.

Force, D. C., and P. S. Messenger. 1964b. Fecundity, reproductivity rates, and innate capacity for increase of three parasites of Therioaphis maculata (Buckton). Ecology 45: 706-715.

Force, D. C., and P. S. Messenger. 1965. Laboratory studies on competition among three parasites of the spotted alfalfa aphid, Therioaphis maculata (Buckton). Ecology 46: 853-859.

Griffiths, K. J. 1969. The importance of coincidence in the functional and numerical responses of two parasites of the European pine sawfly, Neodiprion sertifer. Can. Entomol. 101: 673-713.

Griffiths, K. J., and C. S. Holling. 1969. A competition submodel for parasites and predators. Can. Entomol. 101: 785-818.

Hafez, M., and R. L. Doutt. 1954. Biological evidence of sibling species in Aphytis maculicornis (Masi) (Hymenoptera, Aphelinidae). Can. Entomol. 86: 90-96.

Hagen, K. S. 1950. Fecundity of Chrysopa californica as affected by synthetic foods. J. Econ. Entomol. 43: 101-104.

Hagen, K. S. 1964. Developmental stages of parasites. Chap. 7. In Biological Control of Insect Pests and Weeds, P. DeBach (ed.). Reinhold Publ. Co., N. Y. 844 pp.

Hansberry, R. 1968. Prospects for nonchemical insect control—an industrial view. Bull. Entomol. Soc. Amer. 14: 229-235.

Harris, P., D. Peschken, and J. Milroy. 1969. The status of biological control of the weed Hypericum perforatum in British Columbia. Can. Entomol. 101: 1-15.

Hassell, M. P., and C. B. Huffaker. 1969a. The appraisal of delayed and direct density dependence. Can. Entomol. 101: 353-361.

Hassell, M. P., and C. B. Huffaker. 1969b. Regulatory processes and population cyclicity in laboratory populations of *Anagasta kuhniella* (Zeller) (Lepidoptera: Phycitidae). III. Development of population models. Res. Pop. Ecol. 11: 186-210.

Hassell, M. P., and G. C. Varley. 1969. New inductive population model for insect parasites and its bearing on biological control. Nature 223: 1133-1137.

Haynes, D. L., and P. Sisojevic. 1966. Predatory behavior of *Philodromus rufus* Walckenae (Araneae: Thomisidae). Can. Entomol. 98: 113-133.

Holling, C. S. 1959. The components of predation as revealed by a study of small mammal predation of the European pine sawfly. Can. Entomol. 91: 293-320.

Holling, C. S., and S. Ewing. 1970. Blind-man's buff—Exploring the response space generated by realistic ecological simulation models. Proc. Intern. Symp. Statistc. Ecol., G. P. Patil and W. E. Waters (eds.). Penn. State Univ. Press. Vol. 2, pp. 207-229.

Howard, L. O., and W. F. Fiske. 1911. The importation into the United States of the parasites of the gipsy moth and the brown-tail moth. U.S. Dept. Agr. Bur. Entomol. Bull. 91, 312 pp.

Huffaker, C. B. 1957. Fundamentals of biological control of weeds. Hilgardia 27: 101-157.

Huffaker, C. B. 1958. Experimental studies on predation: (II) Dispersion factors and predator-prey oscillations. Hilgardia 27: 343-383.

Huffaker, C. B. 1964. Fundamentals of biological weed control. Chap. 22. *In* Biological Control of Insect Pests and Weeds, P. DeBach (ed.). Reinhold Publ. Co., N. Y. 844 pp.

Huffaker, C. B. 1966. Competition for food by a phytophagous mite. Hilgardia 37: 533-567.

Huffaker, C. B. 1967. A comparison of the status of biological control of St. Johnswort in California and Australia. Mushi (Suppl.) 39: 51-73.

Huffaker, C. B., and R. L. Doutt. 1965. Establishment of the coccinellid *Chilocorus bipustulatus* L. in California olive groves (Coleoptera: Coccinellidae). Pan-Pac. Entomol. 41: 61-63.

Huffaker, C. B., and C. E. Kennett. 1953. Ecological tests on *Chrysolina gemellata* (Rossi) and *C. hyperici* Forst. in the biological control of Klamath weed. J. Econ. Entomol. 45: 1061-1064.

Huffaker, C. B., and C. E. Kennett. 1966. Studies of two parasites of olive scale, *Parlatoria oleae* (Colvee). IV. Biological control of *Parlatoria oleae* (Colvee) through the compensatory action of two introduced parasites. Hilgardia 37: 283-335.

Huffaker, C. B., and C. E. Kennett. 1969. Some aspects of assessing efficiency of natural enemies. Can. Entomol. 101: 425-447.

Huffaker, C. B., C. E. Kennett, and G. L. Finney. 1962. Biological control of olive scale, *Parlatoria oleae* (Colvee), in California by imported *Aphytis maculicornis* (Masi) (Hymenoptera: Aphelinidae). Hilgardia 32: 541-636.

Huffaker, C. B., C. E. Kennett, B. Matsumoto, and E. G. White. 1968. Some parameters in the role of enemies in the natural control of insect abundance. Pp. 59-75. *In* Insect Abundance, T. R. E. Southwood (ed.). Symposia Roy. Entomol. Soc. London, No. 4.

Huffaker, C. B., and P. S. Messenger. 1964b. The concept and significance of natural control. Chap. 4. *In* Biological Control of Insect Pests and Weeds, P. DeBach (ed.). Reinhold Publ. Co., N. Y. 844 pp.

Huffaker, C. B., K. P. Shea, and S. G. Herman. 1963. Experimental studies on predation: (III) Complex dispersion and levels of food in an acarine predator-prey interaction. Hilgardia 34: 305-330.

Huffaker, C. B., M. van de Vrie, and J. A. McMurtry. 1969. The ecology of tetranychid mites and their natural control. Ann. Rev. Entomol. 14: 125-174.

Janzen, D. H. 1970. The role of herbivores in tropical tree species diversity. Amer. Nat. 104: 501-528.

Kennett, C. E. 1967. Biological control of olive scale, Parlatoria oleae (Colvee) in a dediduous fruit orchard in California. Entomophaga 12: 461-474.

Kennett, C. E., C. B. Huffaker, and G. L. Finney. 1966. Studies of two parasites of olive scale, Parlatoria oleae (Colvee). III. The role of an autoparasitic aphelinid, Coccophagoides utilis Doutt, in the control of Parlatoria oleae (Colvee). Hilgardia 37: 255-282.

MacFadyen, A. 1957. Animal Ecology: Aims and Methods. Pitman and Sons Ltd., London. 255 pp.

Madden, J. L., and D. Pimentel. 1965. Density and spatial relationships between a wasp parasite and its housefly host. Can. Entomol. 97: 1031-1037.

Malthus, T. R. 1803. An Essay on the Principle of Population as it Affects the Future Improvement of Society. J. Johnson, London, 2nd ed. 610 pp.

McLean, L. A. 1967. Pesticides and the environment. BioScience, Sept. 1967, pp. 613-617.

Messenger, P. S. 1964. Use of life tables in a bioclimatic study of an experimental aphid-braconid wasp host-parasite system. Ecology 45: 119-131.

Messenger, P. S. 1969. Bioclimatic studies of the aphid parasite Praon exsoletum. 2. Thermal limits to development and effects of temperature on rate of development and occurrence of diapause. Ann. Entomol. Soc. Amer. 62: 1026-1031.

Messenger, P. S., and D. C. Force. 1963. An experimental host-parasite system: Therioaphis maculata (Buckton) - Praon palitans Musebeck (Homoptera: Aphididae-Hymenoptera: Braconidae). Ecology 44: 532-540.

Milne, A. 1962. On the theory of natural control of insect populations. J. Theoret. Biol. 3: 19-50.

Morris, R. F. 1959. Single-factor analysis in population dynamics. Ecology 40: 580-588.

Morris, R. F. 1963. Predictive population equations based on key factors. Pp. 16-21. In Population Dynamics of Agricultural and Forest Insect Pests. Mem. Entomol. Soc. Canada, No. 32.

Muldrew, J. A. 1953. The natural immunity of the larch sawfly (Pristiphora erichsonii (Htg.) to the introduced parasite Mesoleius tenthredinis Morley in Manitoba and Saskatchewan. Can. J. Zool. 31: 313-332.

Nagel, W. P., and D. Pimentel. 1963. Some ecological attributes of a pteromalid parasite and its housefly host. Can. Entomol. 95: 208-213.

Nicholson, A. J. 1933. The balance of animal populations. J. Anim. Ecol., Suppl. to Vol. 2: 132-178.

Nicholson, A. J. 1954. An outline of the dynamics of animal populations. Austr. J. Zool. 2: 9-65.

Nicholson, A. J., and V. A. Bailey. 1935. The balance of animal populations. Proc. Zool. Soc. London, Part 1. Pp. 551-598.

Odum, E. P. 1959. Fundamentals of Ecology. 2nd ed., W. B. Saunders Co., Philadelphia. 545 pp.

Pimentel, D. 1961a. Animal population regulation by the genetic feed-back mechanism. Amer. Nat. 95: 65-79.

Pimentel, D. 1961*b*. Species diversity and insect population outbreaks. Ann. Entomol. Soc. Amer. 54: 76-86.

Pimentel, D. 1963*a*. Introducing parasites and predators to control native pests. Can. Entomol. 95: 785-792.

Pimentel, D. 1963*b*. Natural population regulation and interspecies evolution. Proc. 16th Intern. Congr. Zool. (Washington, D.C.) 3: 329-336.

Pimentel, D. 1964. Population ecology and the genetic feed-back mechanism. *In* Genetics Today. Proc. XI Intern. Congr. Genetics, The Hague, Netherlands, Sept., 1963. Pergamon Press, London.

Pimentel, D. 1968. Population regulation and genetic feed-back. Science 159: 1432-1437.

Pimentel, D., and R. Al-Hafidh. 1965. Ecological control of a parasite population by genetic evolution in the parasite-host system. Ann. Entomol. Soc. Amer. 58: 1-6.

Pimentel, D., W. P. Nagel, and J. L. Madden. 1963. Space-time structure of the environment and the survival of parasite-host systems. Amer. Nat. 97: 141-167.

Pimentel, D., and F. A. Stone. 1968. Evolution and population ecology of parasite-host systems. Can. Entomol. 100: 655-662.

Putnam, W. L., and D. C. Herne. 1966. The role of predators and other biotic factors in regulating the population density of phytophagous mites in Ontario peach orchards. Can. Entomol. 98: 808-820.

Sandness, J. N., and J. A. McMurtry. 1970. Functional response of three species of Phytoseiids (Acarina) to increased prey density. Can. Entomol. 102: 692-704.

Simmonds, F. J. 1959. Biological control—past, present and future. J. Econ. Entomol. 52: 1099-1102.

Simmonds, F. J. 1967. The economics of biological control. J. Roy. Soc. Arts, Oct., 1967, pp. 880-898.

Smith, H. S. 1929. Multiple parasitism: its relation to the biological control of insect pests. Bull. Entomol. Res. 20: 141-149.

Smith, H. S. 1935. The role of biotic factors in the determination of population densities. J. Econ. Entomol. 28: 873-898.

Solomon, M. E. 1957. Dynamics of insect populations. Ann. Rev. Entomol. 2: 121-142.

Southwood, T. R. E. 1967. The interpretation of population change. J. Anim. Ecol. 36: 519-529.

Southwood, T. R. E., and M. J. Way. 1970. Ecological background to pest management. Pp. 6-29. *In* Concepts of Pest Management, R. L. Rabb and F. E. Guthrie (eds.). North Carolina State Univ. Press.

Steinhaus, E. A. 1954. The effects of disease on insect populations. Hilgardia 23: 197-261.

Tanada, Y. 1964. Epizootiology of insect diseases. Chap. 19. *In* Biological Control of Insect Pests and Weeds, P. DeBach (ed.). Reinhold Publ. Co., N. Y. 844 pp.

Taylor, T. H. C. 1955. Biological control of insect pests. Ann. Appl. Biol. 42: 190-196.

Thompson, W. R. 1924. La theorie mathematique de l'action des parasites entomophages et le facteur du hasard. Ann. Fac. Sci. Marseille, Ser. 2, 2:69-89.

Thompson, W. R. 1939. Biological control and the theories of the interactions of populations. Parasitology 31: 299-388.

Thompson, W. R. 1956. The fundamental theory of natural and biological control. Ann. Rev. Entomol. 1: 379-402.

Tinbergen, L., and H. Klomp. 1960. The natural control of insects in pine woods. II. Conditions for damping of Nicholson oscillations in parasite-host systems. Archs. neel. Zool. 13: 344-379.

Turnbull, A. L. 1967. Population dynamics of exotic insects. Bull. Entomol. Soc. Amer. 13: 333-337.

Turnbull, A. L. 1969. The ecological role of pest populations. *In* Proc. Tall Timbers Conf. Ecol. Anim. Control by Habitat Mgmt. No. 1, Tallahassee Fla., (1969): 219-232.

Turnbull, A. L., and D. A. Chant. 1961. The practice and theory of biological control of insects in Canada. Can. J. Zool. 39: 697-753.

Ullyett, G. C., and D. B. Schonkon. 1940. A fungus disease of *Plutella maculipennis* (Curtis) with notes on the use of entomogenous fungi in insect control. Union S. Africa, Dept. Agr. Sci. Bull. 218, 24 pp.

Uvarov, B. P. 1931. Insects and climate. Trans. Entomol. Soc., London 79: 1-247.

van den Bosch, R. 1964. Encapsulation of the eggs of *Bathyplectes curculionis* (Thomson) (Hymenoptera: Ichneumonidae) in larvae of *Hypera brunneipennis* (Boheman) and *Hypera postica* (Gyllenhal) (Coleoptera: Curculionidae). J. Insect Pathol. 6: 343-367.

van den Bosch, R. 1968. Comments on population dynamics of exotic insects. Bull. Entomol. Soc. Amer. 14: 112-115.

van den Bosch, R., E. I. Schlinger, E. J. Dietrick, K. S. Hagen, and J. K. Holloway. 1959. The colonization and establishment of imported parasites of the spotted alfalfa aphid in California. J. Econ. Entomol. 52: 136-141.

van den Bosch, R., E. I. Schlinger, E. J. Dietrick, J. S. Hall, and B. Puttler. 1964. Studies on succession, distribution, and phenology of imported parasites of *Therioaphis trifolii* (Monell) in southern California. Ecology 45: 602-621.

Varley, G. C. 1959. The biological control of agricultural pests. J. Roy. Soc. Arts, London, 107: 475-490.

Varley, G. C., and G. R. Gradwell. 1958. Balance in insect populations. Proc. X Intern. Congr. Entomol., Montreal (1956) 2: 619-624.

Varley, G. C., and G. R. Gradwell. 1963. The interpretation of insect population changes. Proc. Ceylon Assoc. Adv. Sci. 18: 142-156.

Varley, G. C., and G. R. Gradwell. 1970. Recent advances in insect population dynamics. Ann. Rev. Entomol. 15: 1-24.

Watt, K. E. F. 1965. Community stability and the strategy of biological control. Can. Entomol. 97: 887-895.

Wangersky, P. J., and W. J. Cunningham. 1957. Time lag in prey-predator population models. Ecology 38: 136-139.

Webster, L. T. 1946. Experimental epidemiology. Medicine 25: 77-109.

White, E. G., and C. B. Huffaker. 1969a. Regulatory processes and population cyclicity in laboratory populations of *Anagasta kuhniella* (Zeller) (Lepidoptera: Phycitidae). I. Competition for food and predation. Res. Pop. Ecol. 11: 57-83.

White, E. G., and C. B. Huffaker. 1969b. Regulatory processes and population cyclicity in laboratory populations of *Anagasta kuhniella* (Zeller) (Lepidoptera: Phycitidae). II. Parasitism, predation, competition, and protective cover. Res. Pop. Ecol. 11: 150-185.

Woodworth, C. W. 1908. The theory of the parasitic control of insect pests. Science, N.S. 28: 227-230.

Wynne-Edwards, V. C. 1962. Animal Dispersion in Relation to Social Behaviour. Oliver and Boyd, Edinburgh and London. 653 pp.

Zwolfer, H. 1963. The structure of the parasite complexex of some lepidoptera. Zeit. angew. Entomol. 51: 346-357.

Chapter 3

THE ADAPTABILITY OF INTRODUCED BIOLOGICAL CONTROL AGENTS

P. S. Messenger and R. van den Bosch

Division of Biological Control
University of California, Berkeley

INTRODUCTION

The ecological circumstances necessary for successful establishment of an entomophagous insect in a new environment are complex. The climate must be suitable in its direct effects on development, survival, and reproduction of the new natural enemy; it must also be suitable in its indirect effects on life cycle synchrony between enemy and host[1] (Flanders, 1940; Elton, 1958).

In the new environment, the enemy must be attracted to or be able to find the local habitat of the host or prey, and, once this is done, the host or prey within the habitat. The host must prove acceptable to the parasite, and when attacked must be suitable for the enemy's use (development or consumption). With most parasitic Hymenoptera, environmental circumstances must favor oviposition of fertilized eggs, so that an adequate number of female progeny will be produced in the following generations. All requisites of the adult enemy must be present in adequate supply, including food, water, resting or congregating places, and mating habitats, and its densities must be high enough in each generation to insure meeting of the sexes for mating. When such conditions are met, the introduced species must on occasion contend with competitors; if competitors are present, the new species must be superior in some parts of the target area, or face exclusion (DeBach, 1966; DeBach and Sundby, 1963; Flanders, 1966).

---

[1]References to either "parasite" or "predator," or to "host" or "prey," in general apply to the other also, unless the distinction is obvious.

68

The above requirements, and more, merely favor establishment of an introduced species; they do not insure biological control. While the degree of control obtained involves the nature of the new environment, success, in addition, depends also on the intrinsic properties of the natural enemy and the pest itself. So the circumstances required for successful biological control are much more stringent than those needed for establishment. Successful control, then, is dependent upon the needs and limitations of the natural enemy, which we can call its adaptability, the properties of the host, and the properties of the environment. In this Chapter we are concerned particularly with natural enemy adaptability and the problems this presents in biological control programs.

PROCEDURES RESTRICTING THE GENETIC VARIABILITY OF COLONIZING STOCKS

Species of entomophagous insects, like other animals, exist in nature as groups of populations, with each population differing to a greater or lesser extent from the others. Many of these population differences are the result of local variations in the environment acting upon slightly different allocations of gene alleles in the individuals making up these populations. These population differences, because they are genetically inherited, tend to persist in time.

Existence of such observably different populations may be recognized by the proliferation of terms used to characterize them in the literature. Hence we meet such terms as polymorph, ecotype, biotype, strain, and biological race. In many cases, where evidence has been sought, these populations have been found to be truly conspecific, that is, individuals from respective populations can interbreed to produce viable, interbreeding offspring. However, in some cases these apparent "strains," although morphologically indistinguishable, do not hybridize, in which case they are called "sibling species." Then there are what can be called cryptic species, or species which at first appear to be siblings because of morphological similarity, but which on closer examination are found to have minor but consistent structural differences. In biological control programs, all such population or species entities have been encountered; in some cases their discovery has been ignored or neglected as potential contributors to the solution of pest control problems.

Most significantly, the adaptability of an entomophagous species introduced into a new environment is often determined or limited to a large degree by the qualities of the source population from which the culture was derived and by the way this stock is handled upon receipt in the target area. As background, it is useful to review briefly the steps followed in the importation and colonization of an entomophagous species in biological control programs. The customary procedure for implementing a typical program is to seek in a foreign area an enemy species not yet present in the target area, extract from one of its populations what almost invariably turns out to be a minute sample of the species, ship it to the target area and there receive it in quarantine, identify it, separate it from all other entomophagous and phytophagous species, culture it, transfer it to research laboratories for study and to an insectary for mass production, and eventually try to colonize it in a variety of ways in the field.

We customarily attempt to obtain a wide variety of well-adapted stocks of each natural enemy for use in the target area in order to increase the chances for establishment and effective biological control. However, this goal is limited in a number of ways. The first, and one of the most important steps is the initial search for and sampling of the populations of the species in the source region, for it is this sampling process itself that sets the first limits on the ability of the species to become established in, and adapted to the target environment. It has always been considered proper and desirable to ship as many individuals from as many localities as possible to serve as material for quarantine handling and evaluation, study of life histories, development of culturing techniques, and eventual colonization. But practical problems connected with foreign exploration, coupled with the frequent rarity and cryptic nature of the species, often results in a sample that is quite restricted both in numbers of individuals and in geographical distribution of the source material. In some cases, the entire importation consists of a single, small sample containing only a few individuals; at times, it has comprised only a single female. When one considers that the entomophagous species in question may well exist in large populations distributed over wide geographic areas, sometimes over entire continents (e.g., an estimated two million square miles for the parasite, *Bathyplectes curculionis* (Thomson) , in which climate and other environmental features vary widely, this narrow, man-made selection constitutes an extremely fine genetic sieve.

The second step in this process is the quarantine reception, evaluation, study, and culture of the imported stock. Since it is physically impossible to scrutinize carefully every individual contained in large lots of a natural enemy, and then to maintain such large numbers while a suitable culture technique is devised, quarantine handling tends to restrict the sample size and therefore the potential adaptability of the introduction. Eventually, when the culture technique is perfected, again a relatively small, or certainly narrowed, still more homogeneous sample of the species is transferred to the mass culture laboratory for the necessary numerical build-up of colonization stocks. Hence, in the quarantine procedure there is a good bit of additional (though not deliberate) selection of the original inoculum.

Similar criticism can be leveled at the insectary mass-culture procedures. While numbers are not restricted here, the cultural conditions are usually closely defined and maintained for successive generations. In such a situation there is likely to occur additional loss of genes, due either to unknown selective factors operating under the conditions of culturing, or simply to chance loss of genes (because of the relatively small size of the population). Thus the insectary culture again serves as a narrowing, selective bottleneck for the imported stock. And when, as frequently happens, the stock culture is put on standby routine for maintenance purposes through the off-season, the reduction in numbers of individuals serves to limit the variability of the stocks still further.

And finally, at the third step, the colonization procedure itself serves to limit potential adaptability. In order to give the species an opportunity to establish at a number of places differing ecologically, and for other reasons, we are usually forced to release, at most, only a few thousand individuals at various point localities in the target areas. This, in relation to the many millions of individuals that constitute a natural population, again results in an unintentional selection sieve for the natural enemy as it is finally inserted into each local habitat of a perhaps widely varied new environment.

So, it is possible that the potential adaptability of the entomophagous species becomes much reduced during the process of introduction. The record shows that a substantial fraction (80 per cent in continental United States—Clausen, 1956; 90 per cent worldwide—Turnbull, 1967) of the natural enemies introduced into new environments for biological control fail to establish, and also that of those that do establish, only a portion provide adequate control of the respective host pests (Wilson, 1960; Turnbull, 1967; van den Bosch, 1968). This raises the question of whether the narrow adaptive potential of the typically restricted inoculum (*vide supra*) has contributed to this unsatisfactory establishment record (Clausen, 1936; Smith, 1941; Allen, 1958; Simmonds, 1963). Perhaps, as suggested by Flanders (1959), some of the programs which resulted in outright failure or only partial success should be reattempted in the light of what we now know about natural enemy adaptability. Additionally, in order to take full advantage of the total adaptability of an entomophagous species, perhaps we should revise our technical procedures for carrying out our classical biological control programs.

To understand better the nature of the adaptiveness of biological control agents, and the importance of these limitations in imported stocks, it is useful to analyze in some detail various examples that bear on the matter. First, we shall consider the various adaptive limitations discernable among different entomophagous species attacking the same host. This will give insight into what it is that limits the establishment, spread, and effectiveness of these organisms. Then we shall look at the limiting factors encountered by a given species in different environments. Finally, we shall deal with the different adaptive qualities displayed by races, strains, and forms of a given natural enemy species in the same environment.

DIFFERENT NATURAL ENEMIES CONTEMPORANEOUSLY ATTACKING A HOST IN THE SAME REGION

It is widely assumed that a natural enemy is usually restricted to a narrower range, either geographically or ecologically, than its host. This, of course, is based on the presumption that the natural enemy is host-specific. Since the more effective natural enemies have been found by experience usually to be rather highly host-specific (Chapter 2), the nature and degree of this narrowed range is well worth considering. This may be due, in part, to the fact that at the periphery of the distribution of the host, the host itself becomes an undependable resource for the natural enemy. But more likely, such limitations are the result of narrower climatic tolerances, asynchronies in the reproductive life cycles of the host and enemy, different relative capabilities for going dormant, or the presence of interfering competitors. Often, the geographic range of a natural enemy may be substantially less than that of the host. It may then be found that the given host species will be attacked by a complex of enemy species, sometimes of the same genus, with the range of such a host variously divided among them, each adapted to the conditions of a part of the range, but not all of it.

In the case of the spotted alfalfa aphid, *Therioaphis trifolii* (Monell), in California, the imported parasite *Praon exsoletum* (Nees) is able to persist in effective numbers in the colder, more northerly Great Basin climatic zone where summers are short and not too hot and where a hibernal diapause condition protects it from intense and prolonged winter cold. On the contrary, this parasite is intolerant of the hot, dry, prolonged, summer conditions of the interior desert areas of southern California, and is incapable of surviving there. However, another imported parasite, *Trioxys complanatus* Quilis Perez, thrives in these latter areas because an estival diapause protects it from the intense heat of such summers. On the other hand, it does not occur in effective numbers in the northerly areas of the State, presumably because of the intense, prolonged cold winters. A third parasite, *Aphelinus asychis* Walker, unable to diapause either in summer or winter, is only able to persist in good numbers in coastal habitats where neither summers nor winters are very severe. In fact, because of this lack of diapause, *A. asychis* remains active in the mild winters of coastal California, unhindered by its diapausing but otherwise superior competitors, *P. exsoletum* and *T. complanatus.*

Hence, we can see that these three parasites of the same host species, living in the same region, are differentially distributed geographically, mainly because of differential responses to climate. It is worth while noting an additional rather obvious fact from this illustration. While each of the parasites of the spotted alfalfa aphid is more narrowly adapted to California habitats than the host, the three parasites together give much greater geographic coverage of the aphid than does any one species alone. This is a case where, because of the different adaptive powers of the parasites, multiple introductions (the colonization of more than one natural enemy species against a pest in a given target area) have clearly resulted in better regional control than would have occurred from only a single species introduction (see further Chapter 2, and DeBach, 1964).

There are numerous other examples of differences in adaptiveness of different parasite species among a complex introduced to control a host pest in the same region. An excellent recent illustration concerns two parasites of the olive scale, *Parlatoria oleae* (Colvee), in California (Huffaker *et al.,* 1966). Here, a first parasite, *Aphytis maculicornis* (Masi), introduced in the early 1950's, was found to be very effective during the winter and spring, but rather drastically inhibited by summer heat and dryness. In the late 1950's, a second parasite, *Coccophagoides utilis* Doutt, was discovered, which turned out to be just what was needed to fill in the existing gap. This parasite thrives in the fierce summer heat and aridity of the Central Valley of California because it develops slowly as an internal parasite and does not emerge as an adult during the period of severity. It attacks the spring generation of olive scales that are only little affected by *A. maculicornis* which can only attack them when they are nearly mature, i.e., in the summer. The comparative adaptive powers of these two parasites are strikingly different, yet the two together complement each other in terms of total biological control of the scale. This example particularly shows how parasites, because of adaptive limitations, can be restricted not spatially but seasonally, relative to the host species. (See further Chapters 3 and 7, and Huffaker and Kennett, 1966.)

The series of parasites encountered in the biological control program against the California red scale, *Aonidiella aurantii* (Maskell), in southern California, described most recently by DeBach (1965*b*, 1969) (and see Chapter 7), provide additional insights of the adaptive capabilities of natural enemies. While several different genera of scale parasites are involved, it seems more meaningful from an ecological standpoint to consider these natural enemies in the order of their establishment in California rather than from a taxonomic standpoint. Several different aspects of parasite adaptiveness are involved, including climatic tolerance, competitiveness, host stage preference, and host control effectiveness. California red scale, for decades the key insect pest of citrus in California, has long been the target of a substantial biological control program.

For many years after its accidental introduction into California around 1900, the aphelinid parasite, *Aphytis chrysomphali* (Mercet), proved to be the main enemy associated with California red scale. It was found to be generally well distributed throughout the citrus growing areas of southern California, provided excellent biological control in certain limited, favorable areas, though relatively ineffective in the majority of places. After 1947, competition by other introduced parasites (*vide infra*) restricted *A. chrysomphali* to only a few, small, coastal sites. From this, it can be concluded that *A. chrysomphali* is adequately adapted to climate over a broad range, but is a relatively poor control agent (for unknown reasons), and is a poor competitor in respect to those species introduced later (*vide infra*).

The first parasite to be intentionally introduced into southern California to control the California red scale was the solitary, endoparasitic encyrtid, *Habrolepis rouxi* Compere, imported from South Africa in 1937. It was colonized in all citrus areas, but became established only on the coast near San Diego. Attacking only late stages of the host, it is an ineffective control agent of red scale in the area where it persists. Hence, *H. rouxi* is an example of a parasite which is poorly adapted in almost all respects to the citrus environments of California.

The California red scale strain of the encyrtid, *Comperiella bifasciata* Howard, another solitary endoparasite attacking this scale, was imported from South China in 1941, and became well established in the interior and intermediate citrus zones of southern California. It does not occur in the milder coastal areas, probably because of climatic limitations (DeBach, 1969). It is not an effective control agent. It apparently is not displaced by competing parasites since laboratory studies show it can co-exist with the *Aphytis* species. It is competitively superior to *Prospaltella perniciosi* Tower (*vide infra*). Hence, *C. bifasciata* can be rated as well adapted to the more severe climates of the interior and intermediate zones of the region, competitively self-sufficient, but not well adapted for numerical increase and economic control of its host.

After colonization in 1947, the aphelinid, *Aphytis lingnanensis* Compere, imported from South China, became quite abundant throughout the citrus region of southern California. The results of this release were striking. Relative to its predecessor congener, *A. chrysomphali, (vide supra)* it is an excellent competitor, having displaced *chrysomphali* from all the interior and intermediate zones where the latter had maintained itself for some 50 years (DeBach and Sundby, 1963). *A. lingnanensis* proved to be a moderately good control agent. Its distribution during the years 1947 to 1956 show it to be widely adapted climatically. Since 1956, with the establishment of another closely related species, *Aphytis melinus* DeBach, *A. lingnanensis* has in turn been displaced such that it now appears mainly along the coast where in some localities it is still dominant and a moderately effective control agent.

The California red scale strain (also called the Oriental strain) of the aphelinid, *Prospaltella perniciosi* Tower, imported into southern California from Formosa in 1949, appeared at first to be established at scattered sites throughout the citrus areas, but now is found in effective numbers at only a few coastal sites. Circumstantial evidence suggests its retreat may have involved competitive pressure from other parasites; it is competitively inferior to *Comperiella bifasciata.* In any event, *P. perniciosi* in southern California can also be rated as but poorly adapted climatically, poor competitively, and poor in control effectiveness.

Aphytis melinus, closely related to *A. lingnanensis,* and imported in 1956-1957 from Pakistan, became well distributed in the interior and intermediate citrus zones, where it provides from fair to good biological control of the host. It flourishes as the dominant parasite in these zones, having displaced *A. lingnanensis* and probably also *P. perniciosi.* It is less successful along the coast, where *A. lingnanensis* remains dominant. So *A. melinus* can be rated as well adapted climatically, perhaps the best competitor over most of the citrus region, and certainly the best biological control agent of red scale in southern California.

And, finally, to point up the cryptic nature of parasite adaptability, there is the case of *Aphytis fisheri* DeBach, imported from Burma in 1957. DeBach (1965*a*) classified this parasite and *A. melinus* as sibling species, since they are distinguishable only biologically (i.e., they are reproductively isolated), but not morphologically (except by pupal coloration). Yet in spite of an energetic colonization program throughout the region, and before *A. melinus* became entrenched, *A. fisheri* failed to establish. Debach (1965*a*) concluded from laboratory studies that *A. fisheri* is probably not particularly limited by climate, but rather because it is competitively inferior to *A. melinus* and to *A. lingnanensis,* it failed to establish because these latter two species were introduced earlier.

VARYING PERFORMANCES OF AN ENEMY IN DIFFERENT ECOLOGICAL SITUATIONS

From the foregoing, it is apparent that given natural enemies may flourish in some situations and do poorly in others. There are numerous cases at hand showing limited abilities of natural enemies to attack respective hosts in one region contrasted to their more effective capabilities against the same hosts in some other region(s). These examples include comparisons of parasite performance, either when colonized in two different countries, with varying results, or when colonized in one region relative to performance in the source area.

An excellent example of climatic limitation is that of the citrus black fly parasite, *Eretmocerus serius* Silvestri. Introduced from Malaya into Cuba, Trinidad, Panama, and Costa Rica, it became well established and soon provided very effective control of the pest (Clausen, 1956). This shows the parasite to be climatically well adapted to such sub-tropical and tropical environments, as well as being highly adapted for control effectiveness in such regions. However, when introduced into western Mexico, establishment occurred, but only in scattered areas, and little control was effected (Clausen, 1958). Winters in Mexico proved too severe; summers in many regions were too arid. In Mexico, the parasite clearly is not as well adapted climatically as its host, nor as widely adapted, generally.

A similar example concerns *Bathyplectes curculionis,* a braconid parasite of the alfalfa weevil, *Hypera postica* (Gyllenhal). In California, the parasite occurs wherever its host does, but it is a much more effective control agent in the mild coastal valleys of central California, while less effective in intermediate valleys, and of little effect in the climatically more severe Great Central Valley (Michelbacher, 1943). Hence, the climatic adaptability of *B. curculionis* enables it to persist through the range of climates of this region, but not to have the same biological control effectiveness throughout.

Sometimes the effect of climate is indirect, acting not on the natural enemy itself, but rather on the host (or on some essential alternate host) species. Such is the case with *Metaphycus helvolus* (Compere), which is effective against black scale, *Saissetia oleae* (Olivier), on citrus in coastal and intermediate zones of southern California, but ineffective in the interior zones of southern California and in the Central Valley. Where *M. helvolus* is effective the mild climate allows the host to develop more or less continuously, so that all stages of the life cycle are present at any point in time. In the climatically more severe, interior or northerly areas the life cycle of the host is strongly synchronized so that only one stage occurs at a given time. In the former situation, suitable host stages are present almost continuously, in the latter, particularly in the summer, there is a gap in availability of suitable stages of the host and the parasite is diminished in abundance drastically because of this. So the parasite is able to maintain itself in these various areas, but in the interior it is ineffective because of the climatically induced chronologically discontinuous host population (Clausen, 1956).

Sometimes the adaptive limitations manifest themselves as aspects of host preference. This was found to be the case with the tachinid parasite, *Paratheresia claripalpis* (van der Wulp) on *Diatraea saccharalis* (Fabricius) and *Zeadiatraea lineolata* (Walker) (Simmonds, 1963). In Venezuela, in maize, the parasite commonly attacks both hosts, whereas in Trinidad it mainly attacks *D. saccharalis*, and only rarely *Z. lineolata*. In this particular case it is not known whether this difference in host preference is due to racial differences in the parasite or in the host.

A similar phenomenon concerns the parasite *Palpozenillia palpalis* (Aldrich), which seems to occur in several host-specific strains (Simmonds, 1963). In Mexico it is adapted to *Zeadiatraea muellerella* (Dyar and Heinrich); in Venezuela, it attacks *Eodiatraea centrella* (Moschler); in British Guiana, it prefers *Castnia licoides* Bois. In the laboratory, it can be reared easily on *Z. lineolata*. However, when colonized in several Caribbean Islands against *E. centrella* and *C. licoides*, it failed to establish. Simmonds concluded that the strain of *P. palpalis* used was not adapted to the hosts in question.

One of the more promising parasites introduced against oriental fruit moth, *Grapholitha molesta* (Busck), in eastern United States in the 1930's was the braconid *Agathis diversus* (Muesebeck). Important here is the observation that this species apparently underwent a rather marked adaptive shift in habits after colonization. In Japan it attacks host larvae in peach twigs in spring and summer; in eastern United States it attacks only late-season, fruit-infesting host larvae. It is possible that during spring and summer, contrary to what is observed in Japan, *A. diversus* in eastern United States is associated with an alternate host. In any event, environmental differences (perhaps biotic ones involving alternate hosts) between Japan and the United States caused *A. diversus* to shift its phenological association with its normal host on peach.

Poor synchrony (temporal or spatial) between parasite and host are known to affect parasite success. One case involves a tachinid parasite of Mexican bean beetle, *Epilachna varivestis* Mulsant. In Mexico the parasite, *Paradexodes epilachnae* Aldrich, is well adapted to its host, and maintains a good biological control (Clausen, 1956). However, upon introduction into the United States, it was found to be incapable of hibernating through the winter as the host is able to do. The parasite was unable to find suitable hosts available to it during this season, and thus did not become established.

Another case of asynchronous host-parasite association, described by Clausen (1956), concerns the tachinid parasite, *Hyperecteina aldrichi* Mesnil, which attacks the adult Japanese beetle, *Popillia japonica* Newman. In northern Japan, this is the principal parasite holding the beetle in check. In the United States, however, though established, it provides only sporadic, ineffective control. This is because after hibernation the parasite emerges earlier in the spring than the beetles do, so that most of the fly adults are dead by the time of peak beetle emergence. Parasitization rates are thus very low. The reason for the poorer synchronization in America seems to be related to differences in snow cover in the two environments. Here again, we have a parasite that on first analysis appears to be adapted climatically to the area occupied by the host, but whose life cycle is shifted slightly, relatively to that of the host, to the point of its becoming almost completely ineffective as a control agent (Clausen, 1956).

The nature of the plant on which the host insect lives may affect performance of a natural enemy; i.e., this may pose another challenge to suitable adaptiveness. The California red scale strain of *Comperiella bifasciata* attacks and does well on California red scale occurring on citrus, but apparently is poorly adapted to, and encounters heavy mortality in, this species on sago palm (Clausen, 1956).

DIFFERENT ADAPTABILITIES EXHIBITED AMONG STRAINS OF A NATURAL ENEMY

A number of parasites are known to exist as strains, undifferentiated as to morphology, which exhibit differences in their adaptability to given pest control situations. The evidence that such adaptive differences occur presents important possibilities for turning many failures of the past into successes, for improving upon partially successful programs, and, of course, for the conduct of new programs. The important potential disclosed by the discovery and use of biological races and strains of entomophagous insects, or of sibling species masquerading as strains, has been recognized for many years (Clausen, 1936; Smith, 1941; Flanders, 1950; DeBach, 1953, 1958; Simmonds, 1963; DeBach and Hagen, 1964; Wilson, 1965). Unfortunately, remarkably little has been done about it; this potential remains largely untapped.

The evidence available which demonstrates the occurrence of adaptive differences exhibited by such strains involves aspects of host selection, host immunity, or climatic tolerance.

Host-Specific Strains

Examples of host-specific strains include strains of *Aspidiotiphagus citrinus* (Craw), one of which attacks either oleander scale, *Aspidiotus hederae* (Vallot), or yellow scale, *Aonidiella citrina* (Coquillett), while another only

California red scale; two strains of *Metaphycus luteolus* (Timberlake), one attacking *Saissetia oleae,* the other, *Coccus hesperidum* Linnaeus; and two strains of *Comperiella bifasciata* attacking, respectively, California red scale (*vide supra*) and yellow scale, *Aonidiella citrina* (Coquillett). Regarding this last example, DeBach (1969) indicated that the two strains of *C. bifasciata* hybridize readily, producing normal progeny, though there is evidence that the hybrid may be less fit than either parent. The tachinid, *Trichopoda pennipes* Fabricius, exists in three different strains in North America, an eastern one attacking squash bug, *Anasa tristis* (DeGeer), a southeastern one attacking pentatomids, including the southern green stink bug, *Nezara viridula* (Linnaeus), and a western one attacking the bordered plant bug, *Euryophthalmus cinctus californicus* Van Duzee. In general, such strains either fail to attack or do poorly on hosts other than those normally attacked.

In this last example, and what may well apply to others, crossbreeding studies suggest that perhaps these "strains" are in fact sibling species (see, for example, Hafez and Doutt, 1954, concerning sibling species of the olive scale parasite, *Aphytis maculicornis*). This means that the adaptive variations displayed by their preferences for altogether different host species may actually turn out to be merely the rather more normal adaptations expected of *species* rather than *strains* or *races.* Nevertheless, the value of recognizing that such strains or sibling species exist is clear, as exemplified by *Aphytis mytilaspidis* (Le Baron). This parasite attacks several diaspine scales, including the fig scale, *Lepidosaphes ficus* (Signoret). A strain of this parasite became introduced through some natural channel into California many years ago, but, though associated with fig scale, never produced more than a small amount of parasitism. Another strain of *A. mytilaspidis,* attacking fig scale in Italy, was imported into California in the early 1950's, soon giving much higher percentage parasitism of the scales (Doutt, 1954; Huffaker, 1956). The reasons for this, particularly from the standpoint of enemy adaptability, do not seem to be known, but the significance is clear.

Differential Host Immunity to Strains
of Natural Enemies

Occasionally it is found that a given natural enemy species attacks two (or more) closely related host species. Closer examination, in certain instances, shows that the enemy population associated with one host is different from the enemy population successfully attacking the other. This difference lies mainly in the fact that the host is immune to the one enemy population, or strain, and not to the second.

The immune, or defense, response of host insects to parasitoids has been exhaustively reviewed by Salt (1963), as have also the mechanisms which parasitoids have developed to overcome such defense responses (Salt, 1968). For our purposes, it is the defense response of hosts termed *haemocytic encapsulation* which is important here. Some strains of a parasitic species are able to attack a given host species without inducing an immune response, whereas others, which we would call non-host-adapted strains, do invoke such responses in this host. In one type of case, the response is invariable, and the host is protected from mortal parasitization. Such a parasite strain is non-adapted, and is of no use in biological control of the immune host species. It has not evolved to take in that species as a host or else a former host has become a non-host (but see following). However, there is another type or case where the immune response changes in time, i.e., it does not occur in every host attacked, so that, gradually, through natural selection, a strain of the parasite develops which is adapted to and can successfully attack this host species.

The numbers of such cases are few. A most significant case of this sort concerns the larch sawfly, *Pristiphora erichsonii* (Hartig), and its parasite *Mesoleius tenthredinis* Morley in central Canada (Muldrew, 1953; Turnbull and Chant, 1961). Imported from England into Manitoba during 1910 to 1913, *M. tenthredinis* soon became established and effective, so that until 1938 levels of parasitization reached 75 to 88 per cent. From then on, however, larch sawfly outbreaks became more prevalent, and it was discovered that host larvae were encapsulating and thereby killing parasite eggs. Parasitization rates declined to less than 5 per cent, and in many areas of Manitoba the parasite now cannot be found. The former host has become a non-host.

However, the parasite still remains effective in British Columbia and in the Maritime Provinces (Turnbull and Chant, 1961). Salt (1963) interpreted this as an example of strains of hosts differently adapted for parasitic defense. Recently, Kelleher (1969) indicated that there are, in fact, strains of the parasite differently adapted to the sawfly, a strain from Bavaria (Germany) being able to attack successfully the host form in Manitoba.

Another case of varying power of encapsulating among different host strains is that of *Drosophila melanogaster* Meigen in relation to attack by the endoparasitic cynipid, *Pseudeucoila bochei* Weld (Walker, 1959, cited in Salt, 1963). In some strains of this dipterous host, the parasite eggs and larvae develop without any detectable reaction by the host. In other strains, when the stage attacked was the second or third larval instar, an encapsulation response often resulted.

In respect to a given host-parasite combination, several factors may influence the occurrence (or not) of an encapsulating response (Salt, 1963).

These include health or vigor of the host, its age, and whether or not it is superparasitized (see also van den Bosch, 1964).

The example of *Bathyplectes curculionis,* a larval parasite of the alfalfa weevil, *Hypera postica,* provides a number of insights for biological control endeavors. The effectiveness of *B. curculionis* in controlling its host in California was treated previously. This parasite, imported into Utah from southern Europe in the early part of this century, became a well established and relatively effective control agent. In 1933, it was introduced into central California, with similar results. In 1939, the related Egyptian alfalfa weevil, *Hypera brunneipennis* (Boheman) was discovered for the first time in North America at Yuma, Arizona, from which site it had spread westward into the Imperial and Coachella Valleys of southern California by 1949, and to the coast near San Diego by 1950. *B. curculionis,* derived from the Utah populations, was colonized on *H. brunneipennis* in southwestern Arizona in 1942. The species was found some years later attacking the Egyptian alfalfa weevil in the San Diego area (Dietrick and van den Bosch, 1953). There were, of course, two possible sources for this coastal southern California parasite population, central California and Arizona. The latter was considered more likely.

Larval dissections showed the population of *H. brunneipennis* near San Diego to be partially immune to *B. curculionis,* the immunity being mani-fested by an encapsulation response against parasite eggs and young larvae. The immunity was determined to be of only moderate intensity, not occurring in all cases, and being frequently overcome in superparasitized hosts, small hosts, or hosts inadequately nourished (van den Bosch, 1964). Nevertheless, samples taken in the 1950's showed 35 to 40 per cent destruction of parasite eggs and larvae by this response.

Some fifteen years later, Salt and van den Bosch (1967) examined again this immune response by the Egyptian alfalfa weevil in southern California, finding that samples collected in 1966 showed only about 5 per cent effective encapsulation. They concluded that the normal parasite of *H. postica,* though originally poorly adapted to *H. brunneipennis,* had increased its adaptation as time passed.

This finding is noteworthy. It suggests that, at least in this case, a parasite can become adapted to an initially largely immune host within a surprisingly short time. It is possible, of course, that this facility exists only when the new host is closely related to the original one.

The apparent facility of this adaptive relation in *B. curculionis* has been explored further by Salt and van den Bosch (1967). Comparative studies showed that the central California strain, associated with *H. postica,* elicits a very high proportion of immune responses in southern California *H.*

brunneipennis. This is not too surprising. More interesting is the finding that southern California *B. curculionis,* having become adapted to *H. brunneipennis,* likewise elicited some haemocytic reaction in its formerly normal host, *H. postica,* in central California. This means that the parasite is not only becoming adapted to a new host, but that in so doing it seems to be losing its adaptation to its original host. This again is not too surprising, for the mechanism of adequately duplicating the biochemistry of one host such as to elicit no immune reaction could well preclude the adequate duplication of the biochemistry of the other host, if the frequencies of the immune response in the two host species are rather distinct. This also illustrates Simmonds' (1963) contention that precolonization adaptability (wide adaptive potential) and post colonization adaptation (subsequent change in fitness) are antagonistic, that is, as the latter occurs through selection, some of the former diminishes.

The *B. curculionis* present in North America came from southern Europe (southern France and Italy), where it is associated naturally with *H. postica.* What was believed to be *H. brunneipennis* in Iraq is attacked there by *B. curculionis* also (van den Bosch and Dietrick, 1959). The marked difference in climate between southern Europe and Iraq, coupled with what we know now about host-specific strains of this parasite and others, suggests that this Iraqian *B. curculionis* may be admirably suited to the southeastern desert valleys of California and Arizona where severe climatic conditions have prevented our present strains of this parasite from being effective.

We may conjecture further about this host-parasite system. *H. brunneipennis* has for some years been expanding its range northward in California. One wonders what will happen when *B. curculionis* adapted to *H. brunneipennis* in southern California, and *B. curculionis* adapted to *H. postica* in northern California become intermingled in central California. Will they hybridize? If they hybridize, will there be cyclical shifts in adaptability to one host and then to the other, in which *H. postica* becomes favored with immunity at one time thereby becoming numerically dominant, while *H. brunneipennis* is controlled and becomes less numerous, and *vice versa*?

Climatic Strains

There are several examples of adaptive strains in which the biological or ecological differences between them are known. One of the earliest examples of climatically adapted strains concerns *Trichogramma minutum* Riley, which has been found to contain strains which either develop at different rates, have varying longevities, or take on different sizes, shapes, or colors at different temperature and humidity levels (Flanders, 1931; Lund, 1934). The field value of such adaptive variations has not been demonstrated, perhaps because the practical value of *Trichogramma,* itself, as a useful natural enemy in periodic colonization programs remains debatable.

Aphelinus mali (Haldeman), the famous parasite of the woolly apple aphid, *Eriosoma lanigerum* (Hausmann), exists in several strains derived from different regions, and presumably differently adapted to climate. A so-called "Russian strain" gives better control of the aphid in China than does a locally derived "Tsingtao strain" (Lung *et al.,* 1960, cited in DeBach and Hagen, 1964).

The tachinid parasite *Metagonistylum minense* Townsend, which attacks the sugar cane borer in Brazil, is reported to exist in two climatic strains (Tucker, 1939, cited in DeBach and Hagen, 1964). A "wet race" occurs in the Amazon area, a "dry race" in the Sao Paulo area. These vary also in fecundity and color. They have been introduced successfully into various Caribbean and South American areas, but with no signs that their preadaptive climatic responses have provided any local advantage, since the "wet race" was successfully established in the dry areas of Guadaloupe and Venezuela (Simmonds, 1963).

Other examples of climatically adapted strains or ecotypes of natural enemies include the Korean and Japanese races of *Tiphia popilliavora* Rohwer, which differ importantly in reproductive capacity and adult emergence time (Clausen, 1936).

With these examples as background, current progress in the use of *Trioxys pallidus* (Halliday) for control of the walnut aphid, *Chromaphis juglandicola* (Kaltenbach), in California is of interest. In several areas, this aphid is the key walnut pest. Prior to 1959, very little natural parasitization of this aphid occurred in California. In 1959, the parasite *T. pallidus* was introduced from France and colonized, at first, throughout southern California, and within two years, in central and northern California as well (Schlinger *et al.,* 1960). It readily established and rapidly became an important enemy in the coastal and intermediate zones of southern California (van den Bosch *et al.,* 1962), but even after five or six years of intensive colonization, involving hundreds of thousands of wasps released in dozens of locations, it failed to establish or to persist in any effective numbers in the northern two-thirds of the state (Sluss, 1967). Concluding that the much more severe summer and winter climates of this latter area constituted the principal limiting factors relative to successful parasite establishment (there were no significant competitors, and the parasite seems not to require alternate hosts), it was decided to import this same parasite from Iran, where it was known to attack the walnut aphid. The central plateau of Iran possesses a climate much like the Central

Valley of California, though with somewhat more severe winters. In 1968, a culture of this parasite from Iran was introduced, mass-produced, and colonized in a few sites in central California. It was readily recovered soon after colonization and spread rapidly even in the season of first release. By the next year (1969), it was clear that the northern California environment was favorable. It has overwintered successfully in large numbers, increased its abundance strikingly, and spread promisingly. Results of continued colonizations, coupled with further increases in abundance and further natural spread, confirm the adaptive qualities relative to the northern California climate which proved unsuitable to the parasites from France (*vide supra*). Excellent biological control at some release sites has already been accomplished (van den Bosch *et al.,* in press).

The question remains, are the French and Iranian strains of *T. pallidus* merely ecotypes, or are they true species. In recent correspondence, M. J. P. Mackauer stated that he could not detect even the most minor morphological differences, and indicated that there were only differences in size and coloration. However, laboratory hybridization tests show that the two stocks do not mate, hence, no hybrid offspring have resulted. But these findings come from only 20 attempted crosses, and additional results must be made available before a conclusion can be drawn. However, the status of these parasite stocks notwithstanding, the practical benefits to be derived from utilizing stocks of even the same species from different ecological regions are obvious.

POSSIBILITIES FOR IMPROVING ADAPTATIONS

Several workers have devoted attention to ways in which ecological adaptations of a colonized natural enemy may be improved. The subject has been well reviewed by DeBach and Hagen (1964). Possibilities for improving the fitness of a given enemy to a target environment may be classified as (a) artificial selection, and (b) increase in genetic diversity. In the first situation a stock of the enemy is created, by selection in the laboratory, which is better fitted to cope with some limiting environmental factor. In the second situation, by hybridization or by colonization of greater genetic diversity from source populations, a more plastic or diverse stock is created which, after colonization, will have an increased chance for improvement in fitness through natural selection.

Artificial selection of natural enemies for the purpose of improving their fitness, overcoming certain discerned weaknesses, or general enhancement of their adaptability, has been considered by many, attempted by few, and, unfortunately, proven practicable in terms of improved biological control by no one. Recent reviews on the subject of selective improvement have been made by DeBach (1958), Simmonds (1963), DeBach and Hagen (1964), White *et al.* (1970) and Wilson (1965). Some of the technical difficulties met in such attempts to improve parasites by selection, include lack of knowledge concerning the genetic basis for inheritance of the desired characteristics (e.g., simple Mendelian dominance-recessiveness, polygenic inheritance, pleiotropic influence), prior acquisition of genetic diversity on which to base selection, artificial cultural needs of the species, its power of increase, selective level to be used (which in turn is influenced by the genetic basis for inheritance of the desired features, powers of increase, and cultural efficiency), the reproductive mode of the enemy (e.g., uniparentalism versus biparentalism), and the possibility of unintentional co-selection for detrimental qualities (e.g., loss in general vigor, in fecundity, in survival power, in adaptive behavior, and so on).

Adaptive features selected by past workers include, (1) improved climatic tolerance [Wilkes, 1942—using *Dahlbominus fuscipennis* (Zetterstedt); DeBach (unpubl. data), cited in DeBach and Hagen, 1964 and White *et al.,* 1970—working with *Aphytis lingnanensis*], (2) improvement in the sex ratio of females, so as to produce less male and more female progeny each generation, carried out on *D. fuscipennis* by Wilkes (1947), and on *Aenoplex carpocapsae* (Cushman) by Simmonds (1947), (3) improvement of host finding ability in *Trichogramma minutum* by Urquijo (1951), (4) change of host preference of *Horogenes molestae* (Uchida) by Allen (1954, 1958), and *Chrysopa carnea* Stephens by Meyer and Meyer (1946), and (5) creation of DDT-resistance in *Macrocentrus ancylivorus* Rohwer by Pielou and Glasser (1952).

An attempt at improving natural enemy adaptability through intra-specific crossings was made by Box (1956), who produced improved host-preference behavior using different strains of the tachinid parasite *Paratheresia claripalpis*. The potential for altering the ecological characteristics of an insect, proposed by Sailer (1954) in his work with stinkbugs, has been demonstrated amply by Lewontin and Birch (1966), who followed the improvements in fecundity, longevity, and climatic tolerance resulting from inter-specific crossses between *Dacus tryoni* (Froggatt) and *D. neohumeralis* (Perkins), two fruit flies in Australia. Inter-specific hybridization had already been attempted with a natural enemy with some success by Handschin (1932) who crossed *Spalangia orientalis* Graham and *S. sundaica* Graham to gain a fertile hybrid better adapted to environments of North Australia and possessing better fecundity and longevity than either parent in the area of colonization.

The importation of greater genetic diversity from source populations has long been proposed as a desirable procedure for improving the potential adaptability of colonized natural enemies by such workers as Clausen (1936), Simmonds (1963), and Doutt and DeBach (1964). The idea here, already discussed in an earlier section of this paper, is to introduce a greater sampling of the genetic constitution of the natural enemy species from its home habitats, and thereby create a greater probability for post-colonization selection leading to improved fitness.

Remington (1968) has recently discussed, from the standpoint of population genetics, possible strategies for effecting successful insect introductions. He points out that the strategies available for consideration depend upon the population structure of the species concerned. By population structure is meant the degree to which a species population is broken up into subpopulations and, if so broken up, the degree of gene exchange between them. He visualized four categories of population structure: (1) continuous and numerous over large geographic areas, (2) continuous and geographically widespread but rare, (3) subdivided into subpopulations with moderate gene flow among them, and (4) subdivided into subpopulations with only rare gene flow among them.

Remington concluded that most insect species have a type (3) population structure which includes a relatively large, ecologically central subpopulation, and one or more small, ecologically marginal subpopulations with moderate gene flow between them. Individuals from the central subpopulation will mostly be heterozygous, whereas individuals from marginal subpopulations will mostly be homozygous. The central subpopulation will contain numerous deleterious recessive alleles (large genetic load); marginal ones will contain relatively few (low genetic load). Assuming that this population structure is most common, Remington concluded (a) that the most probably successful establishment will come from the colonization of many founders from a large, central, source population. The least probably successful establishment will come from the colonization of only a few founder individuals from a large, central, source population. The best possibilities for the successful colonization, followed by improved adaptation to the new environment, will come from use of many founders from marginal source populations. This latter strategy applies particularly when the new (target) environment is somewhat different (in climate, flora, et cetera) than the source environment.

Remington went on to consider strategies for introducing beneficial insects (parasites, predators, weed controllers, or pollinators). Here he concluded that if the new environment is similar to the source environment the strategy is to introduce a large, wild sample from a large, central source population (similar to that discussed above). It is even better to introduce several such samples from several such source populations. He cautions against introducing only a few founders from a large, central source population or of using many individuals bred from laboratory rearings started from a few such founders. In considerable measure, because of the economics of foreign exploration work, this is just about what many past importation efforts have done.

Practical considerations will restrict application of Remington's recommended strategy for natural enemy importations. First, he proposes colonization of large samples. Plant quarantine considerations, aimed at reducing to a minimum the possibility for introducing a harmful organism, require that all organisms released from quarantine be individually examined and verified, taxonomically. Second, he proposes direct release of "wild" samples. Plant quarantine procedures demand that, if possible, only propagated material be so released. This means that colonizations will usually involve organisms produced in mass-culture programs in the insectary, hopefully bred from many founders, rather than just a few.

All of Remington's strategies presume prior knowledge of the population structure of the candidate natural enemy, particularly the knowledge of the geographic localities of the "large, central source population" and the "small, marginal source population." Such information is rarely, if ever, available to the collector of many of these cryptic, poorly known, parasites and predators. Many of the imported natural enemies are, in fact, new to science and completely unknown as to geographic distribution, biology, relative abundance, and genetic make-up.

A recent paper by Levins (1969) also discusses possible strategies for colonizing natural enemies, but lack of time prevents us from analyzing in detail his premises about pest populations and natural enemy performance. Suffice it to say, Levins' recommendations are provocative because they are different from the usual practices we follow in biological control, and the reader is referred to the original paper for further consideration.

LESSONS FOR FUTURE STRATEGIES

It is our feeling that the most efficient way to acquire better adapted ecotypes, races, or siblings of a particular natural enemy species or species-complex is to ascertain by post-release evaluation the reasons for lack of suitable performance by the earlier colonized enemy or enemies, and follow this with a renewed search in regions of indigeneity for ecotypes possessing the missing capabilities. Such a procedure implies that careful field evaluation will disclose the reason(s) for initial lack of success, and also that the appropriate improvement in the limiting characters can be found. Such a procedure means that the number of individuals imported need not necessarily be high simply to provide great genetic diversity; rather they should be as suitably pre-adapted as possible in order to fit the new environment adequately for acceptable biological control. These requirements may prove difficult in many cases, but our own experience suggests certain clues which have led to success in some cases.

It will be useful to focus again on what it is we are attempting to do. To do this, we may consider Callan's (1969) three types of colonizers:

I.   *Colonizers with built-in success.* These are well pre-adapted and usually become rapidly established and spread widely within a short time.

II. *Colonizers with delayed success.* They are imperfectly pre-adapted (but may possess the essential genetic diversities), and although establishing a foothold, they have to become genetically adjusted, probably often by genetic recombination, before expanding explosively.

III. *Colonizers predestined to failure.* They are unpre-adapted, and, being incapable of or lacking time for achieving genetic adjustment, never become established.

We are not concerned with Type I; there is no problem here, and although such cases might be analyzed, as DeBach (1964) and others have done, to gain insight into what an effective natural enemy looks like, we feel that any clues so derived will only apply generally, not specifically, to the less-than-successful cases. What we are most interested in are Types II and III. Here, we must either acquire altogether different, untried natural enemies, or else try and improve on the adaptabilities of those enemy species already known to exist or that have already been colonized with less than satisfactory results.

Regarding Callan's Type II colonizer, we would subclassify it into type IIa, those delayed colonizers which eventually become successful in achieving effective control, and type IIb, which though becoming established remain ineffective.

Realizing that the adaptability of natural enemies is variable at the source and changeable (naturally, or by artificial selection) after importation, we can offer certain suggestions or guidelines for natural enemy search and colonization. Three circumstances can be visualized, (a) initial search and importation for the natural enemy in question, where little is known about its ecology and probability for successful control, (b) repeated search after an initial colonization has failed, and (c) repeated search after a previous, successful colonization, but where the enemy provides only partial control success, either geographically or numerically. For the first circumstance, little can be done except to follow the empirical principles of obtaining as large a sample of the gene pool as finances and techniques will allow, selecting as areas for search those places whose climates, plant communities, and host plant associations are as near like those of the target region as possible (Flanders, 1959). These criteria are pretty much followed already in our traditional procedures in biological control. We would only emphasize the need for better sampling coverage of the indigenous natural enemy population than has commonly taken place in the past.

The second circumstance, which is the colonization that has failed, at first thought might seem to be much like the first. However, by now we should know more about these natural enemies in regard to life cycle properties and weaknesses, host relations, cultural requirements, climatic responses, host synchrony, host plant synchrony, and so on. If the colonization was done with care, we may also have an idea as to why it failed, as for example because of excessive heat, inability to overwinter, lack of suitable host stages, host-adaptability, etc.

The third circumstance provides the greater and more interesting challenge. Now the natural enemy is established in the pest-infested area, but for one reason or another is not providing acceptable biological control. A large number of examples apply to this case, a useful summary table being provided by DeBach (1964).

In these and many more cases, the possibility exists that a new sample of the natural enemy, collected from a more appropriate locality, or under different circumstances than the original collection, may provide the pre-adaptive qualities, or the post-colonization adaptability, that will result in more effective biological control. We have several good examples where careful post-colonization study of the ineffective natural enemy has provided clues for subsequent search and eventual control success. Two excellent examples are the California red scale - *Aphytis lingnanensis* - *Aphytis melinus* story (*vide supra*) (see also DeBach, 1969) and our own recent success with *Trioxys pallidus* against the walnut aphid (*vide supra*). In both these cases, opinions about the reasons for lack of effectiveness of the existing enemies were coupled with knowledge of their indigenous distribution, hosts and habitats, such that the appropriately adapted organism, if it existed at all, could be located and imported. In each case this was indeed done.

LITERATURE CITED

Allen, H. W. 1954. Propagation of *Horogenes molestae,* an Asiatic parasite of the oriental fruit moth, on the potato tuberworm. J. Econ. Entomol. 45: 278-281.

Allen, H. W. 1958. Evidence of adaptive races among oriental fruit moth parasites. Proc. X Intern. Congr. Entomol., Montreal (1956) 4: 743-749.

Box, H. E. 1956. The biological control of moth borers (*Diatraea*) in Venezuela. Battle against Venezuela's cane borer. Pt. 1. Preliminary investigations and the launching of a general campaign. Sugar (New York) 51: 25-27, 30, 45.

Callan, E. McC. 1969. Ecology and insect colonization for biological control. Proc. Ecol. Soc. Australia 4: 17-31.

Clausen, C. P. 1936. Insect parasitism and biological control. Ann. Entomol. Soc. Amer. 29: 201-223.

Clausen, C. P. 1956. Biological control of insect pests in the continental United States. U.S.D.A. Tech. Bull. No. 1139, 151 pp.

Clausen, C. P. 1958. Biological control of insect pests. Ann. Rev. Entomol. 3: 291-310.

DeBach, P. 1953. The establishment in California of an oriental strain of *Prospaltella perniciosi* Tower on the California red scale. J. Econ. Entomol. 46: 1103.

DeBach, P. 1958. Selective breeding to improve adaptations of parasitic insects. Proc. X Intern. Congr. Entomol., Montreal (1956) 4: 759-768.

DeBach, P. 1960. The importance of taxonomy to biological control as illustrated by the cryptic history of *Aphytis holoxanthus* n. sp. (Hymenoptera: Aphelinidae), a parasite of *Chrysomphalus aonidum,* and *Aphytis coheni* n. sp., a parasite of *Aonidiella aurantii.* Ann Entomol. Soc. Amer. 53: 701-705.

DeBach, P. 1964. Successes, trends, and future possibilities. Chap. 24. *In* Biological Control of Insect Pests and Weeds, P. DeBach (ed.). Reinhold Publ. Co., N. Y. 844 pp.

DeBach, P. 1965a. Some biological and ecological phenomena associated with colonizing entomophagous insects. Pp. 287-306 *in* Genetics of Colonizing Species, H. G. Baker and G. L. Stebbins (eds.). Academic Press, N. Y. 588 pp.

DeBach, P. 1965b. Weather and the success of parasites in population regulation. Can. Entomol. 97: 848-863.

DeBach, P. 1966. The competitive displacement and coexistence principles. Ann. Rev. Entomol. 11: 183-212.

DeBach, P. 1969. Biological control of diaspine scale insects on citrus in California. Proc. First Intern. Citrus Symp., Riverside (1968) 2: 801-815.

DeBach, P., and K. S. Hagen. 1964. Manipulation of entomaphagous species, Chap. 16. *In* Biological Control of Insect Pests and Weeds, P. DeBach (ed.). Reinhold Publ. Co., N. Y. 844 pp.

DeBach, P., and R. A. Sundby. 1963. Competitive displacement between ecological homologues. Hilgardia 34: 105-166.

Dietrick, E. J., and R. van den Bosch. 1953. Further notes on *Hypera brunneipennis* and its parasite *Bathyplectes curculionis.* J. Econ. Entomol. 46: 1114.

Doutt, R. L. 1954. Biological control of fig scale. Calif. Agr. 8(8): 13.

Doutt, R. L., and P. DeBach. 1964. Some biological control concepts and questions. Chap. 5. *In* Biological Control of Insect Pests and Weeds P. DeBach (ed.). Reinhold Publ. Co., N. Y. 844 pp.

Elton, C. S. 1958. The Ecology of Invasion by Animals and Plants. John Wiley and Sons, N. Y. 181 pp.

Flanders, S. E. 1931. The temperature relationships of *Trichogramma minutum* as a basis for racial segregation. Hilgardia 5: 395-406.

Flanders, S. E. 1940. Environmental resistance to the establishment of parasitic Hymenoptera. Ann. Entomol. Soc. Amer. 33: 245-253.

Flanders, S. E. 1950. Races of apomictic parasitic Hymenoptera introduced into California. J. Econ. Entomol. 43: 719-720.

Flanders, S. E. 1959. The employment of exotic entomophagous insects in pest control. J.

Econ. Ent. 52: 71-75.

Flanders, S. E. 1966. The circumstances of species replacement among parasite Hymenoptera. Can. Entomol. 98: 1009-1024.

Hafez, M., and R. L. Doutt. 1954. Biological evidence of sibling species in *Aphytis maculicornis* (Masi) (Hymenoptera, Aphelinidae). Can. Entomol. 86: 90-96.

Handschin, E. 1932. A preliminary report on investigations on the buffalo fly (*Lyperosia exigua* de Meij.) and its parasites in Java and northern Australia. Australian Council Sci. Ind. Res. Pamph. 31, 24 pp.

Huffaker, C. B. 1956. Spread and activity of imported parasites of fig scale. Proc. X Ann. Calif. Fig Inst., p. 23.

Huffaker, C. B., and C. E. Kennett. 1966. Studies of two parasites of olive scale, *Parlatoria oleae* (Colvee). IV. Biological control of *Parlatoria oleae* (Colvee) through the compensatory action of two introduced parasites. Hilgardia 37: 283-335.

Kelleher, J. S. 1969. Introduction practices—past and present. Bull. Entomol. Soc. Amer. 15: 235-236.

Levins, R. 1969. Some demographic and genetic consequences of environmental heterogeneity for biological control. Bull. Entomol. Soc. Amer. 15: 237-240.

Lewontin, R. C., and L. C. Birch. 1966. Hybridization as a source of variation for adaptation to new environments. Evolution 20: 315-36.

Lund, H. O. 1934. Some temperature and humidity relations of two races of *Trichogramma minutum* Riley (Hym. Chalcididae). Ann. Entomol. Soc. Amer. 27: 324-340.

Lung, C. T., Y. P. Wang, and P. G. Tang. 1960. Investigations on the biology and utilization of *Aphelinus mali* Hald., the specific parasite of the woolly apple aphis, *Eriosoma lanigerum* Hausm. [In Chinese with English summ.] Acta Ent. Sinica 10: 1-39.

Meyer, N. F., and Z. A. Meyer. 1946. The formation of biological forms in *Chrysopa vulgaris* Schr. (Neuroptera, Chrysopidae). Zool. Zhurnal, Moscow 25: 115-120.

Michelbacher, A. E. 1943. The present status of the alfalfa weevil in California. Univ. of Calif. Agr. Expt. Sta. Bull. 677. 24 pp.

Muldrew, J. A. 1953. The natural immunity of the larch sawfly (*Pristiphora erichsonii* (Hartung)) to the introduced parasite *Mesoleius tenthredinis* Morley in Manitoba and Saskatchewan. Can. J. Zool. 31: 313-332.

Pielou, D. P., and R. F. Glasser. 1952. Selection for DDT resistance in a beneficial insect parasite. Science: 115: 117-118.

Remington, C. L. 1968. The population genetics of insect introduction. Ann. Rev. Entomol. 13: 415-426.

Sailer, R. I. 1954. Interspecific hybridization among insects with a report on cross-breeding experiments with stink bugs. J. Econ. Entomol. 47: 377-383.

Salt, G. 1963. The defense reactions of insects to metazoan parasites. Parasitology 53: 527-642.

Salt, G. 1968. The resistance of insect parasitoids to the defense reactions of their hosts. Biol. Rev. 43: 200-232.

Salt, G., and R. van den Bosch. 1967. The defense reactions of three species of *Hypera* (Coleoptera, Curculionidae) to an ichneumon wasp. J. Inv. Pathol. 9: 164-177.

Schlinger, E. I., K. S. Hagen, and R. van den Bosch. 1960. Imported French parasite of walnut aphid. Calif. Agr. 14(11): 3-4.

Simmonds, F. J. 1947. Improvement of the sex-ratio of a parasite by selection. Can. Entomol. 79: 41-44.

Simmonds, F. J. 1963. Genetics and biological control. Can. Entomol. 95: 561-567.

Sluss, R. R. 1967. Population dynamics of the walnut aphid, *Chromaphis juglandicola* (Kalt.) in Northern California. Ecology 48: 41-58.

Smith, H. S. 1941. Racial segregation in insect populations and its significance in applied entomology. J. Econ. Entomol. 34: 1-13.

Tucker, R. W. E. 1939. Introduction of dry area race of *Metagonistylum minense* into Barbados. Barbados Dept. Sci. and Agr., Agr. J. 8: 113-131.

Turnbull, A. L. 1967. Population dynamics of exotic insects. Bull. Entomol. Soc. Amer. 13: 333-337.

Turnbull, A. L., and D. A. Chant. 1961. The practice and theory of biological control of insects in Canada. Can. J. Zool. 39: 697-753.

Urquijo, P. 1951. Aplicacion de la genetica al aumento de la eficacia del *Trichogramma minutum* en la lucha biologica. Bol. Patol. Veg. y Entomol. Agr. (Madrid) 18: 1-12.

van den Bosch, R. 1964. Encapsulation of the eggs of *Bathyplectes curculionis* (Thomson) (Hymenoptera, Ichneumonidae) in larvae of *Hypera brunneipennis* (Boheman) and *Hypera postica* (Gyllenhal) (Coleoptera, Curculionidae). J. Insect. Pathol. 6: 343-367.

van den Bosch, R. 1968. Comments on population dynamics of exotic insects. Bull. Entomol. Soc. Amer. 14: 112-115.

van den Bosch, R., and E. J. Dietrick. 1959. The interrelationships of *Hypera brunneipennis* (Coleoptera, Curculionidae) and *Bathyplectes curculionis* (Hymenoptera, Ichneumonidae) in Southern California. Ann. Entomol. Soc. Amer. 52: 609-616.

van den Bosch, R., E. I. Schlinger, and K. S. Hagen. 1962. Initial field observations in California on *Trioxys pallidus* (Haliday), a recently introduced parasite of the walnut aphid. J. Econ. Entomol. 55: 857-862.

van den Bosch, R., B. D. Frazer, C. S. Davis, P. S. Messenger, and R. C. Hom. In press. An effective walnut aphid parasite from Iran. Calif. Agr. (1970)

Walker, I. 1959. Die Abwehrreaktion des Wirtes *Drosophila melanogaster* gegen die zoophage Cynipide *Pseudeucoila bochei* Weld. Rev. Suisse Zool. 66: 569-632.

White, E. B., P. DeBach, and M. J. Garber. 1970. Artificial selection for genetic adaptation to temperature extremes in *Aphytis lingnanensis* Compere (Hymenoptera: Aphelinidae). Hilgardia 40: 161-192.

Wilkes, A. 1942. The influence of selection on the preferendum of a chalcid (*Microplectron fuscipennis* Zett.) and its significance for biological control of an insect pest. Proc. Roy. Entomol. Soc. (London) Ser. B., 130: 400-415.

Wilkes, A. 1947. The effects of selective breeding on the laboratory propagation of insect parasites. Proc. Roy. Entomol. Soc. (London) Ser. B., 134: 227-245.

Wilson, F. 1960. A review of the biological control of insects and weeds in Australia and Australian New Guinea. Commonwealth Inst. Biol. Control, Tech. Comm. 1. 102 pp.

Wilson, F. 1965. Biological Control and the Genetics of Colonizing Species, H. G. Baker and G. L. Stebbins (eds.). Academic Press, N.Y. 588 pp.

Chapter 4

THE USE OF MODELS AND LIFE TABLES IN ASSESSING
THE ROLE OF NATURAL ENEMIES

G. C. Varley and G. R. Gradwell

Oxford University
Oxford, England

INTRODUCTION

The part played by parasites and predators in the dynamics of natural populations has been one of the most elusive and controversial problems of population ecology. Although natural enemies have been successfully used for the "biological control" of insect pests, the properties of these animals have neither been sufficiently well known beforehand to forecast the degree of success, nor have studies after the introduction revealed the mechanisms involved.

Theoretical ideas based on simple assumptions about the form of the interaction between parasite-host or predator-prey populations have largely been ignored and measurements necessary to test these ideas remain unmade. In our opinion these theoretical ideas are of prime importance, since we can claim to understand a process only when measurements conform to some theoretical model which shows how they might have come about.

TERMINOLOGY FOR COMPONENTS OF POPULATION MODELS

Many of the components we now introduce into population models were discussed by Howard and Fiske (1911) in their study of the gypsy and brown tailed moths. They said that "A natural balance can only be maintained through the operation of facultative agencies which effect the destruc-

93

tion of a greater proportionate number of individuals as the insect in question increases in abundance," and they asserted that parasitism "in the majority of instances, though not in all, is truly facultative." Smith (1935) agreed with Howard and Fiske but renamed their facultative agencies *density dependent factors.* Such has been the persuasiveness of the writings of these authors that it is now commonly assumed that any form of density dependent factor will ensure a natural balance (i.e., will regulate a population) and that parasites and predators necessarily act as density dependent factors. We believe that neither of these assumptions is correct.

We accept the definition that a density dependent mortality is one which kills a "greater proportionate number of individuals as the population density increases." Hence, an appropriate test for a density dependent effect of either a parasite or a predator is to see if the percentage mortality they cause rises with increasing host or prey density. This relationship should be sought between measurements made in a succession of generations. If parasites or predators have a choice of local host or prey concentrations in any one generation, then a series of sub-samples from widely differing densities may show a positive correlation between the percentage mortality and increasing density. Hassell (1966) has pointed out that when such a density dependent relationship is the result of behavioral (functional) responses of parasites or predators within a single host generation, the relationship cannot be used to describe the intergeneration relationships between populations. Only intergeneration relationships are relevant to a discussion of population regulation. (See further Chapter 2.)

Even if a density dependent relationship has been found between measurements made in successive generations, this is no guarantee that it is capable of regulating. A number of relationships are shown in Fig. 1 all of which fit the verbal definition of density dependence, but not all will regulate the population within the limits considered. Curve D is too weak to be effective even at very low rates of increase. Both curve B—which is an example quoted by Milne (1957) of a perfect density dependent mortality—and curve C may or may not regulate, depending on the population's rate of increase. Only some curves in the form of curve A—which approach the 100% mortality asymptotically—are able to regulate a population irrespective of its rate of increase.

Much of the experimental work aimed at investigating whether or not parasites act as density dependent factors has been done by allowing known numbers of hosts and parasites to interact and scoring the percentage of parasitism. Takahashi (1968) used single individuals of the ichneumonid *Nemeritis canescens* (Grav.) searching for larvae of *Ephestia cautella* (Walk.) and showed that parasitism was indeed density dependent. On the other hand,

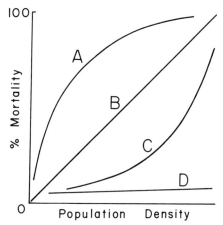

Fig. 1. A number of relationships which fit the verbal definition of a density-dependent mortality.

DeBach and Smith (1941a) used a constant number of the chalcid *Mormoniella vitripennis* Walk. searching for different densities of housefly puparia, and showed the percentage parasitism decreasing with increasing host density; this arises from the long "handling time" (Holling, 1959) which prevents a *Mormoniella* female from attacking more than two hosts in a day. In this case the parasites are acting as *inverse factors* (Smith, 1935).

Although this sort of experimental work may give interesting and important insight into the possible kinds of interactions between parasites and their hosts, the limited objectives of the experiments mean that only rarely can the results be extrapolated to indicate the outcome of population interactions under natural conditions. Our own field work on parasites of winter moth, *Operophtera brumata* (L.) shows that some act as weak inverse factors (Varley and Gradwell, 1968). Most of the parasites we observed kill a very variable proportion of hosts and the proportion is not clearly related in any way to host density; these parasites therefore fit perhaps better with Howard and Fiske's category of "catastrophic agencies" except that their effect is small.

Howard and Fiske said "A very large proportion of the controlling agencies, such as the destruction wrought by storm, low or high temperatures, or other climatic conditions, is to be classed as catastrophic, since they are wholly independent in their activities upon whether the insect which incidentally suffers is rare or abundant." Smith (1935) replaced the term "catastrophic" by density independent factor without any formal change of definition.

He thereby shifted the emphasis from the catastrophic and variable nature of weather and other factors to the idea that a particularly extreme temperature will kill a proportion independent of density.

To illustrate these ideas we first prepared figure 2A and we added the words "Catastrophic = Density Independent Factor." However, although this synonymy has often been quoted, it is obvious that the properties of mortality factors represented by the regression line are entirely different from those represented by the points. The regression line could represent a "legislative factor" of Nicholson (1954) but only the points could represent a key factor.

Next we found that an unexpected relationship appeared in a population model which included a random key factor to represent a catastrophic agency. Instead of the points showing a negligible regression as in Fig. 2A, they came out like Fig. 2B. When plotted either as a k-value against log population, or as

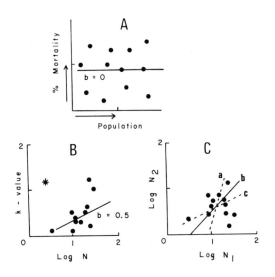

Fig. 2. (A)–Schematic representation of catastrophic mortality, which has been incorrectly synonymised with density-independent mortality, represented by the line. (B)–k-value for a random key factor plotted against the log of the population on which it acts. The calculated regression is spurious because ordinate and abscissa are not independent (see text). (C)–Diagram of statistical test for the figures in B: a is the calculated regression of N_1 on log N_2, b is is the slope b=1, and c is the calculated regression of log N_2 on log N_1. If the mortality had been density-dependent, both lines a and c would have been on the same side of b, and differed significantly from it.

per cent mortality against population density, the points give the appearance expected of a density dependent factor! When a regression was calculated by conventional methods we found a slope of approximately b = 0.5. This is a spurious figure which arises because the ordinate and the abscissa are not independent. The lowest population density *must* have a low mortality and the highest population *must* have a high mortality. If, for instance, the lowest population in Fig. 2B were given a high mortality, as represented by the asterisk (*) this would not destroy the regression, because it would generate a new lowest log population density of $\overline{1}.2$. Choosing to stop plotting at one of the population extremes can reduce the apparent regression, but cannot destroy it. We can distinguish between true density dependent factors and this spurious effect by using the statistical test for density dependency recommended by Varley and Gradwell (1968). This is shown diagrammatically in Fig. 2C. The two independent measurements, the population densities before and after the action of the mortality factors, are plotted against each other. The regression of log N_2 on log N_1 is shown by the line c; the regression of log N_1 on log N_2 is shown by the line a. They are on opposite sides of the line b, which represents b = 1, and neither regression differs significantly from b = 1. A genuine density dependent factor would give calculated regressions both on the same side of line b, and ideally both significantly differing from b = 1.

We know of no field measurements of host density and percentage parasitism for a reasonably long series of successive generations which shows parasitism to act as a density dependent factor. The undoubted ability of parasites to regulate or control the density of the host arises in other ways.

Nicholson (1933) and Nicholson and Bailey (1935) made a great advance in ideas about parasitism and their theory was used by Varley (1947) in population models which interpreted the interaction between parasitism and other mortality factors in determining population levels of the knapweed gall-fly, *Urophora jaceana* Hering. Nicholson apparently did not realize that in both the 1933 and 1935 papers he was, in effect, discussing the effects of two quite different "models." In the first of these he agreed with Howard and Fiske and with Smith that population regulation could be brought about only through the action of density dependent factors (= Nicholson's "controlling agencies"). His second "model" described his theoretical ideas about parasite-host interactions. By means of this model he intended to show how parasites could *regulate* their host's density, but he succeeded in showing only that they could not, because the model was unstable.

Varley (1947, 1958) pointed out that if parasites behaved in the way suggested by Nicholson and Bailey they were behaving in a manner fundamentally very different from that of density dependent factors. Although Nicholson and Bailey's "steady state calculations" suggested that stability between host and parasite was possible, with one exception their population models showed that when displaced from the steady state the interactions were wildly unstable and gave rise to increasing oscillations in host and parasite densities. Their stable model was one in which a parasite was assumed to have two alternative hosts. The density of one was assumed to be constant and unaffected by the parasite. This unnatural device served to stabilize the parasite's density and thus allowed it to stabilize the population density of the second host. Varley (1947, 1953) showed that the Nicholsonian parasite did not act according to the definition of a density dependent factor; in any one host generation the percentage of parasitism was not governed by host density but was solely related to the density of the adult parasites searching for hosts. He therefore proposed a new term *delayed density dependent factor* to cover the action of a Nicholsonian parasite, and his definition was that ". . . a parasite acts as a delayed density dependent factor if its fecundity or its effective rate of increase is strongly correlated with host density." The final criterion for a Nicholsonian parasite is, of course, that the parasite's "area of discovery" should be a constant. In order to test whether or not a parasite is acting as a delayed density dependent factor, estimates must be made either of its rate of increase or of its area of discovery. Both of these estimates require two measurements of parasite density: first, the density of adults searching and, second, the egg or larval density of their progeny. From experimental results, the delayed density dependent relationship can clearly be shown for *Mormoniella* using the figures of DeBach and Smith (1941b) and for *Encarsia,* which parasitizes the greenhouse whitefly, from the data of Burnett (1956).

Most field studies have included some measurement of percentage parasitism, but, because it is more difficult, few have measured adult parasite density. Without this second measurement, critical tests cannot be applied to the data. However, even when this detailed information is not available it may be possible to suggest that the parasites are causing delayed density dependent mortality. When the percentage parasitism is plotted against host density, the points may not lie along one line (as should be the case if the relationship is density dependent), but, when joined together in a time series, they may form a circle or spiral in an anti-clockwise direction. This sort of relationship, however, will be obvious only when parasitism is making a major contribution to population change, in particular when the host population varies cyclically, wholly or partly as a result of parasite action. Examples of this are black-headed budworm *Acleris variana* (Fernald) in Canada (Morris, 1959) and the grey larch tortrix, *Zeiraphera diniana* Gm. in Switzerland (Auer, 1968; Varley and Gradwell, 1970).

Morris (1959) used a model for an interaction between a Nicholsonian parasite and its host and noted that a plot of the host population density in generation n+1 against that in generation n gave an elliptical figure in which the long axis had a positive slope. In developing this model further, Morris (1963a,b) is clearly of the opinion that the linear regression through the spiral of points given by a plot of the log of the population in generation (n+1) against the log of the population in generation (n) has some biological meaning. This view has been contested by Hassell and Huffaker (1969) but reasserted by Morris and Royama (1969). Morris does not clearly distinguish between the action of a density dependent factor and that of a delayed density dependent factor and interprets the regression slope as indicating an amount of density dependency which is contributed by the parasite. We believe this view to be mistaken. We accept, of course, that it may be possible to obtain a statistically significant linear regression through a spiral of points, but are unable to see that any biological meaning can be attached to this line. As we have said, the properties of a density dependent factor and of a delayed density dependent factor are very different, and in a population model the action of a Nicholsonian parasite cannot be said to be equivalent to some amount of density dependency.

We have tested the Morris method of analysis on population models which have included both density dependent and delayed density dependent factors whose properties were precisely defined. The method failed to reveal the relationships which were put into the model and we feel that, at best, Morris' method of measuring density dependent factors lacks precision; also, the approach is misleading in implying that parasite action can always be equated with density dependency.

OBJECTIONS TO THE NICHOLSON AND BAILEY MODEL

The Nicholson and Bailey theory has two consequences which are at variance with what is seen to happen in the field. The first of these is that parasite and host interactions should give rise to increasing oscillations in both their population densities. Nicholson suggested that such oscillations might cause the population to break up into small pockets where such oscillations would continue but not necessarily in phase with one another. However, it has been shown that the Nicholsonian parasite-host model need not give rise to increasing oscillations if there is also a separate density dependent factor operating on the host or the parasite or on both, and thus that the prediction of increasing oscillations is itself not sufficient grounds for rejecting this theoretical idea (Varley and Gradwell, 1963).

In our view, a much more compelling reason for rejecting Nicholson's theory is that it does not allow for the co-existence of two or more specific parasites on one host. Nicholson and Bailey (1935) believed that their model did indeed allow such co-existence and they illustrated a calculation for two parasites on the same host species. However, we have re-examined this claim and find it to be erroneous. Calculation shows that their model is unstable and shows increasing population oscillations. If these oscillations are stabilized by the addition of a density dependent factor acting on the host, one parasite is eliminated by the other (Varley and Gradwell, in press).

The winter moth in England has two parasites which are almost host-specific. The tachinid fly *Cyzenis albicans* (Fall.) attacks the feeding larvae and the ichneumonid *Cratichneumon culex* (Mueller) attacks the pupae in their cocoons in the soil. Simple models describing parasite behavior only by the estimated area of discovery fail to explain their observed co-existence. In our study area, winter moth larvae feed not only on the oak trees but also on bushes of hazel (*Corylus*), hawthorn (*Crataegus*) and blackthorn (*Prunus spinosa*) and on these the winter moth larval densities per unit leaf area are often higher than on oak. Our observations suggest that *Cyzenis* aggregates in areas of higher host larval densities where they cause a higher percentage parasitism (Hassell, 1968). When the winter moth larvae fall to the ground to pupate, the density per square meter (ground area) under oak is higher than that beneath the bushes and *Cratichneumon* concentrates its search under the oaks. A mathematical model (Varley and Gradwell, in press) which is based on this diversity of the habitat of the host and behavioral responses of the parasites is stable and demonstrates one way in which the co-existence of Nicholsonian parasites could be maintained.

PARASITE QUEST THEORY

Recently, Hassell and Varley (1969) (Varley and Gradwell, 1970) have proposed a new "parasite quest theory" in which the mutual interference between parasites, as shown by Huffaker and Kennett (1969) and Hassell and Huffaker (1969), increases as parasite density increases. The relationship between parasite density and its area of discovery as shown by a number of laboratory parasite-host population studies is of the form

$$\log a = \log Q - m \log p$$

where logs are to the base 10 and "a" is the "area of discovery"; "Q" the "quest constant" (the value of a when $p = 1$); "m" the "mutual interference constant" (the slope of the regression of $\log a$ on $\log p$) and "p" is the density of searching adult parasites.

This model for parasite behavior eliminates the main difficulties which arise using the Nicholson and Bailey theory. If the parasite's area of discovery falls as parasite density increases, it is equivalent to a density dependent factor operating on the adult parasite population. Models which represent parasite action in this way show that the parasite-host interaction can be stable, and does not necessarily lead to increasing oscillations. Similarly, stable models can be constructed which allow the co-existence of two or more specific parasites on one host.

Nicholson and Bailey predicted that the introduction of additional parasites into a situation where a host and its parasites were interacting in a stable way would result in an increase in the host's steady density. If this were true, the established practice of attempting multiple introductions of parasites for the "biological control" of a pest could have harmful results.

In contrast, the quest theory predicts that the introduction of further parasites for "biological control" is much more likely to have a beneficial effect and reduce the host's steady density. Only rarely might it have the consequence predicted by Nicholson and Bailey. Hassell and Varley (1969) used the quest theory to model a situation in which a host and two parasites were in equilibrium. They then examined the outcome of introducing a third parasite into this complex and found that, depending on the properties of the third parasite (Fig. 3), there were four possible consequences (Table 1).

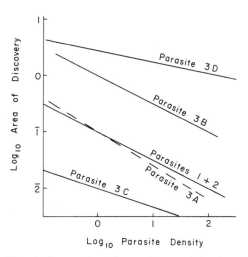

Fig. 3. Properties of parasites used in the calculations for Table 1.

Table 1.   Effect on the host's steady density
of introducing successively acting parasites.   Q =
quest constant.   m = mutual interference constant.
Effective rate of increase of the host = 3.

	Q	m	Host's steady density
Parasite 1.......	0.1	0.5	181
Parasite 1 + parasite 2.......	0.1	0.5	98
Parasite 1+2+ parasite 3C......	0.01	0.33	97
Parasite 1+2+ parasite 3A......	0.1	0.6	72
Parasite 1+2+ parasite 3B......	1.0	0.5	1.8
Parasite 1+2+ parasite 3D.....	2.8	0.2	Unstable

(Data supplied by M. P. Hassell)

The two initial parasites were given the same value of "Q" and "m" and together they produced a host density about half that either could have produced alone. A third parasite with a very much smaller value for "Q" and a smaller "m" (Fig. 3, 3C) has little effect on the host, and in practice would probably fail to become established. A parasite with the same value of "Q" and a slightly larger "m" (3A) co-exists with the others and again causes a considerable reduction in host density. If the third parasite has a value for "Q" ten times that of the first two parasites and the same value for "m" (3B), it may almost replace the first two, but they will hang on at low population densities. The host density will fall to a hundredth. (Note here the difference between the quest theory and Nicholson's theory. A ten-fold increase of the area of discovery reduces host density to a tenth on Nicholson's theory. With the quest theory, the ten-fold rise in "Q" is reinforced by the effect of "m" in relation to changes in parasite population.) If the parasite has a high value of "Q" but a very small "m" of 0.2 or less (Table 1, 3D), it is almost equivalent to a very efficient parasite with a constant area of discovery; in this case the outcome is that predicted by

Nicholson and Bailey. The interaction is unstable and is dominated by the third parasite which causes increasing oscillations in host and parasite densities which in theory could result in extinction of the parasite. In only this last case could the parasite introduction be said to produce a harmful result and, to date, this predicted outcome has never been observed to occur in nature. (We are grateful to Dr. Hassell for providing the information in Table 1.)

How do we make use of these theoretical ideas when attempting to analyze data from a field study? We began a field study of winter moth and its parasites over 20 years ago. Over this period of time we have counted the adult and larval populations of the moth and some of its parasites on five oak trees and have used these figures to construct life tables. By plotting graphs of the successively acting mortality factors expressed as k-values—which measure the logarithmic change in host population (Varley and Gradwell, 1960)—we have a picture of the contribution of each separable mortality factor to the generation mortality as this changes from year to year. This is a key factor analysis (Morris, 1959).

The second stage of our analysis is to seek density relationships; the separately estimated k-values are each plotted against the logarithm of the population density upon which they acted (Varley and Gradwell, 1963, 1968; Southwood, 1966). The mortality occurring between our estimate of the number of eggs laid by the females and subsequent numbers of larvae on the trees is the key factor. It is a "catastrophic" or density independent variable and is the main cause of the year to year changes in winter moth populations. We believe it arises from changes in the degree of synchronization between bud burst and the hatching of the eggs (see also Chapter 9). We have as yet no way of predicting the size of this mortality, nor have we a mathematical model for it.

Mortality caused by *Cyzenis albicans* varies very little from year to year and for practical purposes can be considered as a constant; however, a detailed examination of life tables for this parasite (Hassell, 1969*a,b*) has shown that it is probably acting as a delayed density dependent factor in the way suggested by Nicholson, but because the puparia of *Cyzenis* suffer a strong density dependent mortality, the parasite is a relatively unimportant cause of winter moth mortality.

We have combined the action of those parasites of the winter moth larvae which are neither host-specific nor synchronized with this host. As we said earlier, their total effect is inversely density dependent, but the slope is weak. The k-values for this mortality vary so little that they can be used in a mathematical model as if they were constant. Such an inverse relationship is explicable only if these parasites remain rare, and if their total population varies much less than that of the host. Most of the species lumped in this inverse factor require an alternate host. This relative constancy seems to apply also to the tachinid fly *Lypha dubia*, which like its hosts has a single generation in the year. It is found as a larva in a number of oak-feeding caterpillars, but we are unable to see that the percentage parasitism of any one of these hosts is related in any way to the relative densities of these hosts. Like winter moth and *Cyzenis*, *Lypha* pupates in the soil and we are able to measure the density of emerging adults. *Lypha's* adult density is much less variable than that of *Cyzenis* or those of its several species of host larvae.

Our measurements of the mortality of larvae caused by a microsporidian disease show no relationship with winter moth density. Its action is density independent but it has only killed a very small percentage of larvae. Again, the k-values for this mortality vary little and, at least in simple models, its action can be represented by a constant.

The mortality of winter moth pupae in the soil, due to causes other than the pupal parasite, is very clearly density dependent although, when we join the k-values in a time series, it is also clear that this mortality includes a weak delayed density dependent component. This mortality seems to be mainly caused by predators of which the more important are probably shrews (*Sorex* spp.) and beetles (Carabidae and Staphylinidae). Frank (1967) used a serological technique to investigate the contribution of beetles to this mortality and concluded that the staphylinid beetle *Philonthus decorus* (Gr.) was mainly responsible. Buckner (1969), however, thought that the major part of this mortality was caused by shrews. From the information so far gathered it is not possible to say which of these two views is correct. Shrews may act on winter moth pupae in a density dependent way, but it seems most unlikely that they could be responsible for the observed delayed component in this mortality; an invertebrate predator seems more likely to be responsible for this. Frank has shown that the densities of *Philonthus* have changed in a way which corresponds with the changes in the k-values for this mortality.

We suggest that a predator can cause a direct density dependent mortality only if its density remains much more constant than that of its prey. During the period we have been studying winter moth, its larval populations have varied within a range of about 100-fold. It is most unlikely that the populations of small insectivorous mammals have changed to this extent and, apparently, neither have the populations of many predatory invertebrates. Our measurements show that the population densities of harvestmen (Phalangida), spiders (Araneida), earwigs (*Forficula auricularia* L.) and many species of beetles have varied not more than three-fold during the same period. We can speculate that although many predators, both vertebrate and invertebrate, have definite food preferences they are essentially polyphagous. Perhaps their densities are relatively constant because the change in total abundance of all possible prey is much less than that of any one food item. Such an explanation is, however, likely to be an over-simplification of the interactions involved.

Parasitism of pupae by *Cratichneumon culex* appears to be unrelated to density, but a precise assessment of the role of this parasite is not easy. *Cratichneumon* has at least a partial second generation, and we are able to measure with any degree of accuracy only those adults emerging after overwintering in the soil. We have not been able to measure directly the percentage parasitism of pupae caused by this parasite. The value is deduced from the densities of emerging host and parasite adults. Our attempts to understand the action of this parasite are considered in greater detail in the next section, but briefly our method is to assume that the parasite is acting in one or another of the ways postulated by theory and then to test whether the predicted changes in parasite density and its average level correspond with those we have observed in the field.

The Development of a Model

Our approach to the question of producing a mathematical model for the winter moth has been to commence with something essentially very simple and add to it additional parameters as the consequences of each step are understood. Our first model (Varley and Gradwell, 1963) posed a very simple question. Given the observed potential rate of increase of winter moth and the random mortalities caused by the key-factor, we asked whether the density dependent pupal mortality was sufficient to prevent parasite-host oscillations from increasing, assuming that the parasite was behaving as postulated by Nicholson, with a constant area of discovery as estimated from field data. In order that the answer to this question should be immediately clear, the random changes in host density due to the action of the key-factor (winter disappearance) were removed by using the mean of the value for this mortality as a constant. This model was stable and, moreover, generated a steady host density very similar to the average of our observations. It thus seemed that this model went a considerable way towards mimicking and explaining the level about which fluctuations in winter moth density take place in our area. The model also revealed that, if this correctly described the way in which winter moth populations are regulated, the parasite concerned could not be *Cyzenis*. The parasite densities generated by the model agreed more closely with those measured for *Cratichneumon*.

The next stage in developing the model was to start with the observed values for winter moth and *Cratichneumon* densities in the first year of observation (1950) and use the observed values for winter disappearance (the key-factor) rather than their mean value. The question then posed is: Does this model mimic the observed changes in host and parasite densities? Figure 4 shows that the model closely follows the observed changes in host density. However, although it shows a range of parasite densities which corresponds with that observed, the peaks of the cycle of changes in parasite density differs by two years from that observed. In this model, *Cratichneumon* has been assumed to have only one generation to one of the host; but a model in which it is given two generations does not mimic the observed parasite populations any better, and produces a worse fit with the host population. Clearly, the model suggests that *Cratichneumon* is not behaving as a Nicholsonian parasite.

Can we get a better fit between calculated and observed densities if we model parasite action according to the quest theory of Hassell and Varley?

A plot of the logarithms of the calculated areas of discovery for *Cratichneumon* against the logarithms of our estimates of parasite adult density suggests that the area of discovery decreases as parasite density

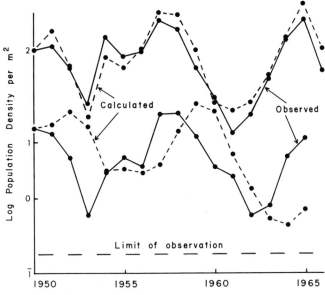

Fig. 4. A model for winter moth and its parasite *Cratichneumon culex* (Mueller). The figures are those for host larvae and parasite adults. The parasite is assumed to have one generation per year and a constant area of discovery.

increases. We are unable to prove this relationship statistically because of considerable errors in our measurements of adult densities, particularly when these are low (for details of the statistical test, see Hassell and Varley, 1969). To test if this theory gives a better description of *Cratichneumon* populations, we have run a series of models with different values of "Q" and "m" such that all slopes pass through the observed mean (log mean parasite density and log mean area of discovery). The model illustrated in Fig. 5 is the best of a small series of such calculations and is probably not the best which can be obtained. Unlike the model illustrated in Fig. 4, this model has two generations of the parasite to each host generation; the points plotted are for the generation which we most accurately measure and the corresponding calculated density. With respect both to the number of generations involved and the parasite densities generated, this model agrees much better with what we have observed in the field. Of course, from this we cannot assert that the parasite is behaving as postulated by the quest theory; nevertheless, it seems

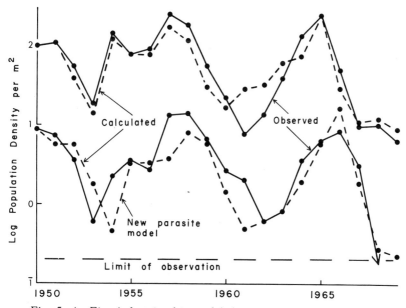

Fig. 5. As Fig. 4, but in this model the parasite is assumed to have two generations a year and the area of discovery falls as parasite density increases.

reasonable to suggest that, whatever the mechanisms involved, the consequences for the parasite population are very similar to those predicted by use of this theory.

Experimental Verification

Having developed a mathematical model of this kind, we would very much like to have some means of testing it. Experimental verification would be ideal, but involves many difficulties of scale, both in space and time. Possible tests fall into four categories:

1) *Prediction in our census area:* We are unable to predict population changes because we have no way of predicting the key factor causing population change. The fact that the model gives the correct value for population levels does not prove that the components of the model correctly represent reality. Errors in one component could be compensated in another.

2) *Experimental modification:* It has often been suggested to us that the experimental use of insecticides could provide a test. DeBach and Huffaker (Chapter 5) have used insecticide check methods in the field and shown that on individual twigs or whole trees that in the absence of parasites or predators a pest can rise in numbers until it severely damages the plant on which it feeds, but this provides proof only that parasites or predators cause mortality. It could provide no check on the validity of the model. Wood (Chapter 19) has ascribed the much larger outbreaks of defoliators of oil palm to the same basic cause. These experiments are so drastic that they eliminate one or more components of the model.

3) *Comparison of two isolated systems:* To a minor degree we have used this method by a comparison between our census area and Wistman's Wood, an isolated oak wood high on the flanks of Dartmoor in Devon. Here the trees are stunted, bear many epiphytes, and appear to be defoliated by winter moth almost every year. The beetle *Philonthus decorus* is present (Frank, 1967) but collections of winter moth larvae and pupae have failed to reveal either *Cyzenis* or *Cratichneumon.* The removal of the effect of *Cratichneumon* from our model results in an average winter moth population density twice as high as that in Wytham; such densities would account for the frequently observed defoliation at Wistman's Wood. A proper comparison between the two areas would require a number of years of detailed study in

both. The idea of an experimental introduction of *Cratichneumon* to Wistman's Wood, which is a Nature Reserve, is not practical. There are no known barriers to the natural spread of this common insect, which is most likely to be absent because conditions are in some way unsuitable.

4) *Biological control of winter moth in Canada:* This seems to provide the best possibility of experimental verification of the model. Winter moth is thought to have been accidentally introduced into Nova Scotia in the twenties, and was identified as a pest in 1949. In 1954 and 1955 various parasites were introduced from Europe, and the ichneumonid *Agrypon flaveolatum* and *Cyzenis albicans* became established. Embree (1965) (see Chapter 9) has provided life table information for some of the years before and after the parasites began to take effect. His information is reproduced in simplified form in Fig. 6. There is nothing in the life table information

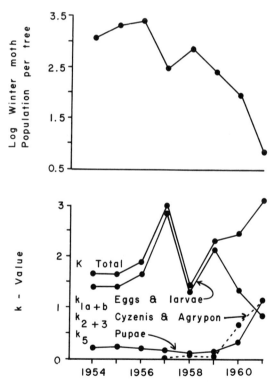

Fig. 6. k-values for mortality and population densities of winter moth in Canada. Data from Embree, 1965.

comparable to the strongly density dependent pupal mortality we observed in England. Although k_5, pupal mortality, in Fig. 6 has risen whilst the population density has been falling, which suggests that pupal mortality is an inverse factor, Embree suggests that this pupal mortality arises from the direct action of *Cyzenis*. In his census the percentage of larval parasitism is estimated from foliage samples, when parasitism is incomplete.

Without the stabilizing influence of the density dependent pupal predation our model predicts that, if *Cyzenis* acts as a Nicholsonian parasite, the interaction between host and parasite will lead to violent population oscillations. We therefore predicted that every 9 or 10 years population outbreaks in Canada would cause defoliation (Varley and Gradwell, 1968). If such outbreaks do not occur, perhaps the large reduction in population density might have allowed some predators of the pupae to cause a significant mortality. Alternatively, if *Cyzenis* behaves in a way describable by the quest theory, this alone might provide enough stabilization. However, Hassell and Varley (1969) found no direct evidence that *Cyzenis* had a significant mutual interference constant. We hope that continuing research by Canadian entomologists will provide further life table information for host and parasites which will enable us to test the model critically.

LITERATURE CITED

Auer, C. 1968. Erste Ergebnisse einfacher stochastischer Modelluntersuchungen uber die Ursachen der Populationsbewegung des grauen Larchenwicklers *Zeiraphera diniana* Gm. (= *Z. griseana* Hb.) im Oberengadin, 1949/66. Zeit. angew. Entomol. 62: 202-235.

Buckner, C. H. 1969. The common shrew (*Sorex araneus*) as a predator of the winter moth (*Operophtera brumata*) near Oxford, England. Can. Entomol. 101: 370-375.

Burnett, T. A. 1958. A model of host-parasite interaction. Proc. X Intern. Congr. Entomol., Montreal (1956) 2: 679-686.

DeBach, P. 1946. An insecticidal check method for measuring the efficacy of entomophagous insects. J. Econ. Entomol. 39: 695-697.

DeBach, P., and H. S. Smith. 1941a. Are population oscillations inherent in the host-parasite relation? Ecology 22: 363-369.

DeBach, P., and H. S. Smith. 1941b. The effect of host density on the rate of reproduction of entomophagous parasites. J. Econ. Entomol. 34: 741-745.

Embree, D. G. 1965. The population dynamics of the winter moth in Nova Scotia, 1954-1962. Mem. Entomol. Soc. Can. 46. 57 pp.

Frank, J. H. 1967. The effect of pupal predators on a population of winter moth *Operophtera brumata* (L.). J. Anim. Ecol. 36: 611-621.

Hassell, M. P. 1966. Evaluation of parasite and predator responses. J. Anim. Ecol. 35: 65-75.

Hassell, M. P. 1968. The behavioural response of a tachinid fly (*Cyzenis albicans* (Fall.)) to its host, the winter moth (*Operophtera brumata* (L.)). J. Anim. Ecol. 37:

627-639.

Hassell, M. P. 1969a. A study of the mortality factors operating upon *Cyzenis albicans* (Fall.) a tachinid parasite of the winter moth (*Operophtera brumata* (L.)). J. Anim. Ecol. 38: 329-339.

Hassell, M. P. 1969b. A population model for the interaction between *Cyzenis albicans* (Fall.) (Tachinidae) and *Operophtera brumata* (L.) (Geometridae) at Wytham, Berkshire. J. Anim. Ecol. 38: 567-576.

Hassell, M. P., and C. B. Huffaker. 1969. The appraisal of delayed and direct density-dependence. Can. Entomol. 101: 353-361.

Hassell, M. P., and G. C. Varley. 1969. New inductive population model for insect parasites and its bearing on biological control. Nature: 223: 1133-1137.

Holling, C. S. 1959. Components of predation as revealed by a study of small mammal predation of the European pine sawfly. Can. Entomol. 91: 239-320.

Howard, L. O., and W. F. Fiske. 1911. The importation into the United States of the parasites of the gypsy moth and the brown-tail moth. U.S. Dept. Agr. Entomol. Bull. 91, 312 pp.

Huffaker, C. B., and C. E. Kennett. 1969. Some aspects of assessing efficiency of natural enemies. Can. Entomol. 101: 425-447.

Milne, A. 1957. Theories of natural control of insect populations. Symp. Cold Spring Harbor Quant. Biol. 22: 253-271.

Morris, R. F. 1959. Single-factor analysis in population dynamics. Ecology 40: 580-588.

Morris, R. F. 1963a. The dynamics of epidemic spruce budworm populations. Mem. Entomol. Soc. Can. 31. 332 pp.

Morris, R. F. 1963b. Predictive population equations based on key factors. Mem. Entomol. Soc. Can. 32: 16-21.

Morris, R. F., and T. Royama. 1969. Logarithmic regression as an index of responses to population density. Can. Entomol. 101: 361-364.

Nicholson, A. J. 1933. The balance of animal populations. J. Anim. Ecol. 2: 132-178.

Nicholson, A. J. 1954. An outline of the dynamics of animal populations. Aust. J. Zool. 2: 9-65.

Nicholson, A. J., and V. A. Bailey. 1935. The balance of animal populations. Part 1. Proc. Zool. Soc. London. Pp. 551-598.

Smith, H. S. 1935. The role of biotic factors in the determination of population densities. J. Econ. Entomol. 28: 873-898.

Southwood, T. R. E. 1966. Ecological Methods. Methuen, London. 391 pp.

Takahashi, F. 1968. Functional response to host density in a parasitic wasp, with reference to population regulation. Res. Pop. Ecol. 10: 54-68.

Varley, G. C. 1947. The natural control of population balance in the knapweed gall-fly (*Urophora jaceana*). J. Anim. Ecol. 16: 139-187.

Varley, G. C. 1953. Ecological aspects of population regulation. Proc. IX Intern. Congr. Entomol., Amsterdam (1951) 2: 210-214.

Varley, G. C. 1958. Meaning of density-dependence and related terms in population dynamics. Nature 181: 1778-1781.

Varley, G. C., and G. R. Gradwell. 1960. Key factors in population studies. J. Anim. Ecol. 29: 399-401.

Varley, G. C., and G. R. Gradwell. 1963. The interpretation of insect population changes. Proc. Ceylon Assoc. Advan. Sci. 18: 142-156.

Varley, G. C., and G. R. Gradwell. 1968. Population models for the winter moth. Pp. 132-142. *In* Insect Abundance, T. R. E. Southwood (ed.). Symp. Roy. Entomol. Soc.

London, No. 4.

Varley, G. C., and G. R. Gradwell. 1970. Recent advances in insect population dynamics. Ann. Rev. Entomol. 15: 1-24.

Varley, G. C., and G. R. Gradwell. In press. Can parasites avoid competitive exclusion? Proc. XIII Intern. Congr. Entomol., Moscow (1968).

Chapter 5

EXPERIMENTAL TECHNIQUES FOR EVALUATION OF THE EFFECTIVENESS OF NATURAL ENEMIES

Paul DeBach

University of California
Riverside, California

and C. B. Huffaker

University of California
Berkeley, California

INTRODUCTION

The role of natural enemies in prey (or host) population regulation has long been debated by ecologists, and various methods of evaluation proposed and attempted. The fact that considerable disagreement remains is sufficient evidence of the need for adequate assessment of the efficiency of enemies. Precise means of evaluation are needed for several reasons: (1) to illuminate basic principles of population ecology, especially the role of biotic and abiotic factors, (2) to furnish a sound ecological basis for manipulation of natural enemies either by demonstrating the need for importation of new ones or by showing that the established ones are rendered ineffective due to interference phenomena, including pesticides, weather or cultural practices, and (3) to provide the proof of effectiveness of natural enemies that is necessary in applied biological control in order to justify needed increased research and development support.

The use of biological control is essentially the antithesis of chemical pest control. Obviously, a lot of skeptics disavow biological control because of their personal interest or experience in the use of pesticides. Additionally, there are still some few ecologists who consider that natural enemies rarely if ever regulate prey populations. Climate often is proposed as the key factor by them, on the basis of circumstantial evidence and poor deductive reasoning.

This leaves no alternative but to answer the skeptics with satisfactory proof.

A clear distinction should be drawn here between the mechanisms involved in regulation of prey (and host) populations by natural enemies and the end results, i.e., the fact and degree of control or regulation by enemies. These concepts involve two distinctly different questions: (1) does prey population regulation by enemies actually occur, and, if so, at what population level, and (2) how does regulation occur?

This chapter is not primarily concerned with the latter question but rather with the former, i.e., with means of proving whether or not regulation by enemies actually does occur, and, if so, at what average population level. The first question, involving proof, is the more important to biological control practice, and possibly theory as well, because, of course, its positive answer is the only reason for asking and trying to explain the second question. Strangely enough, much more has been written about the mechanisms involved in regulation than about the experimental or quantitative demonstration of the actual fact of population regulation and biological control. The mechanisms of regulation are dealt with in Chapter 2.

So much disagreement and argument has taken place among ecologists concerning the academic and semantic complexities of question (2) above, that it has led some of the more theoretical workers to conclude that true regulation by natural enemies either does not occur or cannot be measured. This chapter will emphasize that a variety of comparative experimental techniques may be employed to accurately evaluate and demonstrate the precise contribution of natural enemies in prey (host) population regulation in any particular habitat. Such techniques can be devised to exclude the possible contributing or compensatory effects of other environmental parameters such as weather, competitors, or genetic variation, etc., so as to evaluate the actual total regulatory effect brought about essentially by the action of natural enemies.

By *population regulation* we mean what is commonly referred to as natural balance—the maintenance of an organism's population density over an extended period of time between characteristic upper and lower limits. A regulatory factor is one which is wholly or partially responsible for the observed regulation under the given environmental regime and whose removal or adverse change in efficiency or degree will result in an increase in the average pest population density. Note that this definition is amenable to experimental testing. The ecological term, biological control, connotes prey population regulation by natural enemies. Note that economic qualifications purposely are excluded. The degree of economic achievement must be defined for each case; biological control can occur at very low or very high average densities. Temporary suppression or control by some enemies at times would not satisfy the connotation of "regulation," but in practice, reliable long-term *control* by an effective natural enemy is largely indistinguishable from *regulation* by that enemy. Biological control also is used in another sense, i.e., to refer to this particular ecological field or discipline.

The emphasis herein on experimental means of evaluation does not mean that there are no other approaches. However, we consider that other methods do not provide the rigorous proof of control or regulation by enemies that is needed. Periodic census and life-table data provide much valuable information, but such quantitative methods, including regression and modeling techniques, have weaknesses for the adequate rating of the regulating or controlling power of natural enemies. Detailed discussion and analysis of this subject may be found in articles by DeBach and Bartlett (*in* DeBach, (ed.), 1964, pp. 412-426), Huffaker and Kennett (1969), Hassell and Huffaker (1969), and in Chapter 2 herein. According to Legner (1969), "The ultimate and probably only reliable method for judging a parasite's effectiveness is the reduction in host equilibrium position following liberation." He was, of course, referring to newly imported exotic parasites and an evaluation of their effectiveness by using *before* and *after* density comparisons. The important point in the present context is his (and our own) conclusion that neither precise laboratory studies nor quantitative field studies of parasitization furnish adequate means of evaluating the effectiveness of a parasite in host population regulation and control. Along the same lines, Hassell and Varley (1969), stated, "Laboratory studies can go some way to predict which species [among several introduced] may be successful, but an introduction provides the only real test."

The purpose of this paper is to describe methods of appraising the actual and potential importance of natural enemies, not of their possible utilization in integrated control programs, which subject is considered in other chapters of this book.

SELECTION OF TEST PLOTS

Regardless of the methods employed to evaluate the underlying importance of enemies in population regulation, study plots must be planned and chosen carefully so that cryptic interference phenomena do not inadvertently mask or preclude the ability of enemies to be effective. One cannot learn about inherent biological control potentials in a plot in which, for one man-caused reason or another, natural enemies cannot exist or cannot freely respond to host population increase. An extreme example (but by no means unusual in some circles) would involve the selection of a small plot in the center of a 100-acre cotton field which, in order presumably to study biological control, would be left untreated with chemicals. All the surrounding area would be sprayed and dusted throughout the season, perhaps by airplane. The implausibility of this approach is obvious, yet results from such plots are commonly used as evidence that natural enemies are ineffective and therefore that regular and continuing insecticidal treatment is necessary.

To be ecologically meaningful for the evaluation of the optimal efficiency of natural enemies the biological control study plot must be an adequate representation of the surrounding faunal-complex area. In other words, the results from the plot should enable us to predict the general potential results to be expected over a much larger area if non-treatment were practiced over the larger area. There are many pitfalls in picking a biologically satisfactory plot in an agroecosystem; fewer in a natural habitat of some uniformity. Since most applied biological control research is carried out in agroecosystems, and proper plot selection is most difficult in agroecosystems, we will emphasize this aspect. However, the same general qualifications apply to any study of the regulatory role of parasites, predators or disease in nature.

The size of the study area is of prime importance. The plot must be large enough (including a peripheral buffer zone) to exclude outside influence or practices that would adversely affect the natural enemies. Thus, no consequential drift of pesticides applied elsewhere should penetrate into the study area. Also, abnormal biological drift or migration of organisms, brought about by "population explosions" in nearby pesticide-treated plots should be minimal. For example, spider mites frequently increase greatly in numbers following the chemical treatments and at such high densities some species spin ballooning threads and drift away. A relatively distant plot under biological control at low population levels can suddenly be inundated by such an immigrant mite population, which may then increase rapidly to even higher levels before the resident natural enemies can build up and restore control or equilibrium. Such an occurrence can wrongly convince a grower or casual observer that biological control is inherently ineffective and lead to resumption or continuation of treatments.

Penetration of airborne dust into the plot from outside should be minimized. Dust has been shown (DeBach, 1969) to affect small natural enemies adversely and it is known to commonly cause marked increases in populations of mites and scale insects. Loss of natural enemies by emigration from small plots also is important, especially if they move into adjacent pesticide-treated areas from which there is no return or comparable immigration.

Elimination of adverse within-plot practices or influences is necessary. Various cultural practices come under this category. Again, dust should be minimized to the fullest extent possible. No pesticides or other chemical applications should be applied unless they are absolutely necessary. The harmful effect of certain "non-toxic" materials would not commonly be anticipated. For example, use of certain zinc formulations to correct minor element deficiency in citrus groves is very detrimental to certain natural enemies and commonly leads to large increases in mite populations. The strikingly adverse effects of many commonly used insecticides are too well known to require repeating here. There is no possibility of accurately evaluating the potential effect of enemies when such pesticides are used. Clean cultivation in orchards may be detrimental for enemies as compared to a growth of ground cover, which may provide a source of alternative prey. Ants are known to interfere with or kill parasites and predators of various honeydew secreting insects and their associates, resulting in striking pest increases in some instances. In various crops, ants must be controlled or the potential for biological control will be effectively negated.

Time is critical in biological control studies. A sufficient period must lapse following the last chemical pesticide application for the residue to disappear and for the pest insect and its established enemies to have time to establish a balance. We have found DDT residues to be biologically active against some parasites for at least two years after field application. We have also found chemically-upset insect and mite populations to take three or more years to be brought back under biological control after all treatment had ceased (DeBach, 1969; Flaherty and Huffaker, 1970). Thus, it is obvious that the biological control potential from established enemies cannot be accurately evaluated in study plots which have not gone without major chemical pesticide treatments for significant periods.

In agro-ecosystems, crop production and quality in relation to cost is the final criterion to the grower (not necessarily to the public) regarding his method of pest control. This aspect of the evaluation of biological control needs much more emphasis. A crop area is required that will provide statistically reliable information on production and quality for the biological control (or integrated control) plot as compared with the regularly-treated chemical control program area.

EXPERIMENTAL METHODS OF EVALUATION

There are two related general problems in assessment of natural enemy effectiveness. These involve: (1) measurement of the beneficial results of colonization of newly imported exotic parasites or predators, and (2) measurement of the degree of biological control exerted by already established enemies. Experimental evaluation methods can be classified under the following terms: (1) Addition, (2) Exclusion (or Subtraction), and (3) Interference, according to how natural enemies are manipulated or affected in the experimental procedure used.

The Addition method applies primarily to measurement of the results of importation (including subsequent transfer or relocation) of new natural enemies. The procedure involves "before and after" type comparisons involving initially comparable plots, some having natural enemies colonized, and others not receiving any. Thus, for example, an imported parasite can be colonized in 10 plots and its population increase as well as the host-population decrease measured. Host population trends in 10 plots not receiving parasites would be measured simultaneously. Any differences between the two series can be ascribed to the parasites. Such tests, of course, must be planned well in advance of colonization, because dispersal of some parasites or predators is so rapid as to eliminate any enemy-free plots within a short time. Obviously, plots should be separated initially by a suitable distance. Replication of this type of test can be achieved by conducting the same sort of experiments following transfer of the natural enemy to a new area or country. Results from 10 olive groves where *Aphytis maculicornis* (Masi) alone was present and 10 groves where a second parasite, *Coccophagoides utilis* Doutt, additionally was present, were used by Huffaker and Kennett (1966) to evaluate the net effect of the addition of the second species in the control of olive scale, *Parlatoria oleae* (Colvee) in California. (See further Chapters 2 and 7.)

In many cases of biological control, results are so striking that "before and after" photographs are more spectacular than mere population census figures. Preferably, both types of measurement should be obtained. For the purpose of this chapter, photographs illustrative of each method will be included whenever possible. Although there are now many instances of the use of experimental evaluation techniques reported in the literature, we will rely to a large extent on our own studies for illustration of the general techniques involved. Figure 1, representing the Addition method, shows the degree of biological control achieved in California of the formerly serious Klamath weed, *Hypericum perforatum* L., by colonization of an imported exotic beetle which feeds upon it.

The Exclusion or Subtraction method usually involves elimination and subsequent exclusion of established natural enemies from an appropriate series of plots which are then compared sequentially with an otherwise comparable series of plots having natural enemies undisturbed. Differential "before and after" levels of pest population density, indicating different equilibrium densities for the two series of plots, is a direct meausre of the total regulatory

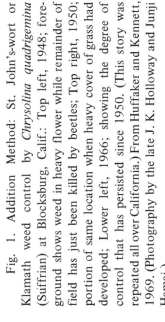

Fig. 1. Addition Method: St. John's-wort or Klamath weed control by *Chrysolina quadrigemina* (Suffrian) at Blocksburg, Calif.: Top left, 1948; foreground shows weed in heavy flower while remainder of field has just been killed by beetles; Top right, 1950; portion of same location when heavy cover of grass had developed; Lower left, 1966; showing the degree of control that has persisted since 1950. (This story was repeated all over California.) From Huffaker and Kennett, 1969. (Photography by the late J. K. Holloway and Junji Hamai.)

and control effectiveness of the natural enemies. The experiments must be so designed as to be biologically realistic, taking into account, for example, behavioral and ecological characteristics, so that the exclusion technique applied is itself not exerting any appreciable influence on the observed results. This is less of a problem with more sessile insects than with those which disperse or migrate actively. The application of ingenuity will doubtless permit evaluation of many cases previously unexplored.

Elimination of natural enemies can be accomplished in various ways, both chemical and mechanical (Smith and DeBach, 1942; DeBach, Dietrick and Fleschner, 1949; Huffaker and Kennett, 1956). The pest may or may not be eliminated simultaneously. If so, it must be subsequently reintroduced into the plot. Following true elimination, the enemies are generally excluded by mechanical means, usually cages, but barriers of various sorts would suffice for non-flying enemies. Predatory ants, for example, may be excluded by hot, electrical resistance-wire barriers. Spatial isolation of the host plant and pest, away from sources of natural enemies, may suffice in certain cases.

Whatever exclusion technique is used, utmost care is required that the comparisons involve only the one major variable, natural enemies. Thus, if cages are used for exclusion, similar cages are used in the non-exclusion, enemy-present plots, with the difference being that openings are present to permit ingress and egress of the enemies. If the method of exclusion is by use of a chemical that might have a stimulative effect on the power of increase of the pest species, this possibility should be checked independently (see below).

Types and results of a series of exclusion tests are shown pictorially in Figs. 2, 3 and 4. Obviously, the exclusion of natural enemies from their prey furnishes a dramatic method of proof of effectiveness. Figure 5 shows a series of insect-tight tree cages, which are being utilized by one of us (DeBach) in current exclusion vs. non-exclusion comparisons. Ambient temperature and humidity are maintained in the cages by large volume air transfer. Various combinations of pests and natural enemies can be tested simultaneously under nearly natural conditions. A single tree serves as a satisfactory universe for the insects being tested in this case.

Although many might be cited, only one example of quantitative differences, as shown by sample counts, will be given in order to conserve space. Census data on percent prey (host) mortality obtained from leaf cell exclusion test comparisons are shown in Fig. 6. These data show that a striking decrease in host (California red scale, *Aonidiella aurantii* (Mask.)) mortality is caused by the absence of parasites. Even more striking results were obtained in similar field tests which involved a different host, the oleander scale, but the same species of parasite, *Aphytis melinus* DeBach. In this case, after some two months, scale mortality was only 8 per cent in

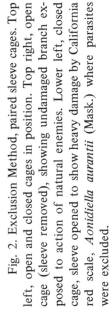

Fig. 2. Exclusion Method, paired sleeve cages. Top left, open and closed cages in position. Top right, open cage (sleeve removed), showing undamaged branch exposed to action of natural enemies. Lower left, closed cage, sleeve opened to show heavy damage by California red scale, *Aonidiella aurantii* (Mask.) where parasites were excluded.

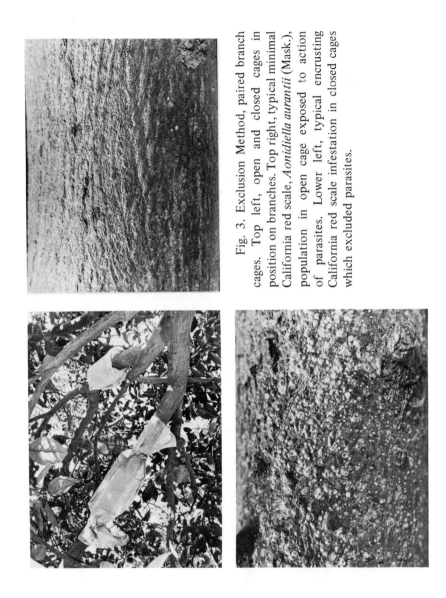

Fig. 3. Exclusion Method, paired branch cages. Top left, open and closed cages in position on branches. Top right, typical minimal California red scale, *Aonidiella aurantii* (Mask.), population in open cage exposed to action of parasites. Lower left, typical encrusting California red scale infestation in closed cages which excluded parasites.

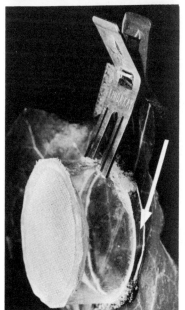

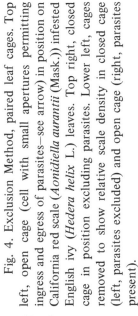

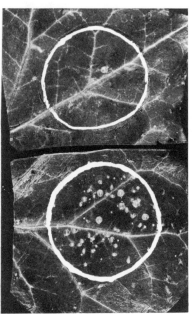

Fig. 4. Exclusion Method, paired leaf cages. Top left, open cage (cell with small apertures permitting ingress and egress of parasites—see arrow) in position on California red scale (*Aonidiella aurantii* (Mask.)) infested English ivy (*Hedera helix* L.) leaves. Top right, closed cage in position excluding parasites. Lower left, cages removed to show relative scale density in closed cage (left, parasites excluded) and open cage (right, parasites present).

Fig. 5. Exclusion Method, tree cages. Two aspects of valencia orange trees enclosed by insect-tight tree cages for evaluation of natural enemies. (Biological Control Grove, University of California, Riverside)

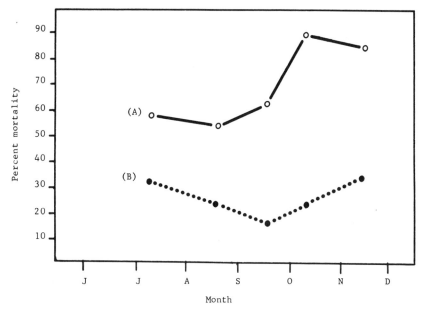

Fig. 6. Exclusion Method. Average California red scale (*Aonidiella aurantii* (Mask.)) mortality on English ivy (*Hedera helix* L.). (A) in open cells permitting entry of parasites and (B) in closed cells excluding parasites.

parasite-excluded leaf cells, whereas it was 92 per cent in cells which naturally occurring parasites were able to enter. The cells of each series were so constructed and situated that any possible differences due to microclimate could be ruled out.

The Interference or Neutralization method involves some means of more or less greatly reducing the efficiency of natural enemies as regulatory agents in one series of plots, as compared to another series having natural enemies operating undisturbed. "Before and after" comparative differences in pest density are measured as in the exclusion method and any increase in population equilibrium density in the interference plots, relative to the normal biological control plots, is a demonstration that natural enemies are responsible for the lower average pest population level observed in the absence of interference. However, such comparisons do not show the total extent of host population reduction, because natural enemies are not completely eliminated or excluded, and so may continue to produce some effect in limiting the host.

In other words, an increased average host density following interference demonstrates that enemies were responsible for the original lower level, but the maximum level the host would attain in their *complete absence* remains unknown, because enemies whose efficiency is reduced by interference may still remain the chief regulating agency at some considerably higher average pest density level, or contribute to the collective regulative process (see Huffaker and Messenger, *in* DeBach, ed., 1964, their Chapter 4), involving perhaps considerable competition for food. When severe damage results (Figs. 2, 7 and 8), the latter becomes the chief regulating agency.

Strange as it may seem, percentage parasitization of the host (i.e., natural enemy caused mortality) may commonly be just as great in the interference plots as in the normal activity plots. The explanation for such a phenomenon has been furnished by Huffaker and Messenger (*in* DeBach, ed., 1964; see their pp. 82-83 and especially Fig. 6). It is based on the theory that the rate at which premature mortality of a pest increases with density determines the density of the pest population at equilibrium. If rate of parasitization by a parasite species increases rapidly as the host population begins to increase, the host population density increase will be rapidly halted and will be low at the equilibrium position; if it increases slowly, the density will be high. This is because the mortality level at which mortality equals natality, i.e., the equilibrium position, is reached rapidly and at low levels in the first instance and not in the latter. This is illustrative of the difference between an effective and an ineffective parasite species and of the improved results accruing from the introduction of an efficient exotic parasite species. It also explains why the application of pesticides—an interference measure— causes pest upsets when effective natural enemies are inhibited. The previously effective enemies are made less efficient by reducing their ability to respond rapidly, with increased parasitization, to the tendency of the prey population to increase. It is important to distinguish between rate of increase of parasitization (i.e., numbers of hosts or prey killed) and percentage parasitiza- tion (percentage mortality). At population equilibrium in a series of different habitats where parasitism is the regulating agency in each case, and provided the pest's reproductive capacity is the same and other mortality is the same, the percentage of parasite-caused mortality will theoretically be the same regardless of the fact that the pest equilibrium densities may differ markedly in the different habitats. Thus, percentage mortality in itself is not an index of the importance of a parasite, predator or pathogen, although it is commonly so used. This is one major reason why we advocate the experimen- tal methods discussed in this chapter.

Several types of interference techniques have been employed, such as the "insecticidal check method," the "biological check method," the "hand-removal method," and the "trap-method" (DeBach and Bartlett, *in* DeBach, ed., 1964). If these techniques were to be applied stringently enough, they might possibly become exclusion methods. As used, all of them either kill or disturb or exclude (often a combination) a large proportion of the natural enemies originally present, so that the net result is a substantial reduction in the average rate of increase of the enemy population with respect to that of the pest. This permits escape of the pest population from a lower to a higher level (if the enemies were indeed regulatory) and thereby demonstrates that enemies were the responsible regulating feature at the original lower level.

The insecticidal (or chemical) check method (DeBach, 1946) employs selective toxic chemicals in one series of plots to kill a large proportion of the natural enemies while, ideally, having little or no adverse effect on the pest population. The development of pest populations then is compared to that of other plots having natural enemies undisturbed. DDT has been a favorite insecticide for this technique because it is very long-lasting and highly toxic to most natural enemies, hence is more useful in long-term field plots. It is considerably less toxic to a substantial number of pest insects, and dosages or timing of application can be varied to maximize the differential toxicity to the natural enemies as compared to the pest. In this method, the chemical is reapplied at intervals as required until sufficient time has passed for any differential population trends to become evident. Any chemical, including such "non-toxic" substances as talc, pyrophyllite, or road dust may be used in this method as long as a strong differential mortality occurs—little or none of the host—but substantial mortality of the natural enemies.

It must be cautioned that materials should be selected that have minimal, or no, effect on the pest's fecundity. Use of an additional form of the check method that does not present this possibility is a desirable safeguard. Huffaker and Kennett (1956) used hand removal as a "check" on the chemical check method they employed to evaluate the role of phytoseiid predators in the regulation of cyclamen mite densities in strawberries. Supplemental information derived from the census data may serve the same purpose. DeBach (1955) and Huffaker, Kennett and Finney (1962) found that DDT used as the check method chemical had no effect on the fecundity of the pest species studied. Progeny production was the same for those females that remained unparasitized on the untreated trees as for those on the DDT-treated trees.

Results achieved by use of the chemical check method are shown pictorially in Figs. 7 and 8. The former shows the damage to a citrus tree that occurred when the California red scale population exploded due to a

Fig. 7. Interference Method: Insecticidal Check. Defoliation and twig die-back caused by California red scale, *Aonidiella aurantii* (Mask.), on a tree sprayed with DDT (left) which inhibited the parasites, as compared to untreated check tree (right) under biological control.

neutralization of parasites by DDT and thus indirectly indicates the tremendous degree of control exerted, principally by only one species of parasite in this case. An idea of the relative differences in population density between trees under biological control and those having biological control of red scale upset by experimental use of DDT is given in Table 1 by the initial and final population figures from tests in several orchards. Initial counts were made in both types of plots just before DDT was applied to one series. Initial population densities were low and quite comparable for a given grove; final counts were made after one or two seasons, depending upon the differences between habitats. A rating of 10 or less indicates that the scale is very scarce and under an excellent degree of economic control, ratings much above 20 are probably not economically acceptable, and those above 50-100 begin to cause visible twig or branch kill.

Figure 8 shows visually the spectacular degree of biological control exerted by a single parasite species in the control of olive scale. This is actual field material, chosen to represent a typical comparison of the minimal extent of infestation occurring under undisturbed parasite activity, with the exploded

Fig. 8. Interference Method: Insecticidal Check. Heavy infestation of olive scale, *Parlatoria oleae* Colvee, on branch of DDT-sprayed olive tree (top) which inhibited parasites, as compared to scale-free branch from untreated tree (bottom) which had normal parasite activity, thus demonstrating the effectiveness of the parasite, *Aphytis maculicornis* (Masi). (Photography by F. E. Skinner)

TABLE 1

THE DEGREE OF BIOLOGICAL CONTROL OF CALIFORNIA RED SCALE,

AS DEMONSTRATED BY THE INSECTICIDAL CHECK METHOD

| Orchard # | Locality | Population Density -- California red scale | | | |
| | | Initial | | Final | |
		DDT-treated	Un-treated	DDT-treated	Un-treated
1	Irvine, Orange Co.	0	2	125	3
2	Sinaloa, Ventura Co.	35*	46*	572	17
3	Sespe, Ventura Co.	1	1	425	7
4	Biological Control Grove UCR, Riverside Co.	8	2	246	8
5	Birdsall, San Bernardino Co.	0	0	67	6
6	Beemer, San Diego Co.	4	5	158	3

*Initially heavy, due to previous upset by ants. Ants were controlled subsequently.

infestation characteristic of DDT-caused depression of natural enemy activity. The ratio of initial to final densities were 100-fold to 1000-fold. (See further Chapter 7.)

DDT is not suitable for use in these methods when it kills a high proportion of the pest. Our attempts to use it for assessment of the extent of biological control of the purple scale, *Lepidosaphes beckii* (Newman) and the chaff scale, *Parlatoria pergandii* Comstock, failed because it greatly reduced the pest population. Tests of various chemicals for suitable selectivity in a method for assessing biological control of purple scale led to the choice of endrin. The great increase in purple scale populations shown in Fig. 9 on endrin-treated trees, as contrasted with untreated trees, constitutes proof that the untreated checks were being held under biological control. In this case, also, it was by only one species of parasite, *Aphytis lepidosaphes* Compere. By varying the timing of the chemical applications, we have also used this method to assess the relative regulatory roles of two different species of parasites attacking the same host population (Huffaker and Kennett, 1966; DeBach *et al.*, Chapter 7, herein).

The biological check method is based on interference with natural enemies by ants (DeBach, Dietrick and Fleschner, 1951; DeBach, Fleschner and Dietrick, 1951). Honeydew-seeking ants constantly interfere with or kill natural enemies, but they do not eliminate or exclude them completely. In the cases we have studied, the ants did not directly favor increases of the pest insects by "cultivating" them as has sometimes been postulated. In fact, the

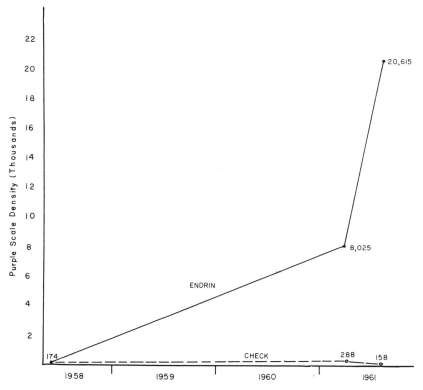

Fig. 9. Interference Method: Insecticidal Check. Increase in population density of purple scale, *Lepidosaphes beckii* (Newm.), over a three year period, resulting from the use of a suitable selective insecticide (endrin) which inhibited activity of the parasite, *Aphytis lepidosaphes* Compere.

interference action of such ants towards parasites and predators is indiscriminate, so that they cause population increases in pests that produce no honeydew and in which ants have no interest whatsoever. The action of ants in producing pest population increases can be explained on the same general basis as is true for the chemical check method.

Experimental comparisons between ant-infested and ant-free plots are conducted in the same general manner as explained for other methods. When starting from "scratch" with comparable initial populations, the biological check method may require from one to two years to achieve maximum differences. When pest populations increase in the presence of ants, as compared to plots where ants are absent, this demonstrates that within the environmental regime the average population density observed in the absence of ants is due to natural enemies. Because the action of ants is not as severe or as constantly maintained as that of DDT, for example, pest populations generally are not upset as much by ants as by the chemical check method. Thus, the total extent of biological control may not be as apparent in this method, but the fact that enemies are responsible for the lower level of control observed in the absence of ants is verified.

Although the use of this method is restricted to studies on crops which have suitable ant species present, such an occurrence is rather common and can sometimes be fostered. It furnishes an excellent complement to other methods and does not have some of the objections which may apply to certain others. Perhaps the most important is that microclimate is not affected by ants, as it may be in certain cage comparisons. More details are mentioned by DeBach and Bartlett (in: DeBach, ed., 1964, pp. 424-426).

The use of this method has furnished some striking visual results, as well as many examples of great quantitative differences determined by census between populations of various pest species occurring in plots infested with ants as contrasted to ones having ants removed or absent. Fig. 10 shows the difference in biological control, expressed by tree damage, which occurred in the presence and absence of ants. The ants interfered with several species of natural enemies and caused at least two species of pests to increase to high levels. The majority of the damage was caused by an enormous population of the California red scale, a species in which the ants have no interest. The red scale populations are very low on ant-free trees.

Fig. 11 shows the typical degree of biological control maintained by natural enemies against the citrus mealybug, *Planococcus citri* (Risso), by comparison with an otherwise comparable plot in which ants interfered with the activity of parasites and predators.

Differences in pest population densities of from a few-fold to 100-fold or more may occur in this method. Table 2 includes some typical quantitative

Fig. 10. Interference Method: Biological Check. Tree damage from California red scale, *Aonidiella aurantii* (Mask.), resulting from ants interfering with the performance of natural enemies (left) as compared to a check tree where ants were eliminated (right). The relative visibility of the researcher in the two photos is an indication of differences in infestation as represented by pest caused defoliation and thus represents the degree of biological control.

Fig. 11. Interference Method: Biological Check. Citrus mealybug, *Plano-coccus citri* (Risso), infestation on orange resulting from interference by ants with natural enemies (top) as compared with typical fruit on a tree having ants excluded (bottom).

results. The higher densities on ant-infested plots show that each pest species was under appreciable biological control when natural enemies were not interfered with.

The hand-removal method for "elimination" of natural enemies from a plot may provide the best approach of all. This is because few technical objections, such as possible influence on microclimate, possible stimulation of pests by chemicals, et cetera, apply to this method. The principal objection is the practical one of providing the trained manpower to monitor one or more rather small plots constantly enough, and over a long enough period (probably one to many months), to permit population changes to become evident. One man can probably only effectively exclude or retard enemies on one shrub or a portion of a tree. Thus, the strategy of this method is superior, but the tactics present formidable problems.

The results from this method can be just as spectacular as those from any of the others. Fleschner, Hall and Ricker (1955) removed natural enemies

TABLE 2

DEMONSTRATION OF THE DEGREE OF BIOLOGICAL CONTROL OF VARIOUS

CITRUS PESTS BY EMPLOYMENT OF THE BIOLOGICAL CHECK METHOD

| Pest | | Final Population Density* | |
Species	Locality	Ants absent	Ants present
California red scale**	Santa Barbara	3	65
Citrus red mite**	Montecito	66	1437
Black scale	Escondido	2	58
Citrus mealybug	Montecito	5	16
Citrus aphids	Escondido	0	5
Brown soft scale	Escondido	0	13

*Initial population densities were comparable. Tests ran from 1 year to 2 years to obtain the differences shown. These are typical examples. Many replicates are available for certain species. The results shown are indicative for the local habitats concerned. They may not apply generally.

**These species do not produce honeydew and are not "tended" or otherwise directly influenced by the ants.

of avocado pests by hand, on a 24-hour basis, for a period of 84 days. From practical necessity, their "enemy-free plots" consisted of individual branches or portions of a tree. These were compared with the rest of the tree, which had natural enemies operating normally. They demonstrated that biological control was responsible for the normally low pest populations in the experimental grove. This included five potential pests in diverse taxonomic groups, namely: the omnivorous looper, *Sabulodes caberata* Gir., the six-spotted mite, *Eotetranychus sexmaculatus* (Riley), the long-tailed mealybug, *Pseudococcus adonidum* (L.), the avocado brown mite, *Oligonychus punicae* (Hirst) and the latania scale, *Hemiberlesia lataniae* (Sign.). Fig. 12 shows the results of hand-removal of predators of the avocado brown mite.

The trap method is a variation of the insecticidal check method that involves a central untreated plot surrounded by a chemically poisoned zone which acts to kill natural enemies as they disperse to or from the central plot. After a period of time, they may become greatly decimated in the untreated central plot, thus permitting differential increase of pests which previously had been held under biological control. Comparison of such a plot with an otherwise comparable, untrapped plot a suitable distance away, provides evidence as to whether natural enemies were exerting biological control. DeBach and Bartlett (1951) reported the first successful use of this method when they demonstrated that the famous vedalia beetle was decimated and its prey, the cottony-cushion scale, increased under this technique.

Later, Bartlett (1957) used the technique to assess biological control of the citrus mealybug by natural enemies. He compared mealybug population trends in an area where natural enemies remained undisturbed with another area where natural enemies were inhibited by surrounding the untreated plot with a barricade zone of DDT-treated trees. He showed that certain natural enemies were severely inhibited, others very little, but the comparisons, which covered two seasons, showed that the mealybugs were held under commercially satisfactory control by natural enemies, except for one month, whereas in the trapped plot mealybug populations exceeded economic limits during eight months.

The advantage of this technique, if carefully planned and applied, is that nothing is modified directly in the plot having natural enemies inhibited. Microclimate is not modified and there is no possibility of chemicals being a confusing factor. Obviously, the method will probably be most effective with fairly sessile pests having rather motile natural enemies, for if the trap effect acts strongly on the pest species, too, the purpose is defeated. Use of several different selective chemicals, applied at intervals as required, could make this a highly sophisticated technique, deserving of more widespread use.

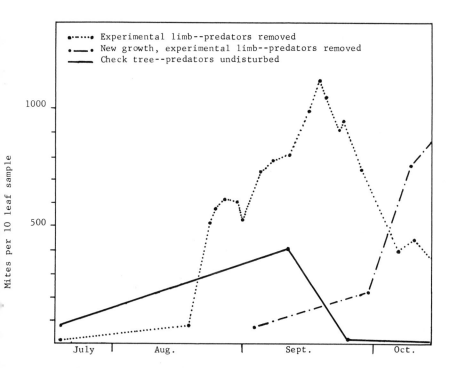

Fig. 12. Interference Method: Hand Removal. Avocado brown mite, *Oligonychus punicae* (Hirst), populations on old and new growth on an experimental avocado limb from which natural enemies were removed, compared to those on the check tree having normal enemy activity (after Fleschner *et al.*, 1955).

DISCUSSION AND CONCLUSIONS

In recent years the premises, principles, practices and future promise of biological control have received considerable discussion. Most of this has been laudatory, but some has raised genuinely constructive questions and a little has been uncritically unfavorable and perhaps purposefully destructive. Authors responsible for the latter, even though lacking in expertise or even minimal experience in the field, need to be answered or their foolishness may be accepted as scientific fact. They have been rebutted in a series of papers by several authors which are summarized in Chapter 2.

The point we wish to emphasize here is that this chapter has illustrated many cases of scientific proof of biological control. Such proof has often been overlooked by skeptics, hopefully because they are unfamiliar with the literature. The application of the various test techniques outlined herein would make it rather easy for those who are skeptical, but openminded, to answer many of their own questions with hard evidence.

Certain of the methods discussed may have individual shortcomings. These are recognized and can generally be circumvented. This has been discussed in the text or in the literature cited. We would stress here that, when possible problems or objections arise with a given method, the simultaneous utilization of two or more methods will provide the confirmation necessary. Thus, we have applied in the field every method discussed in this chapter and virtually all of them at different times and places against one species, the California red scale. Each of them served to confirm others as valid means of evaluation of biological control and to answer critics who, at first, were unwilling to believe that effective natural enemies of the California red scale actually existed anywhere. Similar but less diversified techniques were used with spectacular success in the evaluation of the roles of natural enemies in the control of olive scale. As stated previously, experimental evaluation was also successful in appraising the relative roles of each of two parasite species in olive groves where both were established. We believe that similar assessments can be carried out with most pests in agro-ecosystems and with many species in nature.

It seems apropos here to reaffirm our considered conclusion that *experimental* analysis of the prey-natural enemy population system, involving paired comparisons of plots having enemies present with plots where enemies are absent (or inhibited), furnishes the only satisfactory proof of the control and regulatory power of natural enemies. Such analyses simultaneously permit experimental isolation and evaluation of other parameters, including weather and competitors. Thus, ecologists need not continue to argue from cerebral or laboratory-based information. They can evaluate the role of different parameters in an ecosystem experimentally.

ACKNOWLEDGMENTS

This paper is a contribution, in part, of the IBP Biological Control IRP. It was supported, in part, by NSF grants GB-6776 and GB-14489 and NIH grant A1-1611-13. There have been so many contributors to these studies that we cannot thank them all individually. Mr. Stan Warner and Mr. C. E. Kennett were of especial assistance in many ways.

LITERATURE CITED

Bartlett, B. R. 1957. Biotic factors in natural control of citrus mealybugs in California. J. Econ. Entomol. 50: 753-755.

DeBach, Paul. 1946. An insecticidal check method for measuring the efficacy of entomophagous insects. J. Econ. Entomol. 39: 695-697.

DeBach, Paul. 1955. Validity of the insecticidal check method as a measure of the effectiveness of natural enemies of diaspine scale insects. J. Econ. Entomol. 48: 584-588.

DeBach, Paul. 1964. (Editor). Biological Control of Insect Pests and Weeds. Reinhold Pub. Co., N. Y. 844 pp.

DeBach, Paul. 1969. Biological control of diaspine scale insects on citrus in California. Proc. 1st Intern. Citrus Symp. Riverside, Calif. (1968) 2: 801-815.

DeBach, Paul, and Blair Bartlett. 1951. Effects of insecticides on biological control of insect pests of citrus. J. Econ. Entomol. 44: 373-383.

DeBach, Paul, E. J. Dietrick, and C. A. Fleschner. 1949. A new technique for evaluating the efficiency of entomophagous insects in the field. J. Econ. Entomol. 42: 546.

DeBach, Paul, E. J. Dietrick, and C. A. Fleschner. 1951. Ants vs. biological control of citrus pests. Calif. Citrog. 36: 312, 347-348.

DeBach, Paul, C. A. Fleschner, and E. J. Dietrick. 1951. A biological check method for evaluating the effectiveness of entomophagous insects. J. Econ. Entomol. 44: 763-766.

Flaherty, D. L. and C. B. Huffaker. 1970. Biological control of Pacific mites and Willamette mites in San Joaquin Valley vineyards. Hilgardia 40(10): 267-308.

Fleschner, C. A., J. C. Hall, and D. W. Ricker. 1955. Natural balance of mite pests in an avocado grove. Calif. Avocado Soc. Yearbook 39: 155-162.

Hassell, M. P., and C. B. Huffaker. 1969. The appraisal of delayed and direct density-dependence. Can. Entomol. 101: 353-361.

Hassell, M. P., and G. C. Varley. 1969. New inductive population model for insect parasites and its bearing on biological control. Nature 223: 1133-1137.

Huffaker, C. B., and C. E. Kennett. 1956. Experimental studies on predation: (1) Predation and cyclamen mite populations on strawberries in California. Hilgardia 26: 191-222.

Huffaker, C. B., and C. E. Kennett. 1966. Studies of two parasites of olive scale, *Parlatoria oleae* (Colvee). IV. Biological control of *Parlatoria oleae* (Colvee) through the compensatory action of two introduced parasites. Hilgardia 37: 283-335.

Huffaker, C. B., and C. E. Kennett. 1969. Some aspects of assessing efficiency of natural enemies. Can. Entomol. 101: 425-447.

Huffaker, C. B., C. E. Kennett, and G. L. Finney. 1962. Biological control of olive scale, *Parlatoria oleae* (Colvee), in California by imported *Aphytis maculicornis* (Masi) (Hymenoptera: Aphelinidae). Hilgardia 32: 541-636.

Legner, E. F. 1969. Distribution pattern of hosts and parasitization by *Spalangia*

drosophilae (Hymenoptera: Pteromalidae). Can. Entomol. 101: 551-557.
Smith, H. S., and Paul DeBach. 1942. The measurement of the effect of entomophagous insects on population densities of their hosts. J. Econ. Entomol. 35: 845-849.

SECTION II

OUTSTANDING RECENT EXAMPLES OF
CLASSICAL BIOLOGICAL CONTROL

Chapter 6

THE BIOLOGICAL CONTROL OF WEEDS BY INTRODUCED NATURAL ENEMIES

L. A. Andres

Biological Control of Weeds Investigations
Agricultural Research Service, U.S. Department of Agriculture
Albany, California

and R. D. Goeden

Division of Biological Control
University of California, Riverside

INTRODUCTION

Man's activities on this planet have given rise to a variety of problems—not the least of which stem from his intentional or accidental spread of plants which subsequently become "weeds." In the United States the losses caused by alien and native weeds are believed to equal the combined losses from insects and diseases and to rank second only to those caused by soil erosion (Saunders *et al., in* King, 1966). For the decade ending in 1960, this meant an annual loss of $5.1 billion (U.S. Dept. of Agr., 1965). Cultural and chemical control practices currently are the main approaches to weed control. Both methods are aimed at removing unwanted plants and reducing real and suspected damage as quickly as possible, a short term approach requiring considerable annual expenditures of resources and energies, yet affording only temporary relief, not lasting weed control. The ever-increasing recognition given to weeds as pests and the concomittant increases in expenditures for their control, have focused attention on the need for effective, low-cost, and long-lasting alternative control methods. Biological control provides one such alternative. Although the natural mortality factors in the environment of a weed have long been considered of prime importance in limiting such a plant's distribution and abundance, their practical use has not been exploited. Many

143

of our introduced weeds are probably as abundant as they are primarily because the natural enemies that attack them in their native lands are not found in their new home areas. Thus, these natural control factors can be augmented through the classic biological control method of importing the enemies of such weeds. This approach has received little serious consideration by those currently charged with developing and instituting weed control programs.

Theoretically, any organism that is potentially destructive to a weed might find use as a weed control agent. From the practical standpoint, however, the most effective biological weed control agents used to date are those host-specific phytophagous organisms, like certain species of insects, that limit their attacks to a single or a few closely related weed species. Other types of organisms, such as plant pathogens, nematodes, fish, snails, the manatee, and geese have found only limited use as weed control agents. (Wilson, 1969; Ivannikov, 1969; Hickling, 1965; Seaman and Porterfield, 1964; Mayton et al., 1945). Although work is continuing with a variety of organisms, this discussion is concerned only with the use of host specific insects for weed control.

Weed control with insects is no dangerous, untested, unproven pipe dream, but rather, a well documented accomplishment. Two frequently cited examples are the destruction of over 60 million acres of prickly pear cacti, Opuntia spp., in Australia, primarily by the imported Argentine moth, Cactoblastis cactorum (Berg.) (Dodd, 1959), and the spectacular control of Klamath weed, Hypericum perforatum L., in the western United States, primarily by the beetle Chrysolina quadrigemina Suffrian (Holloway, 1964; Huffaker and Kennett, 1959) (See Chapter 5). Other attempts have been equally successful, if somewhat less spectacular. The status of over 50 weed control projects employing insects throughout the world was recently summarized in a handbook published by the National Academy of Sciences (1968).

Effective weed control implies reduction of the population densities of the weed below its level of economic importance. To achieve long-term control, through biological means, measures are generally undertaken to reduce the weed's competitive ability or by allowing it to dissipate its energy reserves while preventing or curtailing its reproductive growth. Reduction of the plant's competitive ability with insects does not necessarily mean its direct destruction. The gall fly, Procecidochares utilis Stone, imported to Hawaii to control Pamakani, Eupatorium adenophorum (Spreng.), provides a good example of how indirect action by an insect natural enemy can lead to effective weed control. The fly did not directly destroy its host plant; rather, by galling the stems it reduced stem length and foliage production to such an extent that the weed was no longer an effective competitor. Thousands of acres were cleared of pamakani (Bess and Haramoto, 1958). Indeed, heavy destruction of the foliage or aerial parts of a weed by insects or other organisms may not produce the desired control—the plants may survive—unless this destruction occurs at the proper time in the plant's life cycle. The webworms, *Herpetogramma bipunctalis* (Fab.) and *Hymenia fascialis* Cramer, often found in Florida in high numbers on the introduced alligatorweed, *Alternanthera phylloxeroides* (Mart.) Griseb. (Zeiger, 1967) cause heavy defoliation of the weed at a time when its energy reserves are high and thus have little effect on this plant's survival. An effective natural enemy derives its energy from its weed host at a time critical to the latter's survival. The synchronized feeding of the adults and larvae of the *Chrysolina* beetle on the basal foliar rosettes of Klamath weed over a long period in the fall, winter, and spring deprives the root system of its nourishment. Thus the roots largely disintegrate so that the plants cannot secure moisture to survive the long dry summers in California. Irrigated plants commonly recover from such defoliation (Holloway and Huffaker, 1952).

Biological control of weeds with host specific insects does not result in plant eradication. Since the reproduction of the specific insect is tied to the presence of the weed host, an increase in numbers of the insect brings increased pressure to bear on the plant. As the weed is reduced in numbers by this increased feeding, the insects become food-limited and similarly decline, but continue to exert a corresponding feeding pressure on the host plant. Hopefully, a general equilibrium level is reached when the weed's average population density lies below its level of economic importance. Since the insect is host specific, other plants in the community remain unharmed and, in fact, often assist in suppression of the weed through this competitive pressure.

With this brief introduction to the biological control of weeds, we now wish to review several ongoing projects in biological weed control that have met with varying degrees of success.

THE ALLIGATORWEED

The current project on alligatorweed in the southeastern United States is one of the first attempts at controlling an aquatic weed with insects. This plant, *Alternanthera phylloxeroides* (Mart.) Griseb. was first reported in Florida and Alabama just before the turn of the century (Weldon, 1960). It is an emergent or floating perennial that apparently reproduces only by vegetative means in North America. A single stem node can initiate a new infestation.

Figs. 1A and 1B.

   Top, this page (A). Infestation of *Alternanthera phylloxeroides* (Mart.) Griseb. (alligatorweed), Ortega River, Jacksonville, Florida, November, 1965. Original Florida release site of 266 adult *Agasicles* n. sp. beetles, April, 1965.

   Lower, this page (B). Alligatorweed infestation, Jacksonville, Florida, May 26, 1966, showing defoliation due to *Agasicles* n. sp. beetles (photo by C. F. Zeiger).

Figs. 1C and 1D.

Top, this page (C). Alligatorweed infestation, Jacksonville, Florida, May 26, 1966, showing effect of *Agasicles* n. sp. beetles (photo by C. F. Zeiger).

Lower, this page (D). Alligatorweed infestation, Jacksonville, Florida, July, 1966, 15 months after initial release of *Agasicles* beetles.

By 1963, alligatorweed infested an estimated 97,000 acres in eight states from Texas to North Carolina, and was causing navigational, flood control, irrigation, public health and water quality problems (U. S. Army, 1965). Each new dam or waterway is a potential site of infestation and the threat from this weed continues today.

The responsibility for control of alligatorweed in the waterways of the southeastern United States rests with the Army Corps of Engineers. The Corps early recognized that the regenerative powers of the plant and its high tolerance to herbicides made a quick solution, based on mechanical removal and herbicides, impractical. Thus, it broadened the scope of its control program to include a search for and evaluation of biotic agents having a control potential. With funds made available by the Corps in 1960, the Entomology Research Division, U.S. Department of Agriculture, undertook a survey of the insects associated with alligatorweed in its native South America. Initial field exploratory efforts were made in Paraguay, southern Brazil and northern Argentina in areas climatologically homologous to the southeastern United States. The plant is not a pest in its native area—solid stands occurring only in small patches, and these primarily confined to areas disturbed by man (Vogt, 1960). Forty to fifty-three species of insects were found to attack alligatorweed (Vogt, 1961). In contrast, few insects attack it in the United States. This strengthens the supposition that insects might be playing an important role in regulating the plant in its native habitat. Early emphasis centered on the introduction of the leaf and stem feeding *Agasicles* flea beetle.

Following a relatively unsuccessful first release near Savannah, Georgia, in 1964 (Hawkes *et al.*, 1967), a second release of 266 adults on the Ortega River, near Jacksonville, Florida, in April, 1965 was quickly successful. By May, 1966 adults were observed by the hundreds of thousands. By July, 1966 the Ortega release site was essentially free from alligatorweed (Zeiger, 1967) (Fig. 1). Following this rapid build-up and establishment, the beetles were redistributed to eight other states. The beetles have now spread over all watersheds in northeastern Florida and southeastern Georgia and are also established in Alabama and Texas.

Evaluation of the beetle's effectiveness in controlling alligatorweed has been difficult due to the mobility of the plant and its aquatic habit. For example, the compression and expansion of the floating mats due to water movement, and the drifting of mats from one area to another, make the correlation of beetle populations and plant abundance unreliable. Also unexplainable is the rapid build-up of the beetle in some areas and its failure to reproduce in adjacent sites. Preliminary studies indicate that plant nutrition and the rates of mat regeneration influence beetle increase and effectiveness.

Water quality may also influence the attractiveness of the plants to the beetle (D. M. Maddox, unpublished notes).

The success of this beetle in controlling alligatorweed has reduced the need for chemical treatments in some areas. However, as with most weedy situations, treatments were not directed towards alligatorweed alone, but included water hyacinth and several other plant species. Formerly, the herbicide treatments destroyed the less tolerant plants in the community, allowing the more resistant alligatorweed to become dominant. The *Agasicles* beetle has changed this pattern of succession, holding the alligatorweed at low levels. However, it has not reduced the need for treating the other aquatic weeds. There is also some indication that the use of herbicides on alligator-weed enhances the action of the beetles in problem areas, effectively extending the control action of the herbicide (Maddox, unpublished notes).

PUNCTUREVINE

Another current project concerns the biological control of puncturevine, *Tribulus terrestris* L. This weed is a prostrate herb, easily recognized by its hard spiny fruits or burrs. Its natural range extends from the Mediterranean regions of Europe and Africa through the drier regions of Asia and as far north as Siberia (Andres and Angalet, 1963; Squires, 1969). The weed was first introduced into the mid-western United States with livestock from the Mediterranean area. It now occurs from coast to coast but is more abundant in the far western states (Georgia, *in* Johnson, 1932). In the continental United States, puncturevine behaves as a summer annual, its seeds germinate in the spring and it flowers throughout the summer and fall until frost. Seed germination apparently depends on proper conditions of temperature, soil moisture, soil disturbance and possibly light period. Germination may be delayed until mid- to late-summer or the seed may remain dormant in the soil for several years, resulting in wide annual fluctuations in plant abundance.

Two species of beetles, *Microlarinus lareynii* (Jacquelin-Duval), a seed infesting weevil, and *M. lypriformis* (Wollasten), a stem and crown mining weevil, were selected for introduction among several insects found attacking puncturevine in its native habitat.

The adult weevils were tested for host specificity and found to readily feed on a variety of plants under forced conditions, but could reproduce only in the presence of the host. In view of puncturevine's erratic germination pattern, this extended feeding range is an excellent survival mechanism. The weevils were cleared for introduction to the United States on the basis of their restricted reproduction (Andres and Angalet, 1963). Both weevils were

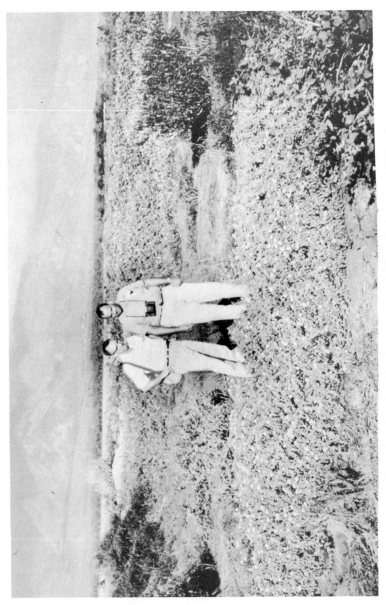

Fig. 2A. *Tribulus cistoides* L. plant, Maui, Hawaii, January, 1964 prior to the release of the seed weevil, *Microlarinus lareynii* Jacquelin-Duval and the stem weevil, *M. lypriformis* (Wollasten) (photo, Hawaii Dept. of Agr., C. J. Davis).

Fig. 2B. *Tribulus cistoides* L. plant, Maui, Hawaii, September, 1965, destroyed by attack of stem weevil, *Microlarinus lypriformis* (Wollasten) (photo, Hawaii Dept. of Agr., C. J. Davis).

introduced to the United States in 1961 (Huffaker, *et al.*, 1961) and are established in the southwest. The following year the weevils were introduced to Hawaii to combat puncturevine and the endemic *Tribulus cistoides* L. (Hawaii Dept. of Agr., 1963; Davis and Krauss, 1966). Despite subsequent collections from a number of economic plants in the United States, neither species has become a pest.

In Hawaii, the weevils are reported to have exerted "tremendous pressure" on *Tribulus* on Maui and Oahu and to have "considerably reduced" puncturevine on Kauai in 1964 (Hawaii Dept. of Agr., 1965). In the most recent assessment of this program, it was stated that "ninety-nine per cent of the puncturevine plants have been destroyed on Kauai by *M. lypriformis* (stem weevil) . . . and a substantial number of seeds were destroyed by the seed weevil, *M. lareynii* . . . There was sporadic reproduction in some localities in 1965 but the weevils moved in rapidly. Puncturevine reproduction was heavy on Oahu and Kauai in 1967, but the young plants were overcome by residual stem weevil populations . . ." (Hawaii Department of Agriculture, 1967). Control was listed as "partial" on the islands of Maui and Molokai (Davis and Krauss, 1966). (Fig. 2.)

Laboratory and field studies in southern California during 1965-1968, on the other hand, indicated that these weevils are only partially effective as biological control agents. *Microlarinus lareynii* annually infested only about half of the seed burrs produced at each of three representative irrigated puncturevine plots, located in different climatic zones. The control potential of both the stem and seed weevils has been measurably curtailed by indigenous predators and parasites (Goeden and Ricker, 1967; in press, and unpublished data). In addition, following an early season increase, a majority of the seed weevils produced at each site dispersed, rather than concentrate in ever-increasing densities sufficient to destroy a large part of the current seed crop and thus effectively reduce the local seed reservoir in the soil (Goeden and Ricker, unpublished data).

However, despite these limitations, the weevils have reportedly reduced the need for roadside treatments of puncturevine in some areas of California and Arizona (Mr. E. L. Dietz, Jr., pers. comm., May and Roney, 1969).

The success of *Microlarinus* spp. in Hawaii is attributed to the tendency of *T. terrestris* to grow as a weak perennial and to the presence, in greater abundance, of the perennial alternate host, *T. cistoides*. The year-round presence of host material allowed the continued reproduction and feeding of both weevils. This has resulted in such large populations that both plants were often killed in great numbers (C. J. Davis, pers. comm.).

The role of parasites and predators in limiting the action of the weevils in Hawaii has not been evaluated. Although the end result of the weevil's introduction to Hawaii has been documented, the population build-up of the weevils was not quantified. While the stem weevil is credited with the major role in controlling the plant in Hawaii, in California the seed weevil, from the time of initial establishment, has been the dominant species.

LANTANA IN HAWAII

The biological control program on the range weed, *Lantana camara* L., in Hawaii has had a prolonged history. Lantana was purposely introduced into Hawaii in 1885 as an ornamental. It has since spread or become a pest over 443,300 acres (93,300 of moderate to heavy infestation) (Hawaii Department of Agriculture, 1962). It occurs at elevations below 3,000 feet and where the rainfall may vary from less than 20 in. to more than 60 in. annually. The Hawaii Department of Agriculture is relying principally on biological control of this pest.

Lantana camara is native to subtropical and tropical Central and South America and is a pest on all continents (Oosthuizen, 1963; Waterhouse, 1966; Greathead, 1967). In 1902, the first significant importation project in biological control of weeds was initiated in Hawaii against this plant when 23 species of phytophagous insects were imported from Mexico. Of these, eight were successfully established, the most important being the leaf-feeding tingid, *Teleonemia scrupulosa* Stal (Perkins, 1903, 1904; Perkins and Swezey, 1924).

According to Imms (1931, p. 2), " . . . The net result of these insect introductions is that the spread of *Lantana* has been appreciably restrained. It no longer readily colonizes lands once cleared and the labor of its eradication has been lightened . . ." However, both Imms and Mr. A. P. Dodd (*in* Fullaway, 1959) agreed that biological control was by no means complete and both recommended that additional natural enemies be introduced.

Accordingly, search for natural enemies in Mexico was resumed in 1952, and during subsequent years was extended to Central and South America, California and Florida, South and East Africa and the West Indies, from which importations continue to date. Several key species of foliage-feeding lepidoptera that were introduced since 1952 include the leaf-feeding noctuid, *Catabena esula* (Druce), introduced from California, and first recovered on Hawaii in 1957 (Weber, 1956; Krauss, 1962); the leaf-feeding pyraustid, *Syngamia haemorrhoidalis* Guenee, introduced from Florida and Cuba, was reported to be well established by Krauss (1962); and the leaf-feeding noctuid, *Hypena strigata* F., imported from Kenya, Rhodesia and the Philippines, which readily became established throughout the islands (Fullaway, 1959; Krauss, 1962). Other introductions have also been reported.

Between 1957 and 1961, following the establishment of these additional lantana insects, improved biological control was reported. This complex of species has somewhat overcome the problem of temporal asynchronization faced by the otherwise effective *Teleonemia scrupulosa* when acting alone. *Teleonemia* caused considerable defoliation of the plants during the summer but allowed them to recover during the remainder of the year. The above three species of leaf-feeding Lepidoptera, on the other hand, were most active during the cooler months, causing winter defoliation and thus complementing the action of *Teleonemia*. This four-species complex now provides excellent control in the formerly troublesome drier areas or where the rainfall is less than 40 in. per year. Intense feeding by *H. strigata, S. haemorrhoidalis* and *T. scrupulosa* reportedly caused the complete dieback of shoots in some localities on Hawaii, and as a result of the continuous defoliation by these and other insects, lantana was reportedly being replaced by forage grasses on portions of Kauai (Hawaii Dept. Agr. and Conserv., 1960). An over-all assessment by Davis (1966) listed lantana as having been brought under "partial" to "substantial" control on the islands of Hawaii, Maui and Molokai, and under "partial" control on Oahu and Kauai.

However, control remains poor in the wetter areas. The introduction of the stem borer, *Plagiohammus spinipennis* (Thoms.) from Mexico may improve this situation. According to Harley and Kunimoto (1969), this beetle requires an annual rainfall of not less than 50 in., the seasonal distribution of this rain being especially critical just prior to and during oviposition. Multiple infestations of the stem borer on single, tall plants have frequently reduced them to mutilated stumps and the prospects for additional lantana control by this natural enemy are excellent.

The leaf-feeding beetles, *Octotoma scabripennis* Guer. and *Uroplata girardi* Pic., are apparently fulfilling a useful role in the lantana insect complex. Their seasonal activity is out of phase with the other imported defoliators, thus their action is considered additive (Harley, 1969).

Though many major successes of biological control of weeds have been credited to the introduction of one or two insect species acting as key control agents, it should not be construed that one or two species are all that will be needed in every case. The course of some projects probably will lie somewhat closer to that which has been demonstrated in the case of lantana control in Hawaii. The wider the range of conditions under which the weed occurs, the greater the number of insect species, or species of wide climatic tolerance, that will be needed for control.

PRICKLY PEAR IN CALIFORNIA

Only one biological control program directed against prickly pear cacti has been conducted in North America, this being against *Opuntia littoralis* (Engelm.) Ckll. and *O. oricola* Philbrick, and their hybrids, infesting rangeland on a 62,000 acre island that lies about 25 miles off the coast of southern California.

This project, now in its final stages, represents the first successful case of biological control of a native weed by intentionally introduced natural enemies. The cacti became pests on this island largely as a result of overgrazing by feral sheep first introduced to the island by early Spanish colonists. These sheep at one time literally denuded the island's grasslands, a situation that eliminated natural plant competition and led to invasion of these lands by the ungrazed, native prickly pears. During the 1940's, for example, it was estimated that 40 percent of the rangeland was rendered useless by the cacti for cattle grazing, which today remains the island's chief industry.

During the last decade, substantial biological control of these cacti has been achieved, principally by the cochineal insect, *Dactylopius opuntiae* Lichtenstein, introduced from Hawaii in 1951. Other species of cactus-feeding insects were introduced, two of which, the coreid bugs, *Chelinidea tabulata* (Burmeister) and *C. vittiger* Uhler, were established. However, *D. opuntiae* is by far the most effective species. *D. opuntiae* subsequently was found to naturally infest these same species of cacti along the southernmost coast of mainland California but here its populations appear to be held at relatively low densities by a complex of insect predators. Once this scale was established on the island, infested pads were redistributed to other areas. This cochineal insect is now established island-wide and has attained high local densities as a result of its abundant food supply and its geographic isolation from some of its principal mainland natural enemies, namely the coccinelid, *Hyperaspis taeniata significanis* Casey and the entomophagous phycitid, *Laetilia coccidivora* (Comstock). (Fig. 3.) By the early 1960's, the ravages of *D. opuntiae* were evident everywhere. Range management practices, begun during the late 1950's, i.e., containment and localized eradication of feral sheep and restrained cattle grazing, have aided these biological control efforts by encouraging the growth and persistence of forage plant species competing with the prickly pears on the formerly overgrazed lands.

Island-wide today, there has been at least a fifty per cent reduction in the prickly pear cacti, particularly if one excludes from consideration certain stretches of rugged, practically inaccessible hillsides that are of little value for cattle grazing and which remain infested with feral sheep. *Dactylopius*-induced

Figs. 3A, 3B and 3C.

Above, this page (A). *Opuntia littoralis* (Engelm.) Ckll. x *O. oricola* Philbrick, Santa Cruz Island, California, October, 1964, at time of initial *Dactylopius* infestation.

Top, opposite page (B). *O. littoralis* x *O. oricola,* Santa Cruz Island, California, September, 1965, showing effect of *Dactylopius* feedings.

Lower, opposite page (C). *O. littoralis* x *O. oricola,* Santa Cruz Island, California, May, 1966, showing effect of *Dactylopius* feeding 19 months after release of this stand.

mortality is still progressing. Regrowth from killed clumps is not surviving. The ultimate benefits from biological control are still to be realized (Goeden, *et al.*, 1967).

TANSY RAGWORT

Another project that has shown promise is the biological control of tansy ragwort, *Senecio jacobaea* L., by the cinnabar moth, *Tyria jacobaeae* (L.) (Lep.: Arctiidae). This poisonous European weed is present in pastures and rangeland of northwestern California, Oregon and Washington (Warren and Freed, 1958) and parts of Canada. It has also been introduced to New Zealand, Tasmania, Australia, South Africa and South America (Frick and Holloway, 1964).

New Zealand was the first to attempt biological control with the cinnabar moth (Cameron, 1935), an effort attended with considerable initial success, but the insects eventually died out (Miller, 1940). Subsequently, the moth was introduced into the United States in 1959. The initial release and establishment was summarized by Frick and Holloway (1964) and the progress of control by Hawkes (1968).

The moth is univoltine, and first appears between mid-April and early May (coastal California). The eggs are usually deposited on the developing second year rosettes, i.e., those plants that will flower during the season. The cinnabar larvae then strip the blossoms and foliage and finally the basal leaves from the plant. Beginning in early August, pupation occurs in the soil where the insect overwinters (Hawkes, 1968).

Larvae released on a dense ragwort stand covering a coastal flat about 200 yards from the ocean at Fort Bragg, Mendocino County, California, gradually increased in numbers and by 1963 were abundant over about two acres. Heavy defoliation occurred within this area. By 1964, the high population expanded to cover an area of five acres; and by 1965, an area of 12 acres. That year, the highest concentrations of larvae were evident on the perimeter of the expanding area with lower numbers behind and in front of the advancing population.

At peak population levels, it was not uncommon to find 100 or more larvae on large plants. The resulting total defoliation, when extended over a period of several seasons, reduced the numbers of flowering stems from 15 to 19 per square yard to less than 1/5 the original abundance (Hawkes, 1973). Striking *before* and *after* photographs of the release site illustrate the control achieved (Fig. 4).

In spite of the high build-up of the *Tyria* population and excellent

Figs. 4A and 4B.
    Top (A). *Senecio jacobaea* L. infestation, Fort Bragg, California, August, 1961.
    Lower (B). Reduction of *S. jacobaea* due to *Tyria jacobaea* (L.) feeding, Fort Bragg, California, August, 1964.

alized destruction of the plant, dispersal of the insect from the original site has been rather slow. From the time of initial large population increase in 1963 through 1967, the effective population had moved only about 500 yards. Hawkes (1968) attributes this slow spread partially to the high density of the plant and the ready availability of ovipositional sites to the newly emerged females. The fecund females fly only haltingly until a good portion of their eggs have been laid. This can result in an over-concentration of eggs near the emergence sites and a severe defoliation of the ragwort by the developing larvae. In several years, the population concentration was so great that, after complete defoliation, the ground was literally covered with migrating larvae in search of food; the majority either starved or were forced into premature pupation.

The potential of the cinnabar moth to control ragwort outside of the Fort Bragg area is as yet unknown. Releases have been made in other areas of California, Oregon and Washington with mixed results (Hawkes, 1968). In most areas, the populations have not as yet increased sufficiently for control of the plant. However, at two or more sites in Oregon, including one near Scio, Linn County, Oregon, in 1960, marked reductions of the infestations have resulted. Little is known of the differences in mortality factors that might account for the different rates of progress.

Attempts are now being made to establish a seed fly, *Hylemya seneciella* Meade, and a crown-feeding fleabeetle, *Longitarsus jacobaea* (Waterhouse) in North America (Frick, 1969, 1970) to supplement the action of the cinnabar moth.

DISCUSSION

Four of the above five projects involve introduced weeds and their biological control through the importation of exotic phytophagous insects. This is the classical approach to biological weed control, and one that has been followed with slight modification with most projects to date. If one considers that almost half of the more than 500 major weeds in the United States are introduced species, and that 13 of the 15 top weeds are exotics, the classic method will continue to offer a fertile approach to biological weed control. The introduction of exotic enemies against native weeds can also be fruitful if suitable enemies of related plants in other regions of the world are available.

Before employment of the classic method can be realized on a large scale, certain problems limiting its application must be overcome. Prior to the movement of phytophagous insects from the native to the introduced area of a weed it must be demonstrated that the candidate insect will not become a pest on other plants. To assure this, an elaborate system of host specificity testing has evolved, aimed at establishing a positive relationship between the insect and its host. The philosophy underlying these test procedures is well summarized by Harris and Zwolfer (1968). Presently, biological and host specificity studies are often conducted for three to five years before clearance for introduction is granted. In view of the large number of introduced weeds for which biological control offers promise, and the number of potential insects that must be screened for each such weed, more rapid and better methods of determining the host relationships must be developed.

Our limited knowledge of how insects suppress plant populations hampers us in selecting the most effective candidate species and developing an introduction strategy. It is obvious from the past record that type of feeding or plant part attacked alone are not sufficient to prejudge the success or failure of a candidate insect. Ultimately, potential effectiveness of candidate insects must be based on detailed, qualitative and quantitative knowledge of the insect's feeding habits, fundamental information on the biology and physiology of the weed, and the contributing roles of prospective competing plants in the invaded region. Although infallibility cannot be guaranteed, basic information of this nature will not only improve the chance of selecting the best insect and therefore of obtaining biological control by the classical method, but it may eventually lead to development of supplemental approaches.

Good documentation of a successful example of biological control of a weed requires exacting and continuing effort over a number of years. Unfortunately, the slow successional changes that occur in a plant community attendant with the removal of the weed are often not recorded. The documentation covering the biological control of Klamath weed in California is a notable exception (e.g., Huffaker and Kennett, 1959; Huffaker, 1967). However, failure to properly document the benefits of biological control do not diminish the effectiveness of the method, but do hinder the recognition of its usefulness.

In discussing the economics of biological control, Simmonds (1967) cited several examples of cost-benefit ratios that can be expected from biological control. Two weed projects, the control of *Cordia macrostachya* (Jacquin) Roemer and Schultes on Mauritius and *Opuntia* spp. on the island of Nevis in the West Indies are producing per annum benefits of 1,000 per cent and 2,000 per cent, respectively, on the money invested. Paradoxically, success may be the nemesis of biological control. Simmonds noted that once the problem is solved through biological control, it is then forgotten and the benefits no longer tallied. Problems that require annual treatment, on the other hand, provide a continuing cost-benefit reminder.

The introduction of exotic weed feeding insects has certain distinct advantages over current weed control practices. At the least, it is a useful adjunct to present control methods; at best, it can supplant the need for artificial controls entirely. Its potential equals that of other control methods, without the spectre of undesirable side effects. However, successful biological control cannot be achieved overnight, nor is every weed likely to be controlled in this manner. Contrary to the impressions conveyed by Holm *et al.* (1969), the use of biological control requires a high degree of skill and an amalgamation of many talents, disciplines and ecological insights. Sensibly, the weed scientist should approach his problems with a variety of solutions in mind. To date, certainly, too little emphasis has been given to the biological control method.

LITERATURE CITED

Andres, L. A., and G. W. Angalet. 1963. Notes on the ecology and host specificity of *Microlarinus lareynii*, and *M. lypriformis* (Coleoptera: Curculionidae) and the biological control of puncturevine, *Tribulus terrestris.* J. Econ. Entomol. 56: 333-340.

Anonymous. 1968. Weed Control. Principles of Plant and Animal Pest Control. Vol. 2. Publ. 1597, Nat. Acad. Sci., Wash., D. C. 471 pp.

Bess, H. A., and F. H. Haramoto. 1958. Biological control of pamakani *Eupatorium adenophorum*, in Hawaii by a tephritid gall fly, *Procecidochares utilis.* 1. The life history of the fly and its effectiveness in the control of the weed. Proc. X Intern. Congr. Entomol., Montreal (1956) 4: 543-548.

Cameron, E. 1935. A study of the natural control of ragwort (*Senecio jacobaea* L.). J. Ecol. 23: 265-322.

Davis, C. J. 1966. Progress report: Biological control status of noxious weed pests in Hawaii–1965-66. Hawaiian Dept. of Agr. 4 pp. (processed).

Davis, C. J., and N. L. H. Krauss. 1966. Recent introductions for biological control. Proc. Hawaiian Entomol. Soc. (1965) 19: 201-207.

Dodd, A. P. 1959. The biological control of prickly pear in Australia. Pp. 565-577. *In* Biogeography and Ecology in Australia (Monogr. Biol., Vol. VIII). Dr. W. Junk, Publ., The Hague, Netherlands.

Frick, K. E. 1969. Attempt to establish the ragwort seed fly in the United States. J. Econ. Entomol. 62: 1135-1138.

Frick, K. E. 1970. Ragwort flea beetle established for biological control of tansy ragwort in northern California. Calif. Agr. 24(4): 12-13.

Frick, K. E., and J. K. Holloway, 1964. Establishment of the cinnabar moth, *Tyria jacobaeae*, on tansy ragwort in the western United States. J. Econ. Entomol. 61: 499-501.

Fullaway, D. T. 1959. Biological control of lantana in Hawaii. Pp. 70-75. *In* Rept. Bd. Agr. and Forest., Hawaii, Bien. ending June 30, 1958.

Goeden, R. D., C. A. Fleschner, and D. W. Ricker. 1967. Biological control of prickly pear cacti on Santa Cruz Island, California. Hilgardia 38: 579-606.

Goeden, R. D., and D. W. Ricker. 1967. *Geocoris pallens* found to be predaceous on *Microlarinus* spp. introduced to California for the biological control of puncturevine, *Tribulus terrestris*. J. Econ. Entomol. 60: 726-729.

Greathead, D. J. 1967. A list of the more important weeds of East Africa. Unpublished mimeo. 7 pp.

Harley, K. L. S. 1969. The suitability of *Octotoma scabripennis* Guer. and *Uroplata girardi* Pic. (Col.: Chrysomelidae) for the control of Lantana (Verbenaceae) in Australia. Bull. Entomol. Res. 58: 835-843.

Harley, K. L. S., and R. K. Kunimoto. 1969. Assessment of the suitability of *Plagiohammus spinipennis* (Thoms.) (Col.: Cerambycidae) as an agent for control of weeds of the genus *Lantana* (Verbenaceae). I. Life history and capacity to damage *L. camara* in Hawaii. Bull. Entomol. Res. 58: 567-574.

Harris, P., and H. Zwolfer. 1968. Screening of phytophagous insects for biological control of weeds. Can. Entomol. 100: 295-303.

Hawaii Dept. of Agr. 1962. Noxious weeds of Hawaii. 89 pp. (processed).

Hawaii Dept. of Agr. 1963. Ann. Rept. 1962-63, Div. Plant Indust. Pp. 49-69.

Hawaii Dept. of Agr. 1965. Ann. Rept. 1964-65, Div. Plant Indust. Pp. 10-12, App. V: 39-52.

Hawaii Dept. of Agr. and Conservation. 1960. Ann. Rept. 1960, Div. Entomol. and Market. S-45.

Hawkes, R. B. 1968. The cinnabar moth, *Tyria jacobaeae*, for control of tansy ragwort. J. Econ. Entomol. 61: 499-501.

Hawkes, R. B., 1973. Natural mortality of cinnabar moth in California. Ann. Entomol. Soc. Amer. 66: 137-146.

Hawkes, R. B., L. A. Andres, and W. H. Anderson. 1967. Release and progress of an introduced flea beetle, *Agasicles* n. sp. to control alligatorweed. J. Econ. Entomol. 60: 1476-1477.

Hickling, C. F. 1965. Biological control of aquatic vegetation. Pest Articles and News Summaries (C). 11: 237-244.

Holloway, J. K. 1964. Projects in biological control of weeds. Pp. 650-670. *In* Biological Control of Insect Pests and Weeds, P. DeBach (ed.). Reinhold Publ. Co., N. Y. 844 pp.

Holloway, J. K., and C. B. Huffaker. 1952. Insects to control a weed. Pp. 135-140. *In* Insects—The Yearbook of Agriculture, 1952. U.S. Govt. Printing Office.

Holm, L. G., L. W. Weldon, and R. D. Blackburn. 1969. Aquatic weeds. Science 166(3906): 699-709.

Huffaker, C. B. 1967. A comparison of the status of biological control of St. Johnswort in California and Australia. Mushi 39 (Suppl.): 51-73.

Huffaker, C. B., D. W. Ricker, and C. E. Kennett. 1961. Biological control of puncturevine with imported weevils. Calif. Agr. 15: 11-12.

Huffaker, C. B., and C. E. Kennett. 1959. A ten-year study of vegetational changes associated with biological control of Klamath weed. J. Range Management 12: 69-82.

Imms, A. D. 1931. Biological control. II. Noxious weeds. Trop. Agr. (Trinidad) 8: 124-127.

Ivannikov, A. I. 1969. A nematode controlling *Acroptilon picris*. Zashita Rasteniya, January. Pp. 54-55. (Trans. by C. C. Nikiforoff, U.S. Dept. Agr., Agr. Res. Ser.)

Johnson, E. 1932. The puncturevine in California. Calif. Agr. Exp. Sta. Bull. 528: 1-42.

King, L. J. 1966. Weeds of the World, Biology and Control. Intersci. Publ. Inc., N. Y. 526 pp.

Krauss, N. L. H. 1962. Biological control investigations on lantana. Proc. Hawaiian Entomol. Soc. 18: 134-136.

May, Judson, and J. N. Roney. 1969. Arizona Cooperative Insect Survey. Rept. No. 33, Aug. 1969.

Mayton, E. L., E. V. Smith, and D. King. 1945. Nutgrass eradication studies. IV. Use of chickens and geese in the control of nutgrass, *C. rotundus* L. J. Amer. Soc. Agron. 37: 785-791.

Miller, D. 1940. Biological control of noxious weeds of New Zealand. Herb. Publ. Ser. Bull. 27: 153-157.

Oosthuizen, M. J. 1963. The biological control of *Lantana camara* L. in Natal. J. Entomol. Soc. South Africa 27: 3-16.

Perkins, R. C. L. 1903. Enemies of lantana. Hawaii Livestock Breeder's Assoc., First Ann. Mtg. Proc., Nov. 17-18, 1902: 28-35.

Perkins, R. C. L. 1904. Later notes on lantana insects. Hawaii Livestock Breeder's Assoc., Second Ann. Mtg. Proc., 1903: 58-61.

Perkins, R. C. L., and O. H. Swezey. 1924. The introduction into Hawaii of insects that attack Lantana. Hawaiian Sugar Planters' Assoc., Entomol. Ser. Bull. 16: 1-53.

Seaman, D. E., and W. A. Porterfield. 1964. Control of aquatic weeds by the snail *Marisa cornuarietis.* Weeds 12: 87-92.

Simmonds, F. J. 1967. The economics of biological control. J. Roy. Soc. Arts London 115(5135): 880-898.

Squires, V. R. 1969. Distribution and polymorphism of *Tribulus terrestris* sens. lat. in Australia. Victorian Naturalist: 86: 328-334.

U. S. Army. 1965. Expanded project for aquatic plant control. 89th Congress, 1st Sess., House Document No. 251. 145 pp.

U. S. Department of Agriculture. 1965. A survey of extent and cost of weed control and specific weed problems. Agr. Res. Serv., ARS 34-23-1, August. 78 pp.

Vogt, G. B. 1960. Exploration for natural enemies of alligatorweed and related plants in South America. Special Report PI-4. U.S. Dept. Agr., Agr. Res. Ser., Ent. Res. Div. Mimeo. 58 pp.

Vogt, G. B. 1961. Exploration for natural enemies of alligatorweed and related plants in South America. Special Report PI-5. U.S. Dept. Agr., Agr. Res. Ser., Ent. Res. Div. Mimeo. 50 pp.

Warren, R., and V. Freed. 1958. Tansy ragwort—a poisonous weed. Oregon State Coll. Ext. Bull. 717: 3-5.

Waterhouse, D. F. 1966. The entomological control of weeds in Australia. Mushi 39 (Suppl.): 109-118.

Weber, P. W. 1956. Recent introductions for biological control in Hawaii. Proc. Hawaiian Entomol. Soc. 16: 162-164.

Weldon, L. W. 1960. A summary review of investigations on alligatorweed and its control. U.S. Dept. Agr., Agr. Res. Ser. CR-33-60. 41 pp.

Wilson, L. W. 1969. Use of plant pathogens in weed control. Ann. Rev. Phytopathol. 7: 411-434.

Zeiger, C. F. 1967. Biological control of alligatorweed with *Agasicles* n. sp. in Florida. Hyacinth Control Jour. 6: 31-34.

Chapter 7

BIOLOGICAL CONTROL OF COCCIDS BY INTRODUCED NATURAL ENEMIES

Paul DeBach

Division of Biological Control
University of California, Riverside

David Rosen

Hebrew University of Jerusalem
Rehovot, Israel

and C. E. Kennett

Division of Biological Control
University of California, Berkeley

INTRODUCTION

Scale insects and mealybugs (Homoptera: Coccoidea) have been the targets of numerous successful biological control projects. The first spectacular success in the history of biological control was achieved against a coccid pest, the cottony cushion scale, *Icerya purchasi* Maskell, in California in the late 1880's. A predatory ladybeetle, *Rodolia cardinalis* (Mulsant), introduced from Australia, was credited with the rapid, complete and permanent control of this serious pest in California. Subsequently, the same predator achieved complete biological control of the cottony cushion scale in 25 additional countries, and a substantial degree of control in four other countries (see DeBach, ed., 1964, Chapter 24).

Many outstanding successes in the biological control of serious coccid pests have since been recorded. Some notable cases include the complete biological control of the citrophilus mealybug, *Pseudococcus fragilis* Brain, in California; the coconut scale, *Aspidiotus destructor* Signoret, in the islands of Fiji, Mauritius, and Principe; Green's mealybug, *Pseudococcus citriculus* Green, in Israel; the red wax scale, *Ceroplastes rubens* Maskell, in Japan; and the coffee mealybug, *Planococcus kenyae* (LePelley), in Kenya. Complete,[1] substantial or partial success was attained in many other projects directed against coccid pests. In fact, coccids have largely dominated the scene in biological control of insect pests, accounting for nearly half of all projects against insects in which some degree of sucess has been attained, including about 67 per cent of all complete successes, 31 per cent of all substantial successes, and 43 per cent of all partial successes. Together with related sternorrhynchous Homoptera (mainly aphids and whiteflies), they account for about 2/3 of all successes in biological control, and for more than 4/5 of all projects in which complete success has been achieved. Armored scales (Coccoidea: Diaspididae) alone account for about 1/5 of all complete successes in biological control. (Data compiled from Clausen, in press.)

The proportion of successful projects has been considerably higher in the Coccoidea than in any other group of animal pests. Altogether, efforts have been made to bring about the biological control of 223 species of insect pests; some degree of practical success has been attained with over half of these species. With coccids, attempts have been made with 64 species and some degree of success attained with 50 species, or 78 per cent. Taking all other insect groups, attempts have been made with 159 species, and some degree of success has been attained with 70 species, or 44 per cent. Although this higher proportion for the coccids may, in part, reflect the greater amount of effort directed towards the biological control of these pests, it also appears to indicate that coccids may be somewhat more amenable to biological control than other groups of arthropod pests.

Perhaps the frequency with which coccids invade new countries and become serious pests, difficult to control with insecticides, is a major part of the explanation of the observed successes. Being sedentary and often rather cryptic, coccids are readily transferred accidentally with their host plants, and have thus often gained entrance into new habitats and countries where, free from the restraining influence of their natural enemies, they have become serious pests, often threatening the very existence of an industry. Such introduced, rampant pests are, of course, prime targets for biological control efforts, hence they may have attracted more research emphasis.

---

[1]"Complete success" refers to complete biological control being obtained and maintained over a fairly extensive area so that insecticidal treatment becomes rarely, if ever, necessary. "Substantial successes" includes cases where economic savings are somewhat less pronounced by reason of the crop area being restricted, or by the control being such that occasional insecticidal treatment is indicated. "Partial successes" are those where chemical control measures remain commonly necessary but less frequent, or cases where complete biological control is obtained only in a minor portion of the pest-infested area.

The sedentary habits, colonial distribution, general chronological continuity of all life stages in a population throughout the year, and the population stability conferred by perennial host plants, are characteristic of many coccids and may enhance their susceptibility to control by natural enemies. Parasites and predators are more likely to approach their full potential of effectiveness on this type of host or prey population.

Although the aforestated statistics for successful cases in biological control are impressive enough, we do not consider that a 50 per cent level of success, or even a 75 per cent level, represents an expected ceiling for biological control projects. A current lack of success with any given species does not mean that favorable results will not be obtained in the future. This is illustrated later in this paper by data from the California red scale project, in which the two most effective parasites were found some 55 years and 65 years, respectively, after the first work was started. Numerous attempts at biological control have been casual or incidental and certain others have lacked scientific sophistication. Many promising projects remain dormant and uninvestigated because of lack of trained personnel or ecologically oriented administrators.

In California alone it is estimated that successful biological control projects have saved the agricultural industry some $200 million during the past 45 years due to reduction in insect pest-caused crop losses and diminished need for chemical control (see DeBach, ed., 1964, Chapter 1). Additionally, biological control offers the best alternative to environmental pollution by hazardous chemical pesticides.

COTTONY CUSHION SCALE IN CALIFORNIA

The pioneering, highly successful biological control of cottony cushion scale in California serves to illustrate some of the basic principles involved in the biological control of coccid and other pests. The cottony cushion scale, *Icerya purchasi* Mask., was apparently accidentally introduced into California around 1868. By 1887 it had increased to such serious proportions as a pest of citrus, that the entire citrus industry of California was threatened with destruction. Since the scale was then regarded as a serious pest throughout its extensive range of distribution except for Australia, the latter country was assumed to be its native home, and a search for natural enemies was made there. Two natural enemies were introduced from Australia into California during 1888-1889: the vedalia ladybeetle, *Rodolia cardinalis* (Mulsant), and a parasitic fly, *Cryptochaetum iceryae* (Williston). Both were released and became established in California. The *Cryptochaetum* was initially regarded as the more promising species, and it has indeed proved to be the most effective natural enemy of cottony cushion scale in certain areas. However, it was largely overshadowed by the spectacular performance of the vedalia in the commercial citrus of southern California. The ladybeetle increased at a fantastic rate and brought about the complete suppression of the scale there by the end of 1889 (Doutt, 1958). This and many other similar examples indicate the inadvisability and futility of trying to predict the success of introduced natural enemies before they are actually tested in the field, and employing such predictions as a basis for preselection of the enemies to be introduced, as has been suggested by certain writers about biological control.

Recent studies by J. Quezada and P. DeBach (unpublished data) have shown that both natural enemies are still effective in maintaining the populations of cottony cushion scale at very low levels throughout California. In desert areas, the vedalia is dominant and displaces the *Cryptochaetum* whenever competition occurs, whereas in coastal areas the *Cryptochaetum* is dominant and tends to displace the vedalia (see Fig. 1). The two species appear to coexist in interior areas, but this is due to dispersal from their respective areas of dominance. The competition between vedalia and *Cryptochaetum* in the interior areas of California does not affect the excellent biological control of cottony cushion scale in those areas. This interesting outcome lends further support to the principle of multiple introduction of natural enemies. As many potentially effective enemies of a pest as are available should be introduced and released; the ones that are most efficient in a given habitat, unless they compete only partially, will eventually displace the others and produce better control of the pest. Contrary to certain suppositions, competition between natural enemies is not known to reduce their over-all effectiveness in biological control. (See further Chapters 2 and 4.)

The adverse effects of modern pesticides on natural enemies were demonstrated with the cottony cushion scale/vedalia complex in California soon after the advent of DDT. Decimation of vedalia populations by DDT was shown to be the cause of serious outbreaks of the scale in citrus groves (DeBach and Bartlett, 1951).

SOME RECENT EXAMPLES OF BIOLOGICAL CONTROL
OF SCALE INSECTS

Because three recent projects illustrate many of the most important principles and problems of biological control unusually well, they are discussed here in some detail. These include the complete biological control of the Florida red scale, *Chrysomphalus ficus* Ashmead, in Israel and of the olive

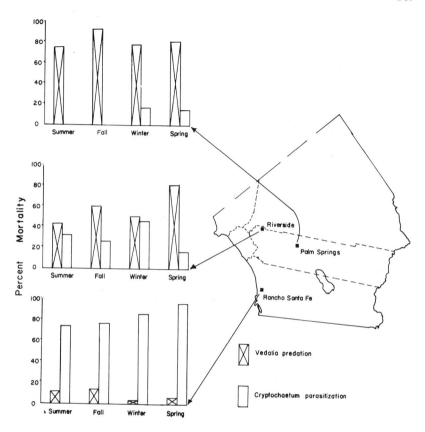

Fig. 1. Mortality of cottony cushion scale caused by vedalia and *Cryptochaetum iceryae* (Williston) on citrus in coastal, interior and desert areas of Southern California (from unpublished data of J. Quezada and P. DeBach).

scale, *Parlatoria oleae* (Colvee), in California, as well as substantial biological control of the California red scale, *Aonidiella aurantii* (Maskell), in California. The striking results in the use of airplane releases of a parasite of limited dispersal powers for control of Rhodes-grass scale, *Antonina graminis* (Maskell), are treated separately in Chapter 10.

Florida Red Scale

The Florida red scale invaded Israel around 1910. Notwithstanding efforts to quarantine it in the northern parts of the country, it eventually spread throughout the citrus-growing areas and soon became a major citrus pest. An indigenous ectoparasite, *Aphytis chrysomphali* (Mercet), was the only parasite commonly recorded from the scale in Israel. This species' attacks were limited mainly to male scale pupae, and its effect on the scale populations was negligible.

Another species of *Aphytis,* then believed to be *A. lingnanensis* Compere, and the endoparasite *Pteroptrix* (=*Casca*) *smithi* (Compere) were simultaneously introduced into Israel from Hong Kong during 1956-1957 against the Florida red scale. The project was remarkably unsophisticated and relatively inexpensive. Adult parasites were collected, mainly by beating, on Florida red scale-infested citrus trees in Hong Kong and were airmailed to Israel. Numerous shipments were made, but they yielded only 150 live specimens of *Aphytis* and 120 of *Pteroptrix,* and these were immediately released. Establishment of the *Aphytis* was soon evident, and its introduction proved to be a remarkable success. In less than two years, it dispersed throughout the infested citrus areas. Its dispersal was aided by transfer of parasitized material to infested groves in 1957, and by release of insectary-reared parasites during 1958-1959. Complete biological control was achieved throughout the coastal plain, which is the main citrus-growing area of Israel, during the first two or three years following the introduction of the parasite, and the Florida red scale was virtually eliminated from that area as a pest of economic importance. A few localized outbreaks of the scale still occur in the hot Jordan Valley, where the parasite proved to be only partly efficient even when periodic releases were made (see Rosen, 1965, 1967). A conservative estimate places the annual savings accrued by this successful project at well over $1 million (see also Fig. 1 in Chapter 20).

Only after the successful termination of the project was it discovered that the introduced species was not *A. lingnanensis,* which is a parasite of California red scale, but the closely related, theretofore undescribed *Aphytis holoxanthus* DeBach, a rather specific parasite of Florida red scale (DeBach, 1960). The initial confusion was later explained when the cryptic history of *A. lingnanensis* was revealed. When that species was first introduced into California from southern China in 1947, it was obtained from a shipment of Florida red scale, and was successfully propagated on California red scale. It was, therefore, believed to be a parasite of both scales. In the light of present knowledge, it appears very likely that the original shipment included a few California red scales, unnoticed at the time, from which several specimens of *A. lingnanensis* emerged. Since the parasites obtained from that shipment were exposed to the California red scale, those that issued from the Florida red scale—probably *A. holoxanthus*—were unable to reproduce, whereas the few *A.*

lingnanensis were able to survive and multiply. This example clearly illustrates the importance of biosystematic research to practical biological control. Had the true identity of *A. holoxanthus* been known, it would have been available for the biological control of Florida red scale in various countries long ago.

Relative to this, a recent study by Rosen of the *Aphytis* material preserved in the collection of the U. S. National Museum has brought to light several specimens, reared by Alfred M. Koebele from Florida red scale on citrus in Hong Kong, apparently sometime near the end of the last century. They were identified as *A. mytilaspidis* (Le Baron), and were subsequently ignored, but they appear to be *A. holoxanthus*. Likewise, it is probable that the species of *Aphytis* described by Taylor (1935) as a "form" of *A. chrysomphali* parasitizing Florida red scale in Java was actually *A. holoxanthus*.

Since the species of *Aphytis* are almost indistinguishable one from the other without careful microscopic examination of properly cleared specimens, it is possible that more than one species was included in the shipments of field-collected adult parasites sent from Hong Kong to Israel. This might explain the almost simultaneous discovery in 1960 of *Aphytis coheni* DeBach, a parasite of California red scale, and *A. lepidosaphes* Compere, a parasite of purple scale, *Lepidosaphes beckii* (Newm.), on citrus in Israel (Rosen, 1967). Accidental ecesis from other parts of the world is, of course, another possible explanation.

Pteroptrix smithi had no appreciable effect in the initial biological control of Florida red scale in Israel. Not until several years later was it recovered near the original release site. It has since dispersed along the coastal plain, and appears to persist at very low host densities. There is no evidence that it has detracted from the effectiveness of *A. holoxanthus;* if anything, it may be regarded as a complementary factor in the excellent control of the scale. The same may be said for another endoparasite, *Habrolepis fanari* Delucchi and Traboulsi, which has recently reached Israel accidentally. On the other hand, *Aphytis chrysomphali* was entirely displaced by *A. holoxanthus* from the populations of the scale. Thus, the competition between the several natural enemies had no adverse effects on the successful biological control of Florida red scale in Israel—another illustration of the basic principle of multiple introduction being either beneficial, or possibly of no effect, but never detrimental to biological control.

Disregarding political boundaries, *Aphytis holoxanthus* soon dispersed into Lebanon, where similar results occurred. It was subsequently introduced into Florida (Muma, 1969), Mexico (Maltby, Jimenez-Jimenez and DeBach, 1968) and South Africa (Bedford, 1968) where it has been responsible for substantial successes in the biological control of Florida red scale. Similar results were recently observed by DeBach in Brazil and Northern Peru. In Israel, however, the remarkable success in the biological control of Florida red scale has been partly offset subsequently by an increase in the severity of the California red scale. In the past, the Florida red scale has been credited with the competitive displacement of the California red scale from mature citrus groves in Israel. The suppression of the former by *A. holoxanthus*, which does not attack the California red scale, has apparently enabled the latter to invade that ecological niche more successfully than before. Thus we must bear in mind that a simultaneous multiple-project approach to biological control may be necessary in certain cases.

Olive Scale

The olive scale became the major pest of olive in California following its establishment near Fresno about 1934. It also became a major pest of many deciduous fruit crops and of ornamental shrubs and trees. It spread rapidly over large areas of the State, embracing, by 1961, the entire San Joaquin and Sacramento Valleys and scattered sites in southern California.

In California, the scale exhibits two generations per year. Young scales (crawlers) of the spring generation hatch, during May, from eggs produced by overwintered females of the fall generation. The crawlers settle and begin developing on the twigs and leaves, reaching maturity during early July (males) and late July or early August (females). Females of the spring generation produce eggs during August and September, the crawlers settling on the twigs, leaves and fruits. These mature by about November and the females overwinter (Huffaker, Kennett and Finney, 1962).

On olive, crawlers of the fall generation (produced in August-September) exhibit a propensity to settle and develop on the fruit. Thus, relatively low scale densities in summer may result in considerable numbers of infested fruits during the early fall. Since scale-discolored fruits are subject to cullage, olive scale is considered to be a "direct pest" of olive. Hence, the development of a successful program of biological control was contingent on the establishment of natural enemies capable of greatly reducing and then maintaining scale densities at extremely low levels.

An intensive program of biological control was started in 1949, when a "strain" of *Aphytis maculicornis* (Masi) was introduced from Egypt. In 1951, great efforts were made to obtain natural enemies of olive scale throughout its range of distribution, including its supposed native home in northern India, West Pakistan and the Middle East. Numerous consignments of olive scale material were shipped to California during 1951 from India, Pakistan,

Afghanistan, Iran, Iraq, Syria, Lebanon, Israel, Cyprus, Egypt, Greece, and Spain, and additional shipments were made from India and Pakistan in 1952-1953. Several species of parasites were obtained from these shipments and released, including four so-called "biological strains" or sibling species of *Aphytis,* all indistinguishable morphologically from *A. maculicornis* but possessing distinct biological attributes. These were reared and released separately (Doutt, 1954; Hafez and Doutt, 1954).

It early became evident that the "Persian strain" of *A. maculicornis,* obtained from Iran and Iraq and first colonized during 1952, was the only natural enemy giving some degree of control in field plots. Consequently, insectary production, colonization, and field study plots were concentrated on this parasite. Over 27 million *A. maculicornis* were colonized during 1952-1960 at several hundred sites in 24 counties in California (Huffaker, Kennett and Finney, 1962).

This aspect of the biological control program illustrates the importance of searching for natural enemies throughout the entire range of the pest's distribution, as well as the value of recognizing biological differences between morphologically similar natural enemies. Had morphological taxonomy been used as the sole criterion, the distinctness and superiority of the Persian *Aphytis* would not likely have been realized.

Early results were very promising. The Persian *A. maculicornis* became readily established, and proved capable of attaining levels of parasitization of 90 per cent or more even after host densities had been reduced to low levels. Scale populations were generally lowered by more than 90 per cent. However, this drastic reduction proved to be economically unsatisfactory in many cases, because if even one scale occurs on a fruit it may be culled. Greater effectiveness was prevented because the parasite proved intolerant of the hot, dry summers of the Central Valley. It reaches its peak of activity, up to 95 per cent parasitization, during the spring, then declines drastically in numbers by early summer. Extremely low levels of survival continue from early June until late August or September, and summertime parasitization of the female scales of the spring generation is very slight. The parasite populations recover to a certain extent in the fall and parasitization may reach a level of 5 to 25 per cent, but winter is, again, too cold for the normal reproductive activity of the adult wasps, and parasitization remains very low until the spring, when a rapid and effective increase usually takes place. Unusually low fall or spring temperatures may retard the parasite's activity and reduce its overall effectiveness (Huffaker, Kennett, and Finney, 1962).

In certain favorable locations, *A. maculicornis* alone produced good control. This was the case in groves where summer parasitization by this species remained as high as 1 to 5 per cent and where annual or biennial pruning, good fertilization and irrigation were practiced, producing vigorous growth and bearing of fruits distant from parent scales on the older wood. However, full economic control was not achieved with this parasite alone and the results of using it were too unreliable for general acceptance (see Tables 1 and 2, 1956 through 1959). The reasons for this limitation in its performance include the presence of occasional scales on the fruits even under a good level of control (less than 5 per cent infested fruits constitutes a tolerable level), disturbances from drift of insecticides, the decimation of the parasite population due to unavailability of the hosts in suitable stages for parasitization in early summer, and, most important, the severe effect of the characteristic hot,

Table 1.   Parasitization and   degree of control[†] of *Parlatoria oleae* (Colvée) by *Aphytis maculicornis* (Masi) and *Coccophagoides utilis* Doutt at the Duncan Grove, Herndon, California, 1956-1968.   (Adapted in part from Huffaker and Kennett, 1966)

Year	Month	No. mature female scales in sample	Av. mature female scales per twig‡	Per cent parasitization			Per cent commercial fruit cullage
				A. macu-licornis	*C. utilis*	Combined net	
1956	April	19	0.5	79.2		79.2	
	Aug.	9	0.2	0.0		0.0	5.4
1957	April	48	1.0	42.6		42.6	
	Aug.	72	0.8	0.0		0.0	4.4
1958	April	313	3.3	37.2		37.2	
	Aug.	190	1.9	3.1		3.1	27.4
1959	April	177	3.7	100.0		100.0	
	Aug.	22	0.4	0.0		0.0	7.4
1960	April	4	0.1	75.0	0.0	75.0	
	Aug.	15	0.3	0.0	35.9	35.9	0.4
1961	April	42	0.7	40.5	59.5	100.0	
	Aug.	19	1.6	0.0	42.1	42.1	0.5
1962	April	38	0.8	40.0	50.0	90.0	
	Aug.	42	0.9	2.4	78.5	80.9	0.0
1963	April	9	0.2	55.6	44.4	100.0	
	Aug.	3	0.1	0.0	33.3	33.3	0.0
1964	April	15	0.3	60.0	40.0	100.0	
	Aug.	4	0.1	0.0	75.0	75.0	0.0
1965	May	14	0.2	7.1	78.6	85.7	
	Aug.	6	0.1	0.0	16.7	16.7	0.5
1966	April	3	0.0	33.3	33.3	66.6	
	Aug.	—	—	—	—	— ·	0.0
1967	Aug.	—	—	—	—	—	0.0
1968	Aug.	—	—	—	—	—	0.5

[†]As represented by per cent cullage.
‡Sample size was varied considerably--from a minimum of 12 twigs for the August 1961 sample to a maximum of 96 twigs when scale densities were expected to be low.

Table 2. Parasitization and degree of control by *Parlatoria oleae* (Colvée) by *Aphytis maculicornis* (Masi) and *Coccophagoides utilis* Doutt at the Oberti Grove, Madera, California, 1957-1968. (Adapted in part from Huffaker and Kennett, 1966)

Year	Month	No. mature female scales in sample†	Av. mature female scales per twig	Per cent parasitization			Per cent commercial fruit cullage
				A. maculicornis	*C. utilis*	Combined net	
1957	March	1,856	30.9	6.1		6.1	
	Aug.	376	6.3	0.0		0.0	22.2
1958	April	1,186	19.8	5.6		5.6	
	Aug.	697	11.6	26.7‡		26.7	60.2
1959	April	204	3.4	79.4		79.4	
	Aug.	23	0.4	0.0		0.0	4.4
1960	April	39	0.7	89.7	5.1	94.8	
	Aug.	30	0.5	0.0	43.3	43.3	0.2
1961	April	22	0.4	50.0	40.9	90.9	
	Aug.	—	—	—	—	—	0.5
1962	Jan.	119	2.5	5.9	29.4	35.3	
	Aug.	21	0.4	0.0	38.1	38.1	0.6
1963	May	13	0.3	61.5	38.5	100.0	
	Aug.	1	0.0	—	—	—	0.4
1964	May	19	0.4	52.6	36.8	89.4	
	Aug.	4	0.1	0.0	25.0	25.0	0.3
1965	April	22	0.5	45.5	50.0	95.5	
	Aug.	6	0.1	0.0	0.0	0.0	1.0
1966	April	15	0.2	20.0	33.3	53.3	
	Aug.	—	—	—	—	—	0.5
1967	Aug.	—	—	—	—	—	0.0
1968	Aug.	—	—	—	—	—	0.5

†Sample size was 60 twigs from 1957-1961 inclusive, 48 twigs from 1962-1964 inclusive, and 96 twigs in 1965 and 1966.

‡Extraordinary summer survival of *A. maculicornis* possibly associated with industrial sprinkling operations adjacent to test trees.

dry summers on the parasite (Huffaker and Kennett, 1966).

The degree to which *A. maculicornis* is capable of controlling the olive scale was demonstrated by the insecticidal check method (see Chapter 5). In less than 3 years, scale populations on trees sprayed twice a year with DDT to the virtual elimination of the parasites, increased tremendously and reached levels 75 to nearly 1,000 times higher than on untreated trees (see Table 3). However, while this contrast was very impressive and attested to the general efficacy of *A. maculicornis,* the level at which the scale persisted in the presence of *Aphytis* was not always low enough to preclude economic injury (see Huffaker, Kennett,and Finney, 1962).

Table 3. Average number of live mature female *Parlatoria oleae* (Colvée) per twig on DDT-treated, *Aphytis*-inhibited trees, contrasted with densities in surrounding biological control block of trees. (Adapted from Huffaker, Kennett, and Finney, 1962.)

| Location | 1958 Fall | | 1959 Spring | | 1959 Fall | | 1960 Spring | | 1960 Fall | | Net change by Fall 1960 as fold increase of densities in Fall 1958 | | Fall 1960 ratio of densities |
	DDT trees	*Aphytis* trees	DDT trees	*Aphytis* trees	DDT trees	*Aphytis* trees	DDT trees	*Aphytis* trees	DDT trees	*Aphytis* trees	DDT trees	*Aphytis* trees	DDT:*Aphytis*
Kirkpatrick (Lindsay)	0.0+	0.0+	0.3	0.3	5.5	1.2	25.5	1.9	67.6	0.9	+676:1	+ 2:1	75:1
Long (Seville)	2.0†	1.6	4.2	9.0	2.7	0.0	12.6	0.0	29.8	0.03	+15:1	- 5:1	993:1
Bell (Hills Valley)	0.8	1.3	10.6	0.1	12.1	0.4	25.6	0.2	90.7	0.1	+113:1	-0:1	907:1
Bell (Clovis)	1.2	3.0	12.4	0.2-	43.9	0.5	55.6	1.9	287.8	2.5	+ 85:1	- 3:1	140:1
Duncan (Herndon)	2.0	3.8	38.1	3.2	134.0	0.7	60.4	0.0+	169.8	1.5	+240:1	1:1	113:1
Oberti (Madera)	16.9	11.6	23.9	3.4	43.5	0.4	120.3	0.6	204.2	0.5	+ 12:1	-23:1	408:1

†Treated with oil through error of grower in June, 1958 killing many scales. The treatment was incorporated with a hormone spray and was unnecessary for olive scale control as the very high parasitization had decimated the scale population during April and May, prior to oil application. The DDT treatments in these trees were begun one scale generation later than in the other groves (original test trees proved inaccessible in winter).

Thus, in spite of the demonstrated effectiveness of *A. maculicornis* in reducing and maintaining olive scale densities much below those prevalent prior to its establishment, its failure to achieve highly reliable economic control meant that the establishment of additional suppressive agents would be required.

In 1957, two additional species of parasites of olive scale were collected in Pakistan. These parasites, *Coccophagoides utilis* Doutt and *Anthemus inconspicuus* Doutt, were initially reared and colonized in California during 1957-1958. Early in 1961, *C. utilis* was found to be well established at two release sites, whereas *A. inconspicuus* had generally disappeared by 1961. Since *C. utilis* showed great promise of improving the control of olive scale in the two groves where it first became established, it was mass-reared and widely colonized. Over four million *C. utilis* were released during 1962-1964 at more than 170 sites in 24 counties (Kennett, Huffaker and Finney, 1966). Large numbers were also collected in the field from scattered, heavily infested "field insectary" trees that had been previously and deliberately "upset" by light DDT treatments for this purpose, and were distributed to many localities (Huffaker, Kennett and Finney, 1962; Kennett, Huffaker and Opitz, 1965; Kennett, Huffaker and Finney, 1966). The production of parasites by this means is most economical and the use of such "insectary" trees should be considered in similar projects.

It is interesting to note that the collection and shipment of *C. utilis* from W. Pakistan in 1957 was made incidentally during a search for parasites of California red scale. Obviously, no amount of planning and preliminary research on their "suitability" can replace the actual empirical search for natural enemies in the field and their trial by release in the new environment.

In contrast to *A. maculicornis*, which is a primary ectoparasite in both sexes, *C. utilis* is an endoparasitic species exhibiting both primary and hyperparasitic habits. The female (fertilized) egg is deposited internally within the scale body, and the female develops as a primary parasite. The male (unfertilized) egg is deposited externally only on developing female prepupae or pupae of its own species within the host scale, and the male develops as a hyperparasite. Unmated female *C. utilis* lay eggs which give rise to male progeny only, whereas the mated female lays eggs which give rise to female progeny only (Broodryk and Doutt, 1966).

In spite of its complicated life cycle, *C. utilis* proved to be an effective addition, and its introduction into California resulted in complete control of the olive scale. Incidentally, *A. inconspicuus* had been judged the better prospect by its performance in the laboratory and *C. utilis* had been largely "shelved" for a while, until the latter proved its superiority in the field. The development of this species is closely synchronized with that of the scale, and it succeeds in attaining moderately high levels of parasitization (20 to 60 per cent, averaging 40 per cent) on each of the two host generations each year. It apparently would not have controlled the scale alone, but it substantially occupies the niche left nearly open by *A. maculicornis* during the summer period, and contributes additional mortality of the fall scale generation. Thus, *C. utilis* acts as a complementary factor in the biological control of olive scale. Although a considerable degree of multiple parasitism may occur, mainly when *A. maculicornis* parasitizes a host already inhabited by an early-stage larva of *C. utilis,* this is usually not extensive because the two parasites are not strict ecological homologues, i.e., *C. utilis* oviposits in the early instars of the host, much prior to the stage which is susceptible to attack by *A. maculicornis;* on the other hand, *A. maculicornis* rejects hosts showing the marked physiological changes caused by the developing larva of *C. utilis.* The two parasites perform in a compensatory manner, and the total parasitization produced by their combined action is higher on an annual basis than that produced by *A. maculicornis* alone and considerably higher than that produced by *C. utilis* alone (Huffaker and Kennett, 1966). However, whereas the role of *C. utilis* in compensating for the poor performance of *A. maculicornis* during summer is obvious, the possible interactions between the two species in the fall require some elaboration. Poor summer survival by *A. maculicornis* results in reduced populations of this species in the fall and in low levels of multiple parasitism, permitting above-normal performance of *C. utilis.* Thus, *C. utilis* compensates partially for the reduced performance by *A. maculicornis.* Good summer carry-over by *A. maculicornis* results in higher levels of parasitization during the fall and spring periods, and although a higher degree of multiple parasitism may eliminate more *C. utilis, A. maculicornis* more than compensates for this by its greater parasitization, which is considerably higher than the maximum ever reached by *C. utilis.*

To a large extent, parasitism by *C. utilis* is merely replacing mortality of olive scale which would have been caused by *A. maculicornis.* Thus, since *A. maculicornis* searches well enough to find 80 per cent of the susceptible hosts, e.g., those not already parasitized by *C. utilis,* it also has searched well enough and presumably found 80 per cent of those already rendered unacceptable by *C. utilis* but refused them. Thus 80 per cent of the scale hosts already rendered unacceptable by *C. utilis* would have been attacked by *A. maculicornis* in any event. This is illustrated by an example representing 10 groves having both species of parasites and 10 groves having only *A. maculicornis* (see Table 4) (see also, Huffaker and Kennett, 1966, Table 3). Moreover, other evidence shows that *A. maculicornis* would not be confronted by a shortage of eggs for parasitizing these additional hosts. On this basis, the indispensable contribution to mortality by *C. utilis* would be only 4 per cent

Table 4. Parasitization of *Parlatoria oleae* (Colvée) by *Aphytis maculicornis* (Masi) and/or *Coccophagoides utilis* Doutt at 10 locations in California where *C. utilis* was established and at 10 locations where it was absent or nearly so. Late April - early May, 1964.

C. utilis present				*C. utilis* absent			
Net per cent parasitization by:				Net per cent parasitization by:			
A. maculicornis	*C. utilis*	2 species combined	*A. maculicornis* on *C. utilis*-free hosts	*A. maculicornis*	*C. utilis*	2 species combined	*A. maculicornis* on *C. utilis*-free hosts
50.3	38.1	88.4	80.9	82.6	1.0	83.6	83.5

for this host generation, rather than the observed 38 per cent. However, it should be emphasized that this "indispensable parasitism" by *C. utilis*, although accounting for a relatively small proportion of total mortality, nonetheless represents an addition which, combined with its action on the spring generation, assures reliable, permanent biological control of olive scale at extremely low population levels (Huffaker and Kennett, 1966).

The olive scale project thus serves as an outstanding demonstration of the principle of multiple introduction of natural enemies.

California Red Scale

The California red scale invaded California between 1868 and 1875 and rapidly became a major pest. In spite of continuing localized quarantine and eradication efforts, it has since spread over virtually all of the citrus-growing areas of the State. The first attempt to introduce its natural enemies into California was made as early as 1889, and intensive efforts have been made more or less continuously ever since. It is of interest to note that the two most important enemies were not discovered and colonized until 1948 and 1957, respectively—over 50 years after the work started. There are many reasons for this, various of which we will discuss later. Suffice it to say for the moment that attempts at biological control of California red scale were considered a complete failure for those 50-odd years. Meanwhile, potentially effective natural enemies remained undiscovered or unrecognized. How many other so-called "failures" can be turned into successes by adequate research remains to be seen, but the possibility should not be discounted even for those currently unsuccessful projects which have had most emphasis.

The first importations mainly comprised predacious coccinellids, a few of which became established in California. These, however, proved to be high-density feeders and made no significant contribution to the control of the scale. It was not until many years later that the development of suitable ecological studies and techniques showed that parasitic species of *Aphytis* were much more effective than the predatory ladybeetles. In fact today the predators have virtually vanished because the parasites have reduced scale densities below levels sufficient for survival of the ladybeetles.

Early in the present century, a species of *Aphytis*—in all probability *A. lingnanensis*— was recognized as a common parasite of California red scale in southern China, but no effort was made to introduce it, due to its misidentification as a species already present in California. Half a century later, *A. lingnanensis* was recognized as distinct and introduced into California. It soon proved to be an effective natural enemy of the scale, far superior to the local *A. chrysomphali*, with which it was previously confused. Thus, the inadequate state of the taxonomy of the parasitic Hymenoptera caused a delay of some 50 years in the introduction of a valuable parasite into California and in other places as well (Compere, 1955, 1961; DeBach, 1969*b*).

A long series of failures to establish imported natural enemies was caused by inadequate taxonomic knowledge of the host scales. Several parasites were repeatedly obtained in the Orient from various scales that had been misidentified as the California red scale, and were introduced into California against that scale, without success because they were not adapted to it. Other importation failures were due to the ignorance of some cryptic effects of the host plant on certain endoparasites of the scale. *Cycas* plants, infested with California red scale, were taken to the Orient on several occasions and exposed in the field in the hope of attracting the natural enemies of the scale. No parasites were obtained, and only later was it discovered that certain endoparasites are incapable of completing their development in California red scale when the latter is growing on *Cycas,* whereas the same scale on citrus is suitable for their development. Ironically, when a species of *Aphytis*—apparently *A. melinus* DeBach—accidentally "contaminated" some of those scale colonies on *Cycas* in Hong Kong, it was destroyed under the assumption that it was identical with *A. chrysomphali. A. melinus,* currently the most effective natural enemy of California red scale, was not "discovered" until many years later. These repeated importation failures led to the erroneous conclusion almost 40 years ago that no effective parasites of California red scale were present in the Orient.

The misclassification of California red scale as a species of *Chrysomphalus*, a genus believed to be of South American origin, then directed the search for natural enemies toward that continent. An unsuccessful search was therefore made in South America in 1934-1935 (Compere, 1961).

When the reasons for some of these importation failures were finally understood, the search for natural enemies was again directed to the Orient, where the scale apparently originated. Two endoparasites were then successfully introduced from China and established in California: *Comperiella bifasciata* Howard in 1941 and *Prospaltella perniciosi* Tower in 1947.

Repeated efforts to introduce another endoparasite, *Pteroptrix* (=*Casca*) *chinensis* (Howard), from China failed, presumably due to insufficient knowledge of its biology. The development of the male of this species is still unknown (Flanders, Gressitt and Fisher, 1958).

By 1947, as knowledge accumulated, the emphasis of the project was shifted toward ectoparasites of the genus *Aphytis,* which are now generally recognized as the most effective natural enemies of armored scales. All the known species of *Aphytis* develop externally on the body of armored scale insects, beneath the scale covering. They do not oviposit in a host unless its body is free beneath the scale, and armored scale insects are not susceptible to their attack during molt periods, when their body becomes attached to the scale covering. The California red scale, in contrast to many other scale species, presents a special problem in this respect, since in this species the body of the maturing female becomes firmly affixed to the scale covering, thus rendering adult females immune to attack by the species of *Aphytis.* The California red scale is, therefore, susceptible to *Aphytis* for relatively short periods during its life cycle (see DeBach, 1969*a*). Nevertheless, *Aphytis* species have in the last two decades proved to be the most effective natural enemies of the California red scale and can control it at very low levels if other adverse factors do not interfere.

Aphytis lingnanensis from southern China was first released in California in 1948, and establishment was evident by the end of that year. It was mass-reared and distributed throughout the infested citrus areas of the state during 1948-1954. It soon became the dominant parasite of the scale, gradually displacing *A. chrysomphali* in interior and intermediate climatic areas. By 1959, *A. chrysomphali* was virtually eliminated from all of southern California, except for a few small coastal pockets (see Figures 2-4) (DeBach and Sundby, 1963; DeBach, 1966). Overall control of the scale was improved, but *A. lingnanensis* was not sufficiently effective in the interior citrus areas, where climatic extremes caused mortality of the parasites and prevented them

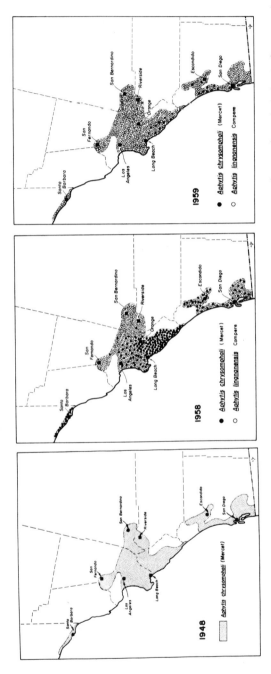

Figs. 2, 3 and 4. Distribution and relative abundance of *Aphytis chrysomphali* (Mercet) and *Aphytis linganensis* Compere in California red scale populations on citrus in Southern California. Semi-schematic. (From DeBach and Sundby, 1963)

Fig. 2 shows the distribution of *A. chrysomphali* in 1948, prior to the introduction of *A. lingnanensis*; Fig. 3, based on population counts made in 1958, shows the interim spread of *A. lingnanensis* in interior areas; and Fig. 4 shows the almost entire displacement of *A. chrysompahli* by *A. lingnanensis* by 1959.

from asserting their full biotic potential. Efforts were therefore made to augment this species by periodic releases (DeBach *et al.,* 1950), and also by the selection of cold-and-heat-resistant strains (White, DeBach and Garber, 1970).

Basic ecological studies in the field and laboratory led to the above-mentioned conclusions regarding the adverse effect of climate on *A. lingnanensis* and to the attempts to manipulate this otherwise effective species to overcome these effects. These studies include one of the most clear-cut examples of the negation of biological control by weather and illustrate some important ecological concepts, principles and practices. Early in this phase of investigation, appropriate experimental field tests showed that average population densities of the host, the California red scale, were not regulated by climate. When parasites were excluded from the scale populations, host densities always increased to maximal levels that the tree could support, regardless of the local climate. This was demonstrated time and again in all portions of southern California where the California red scale occurred (see DeBach, 1965, and Chapter 5). The same type of tests showed that *A. lingnanensis* could successfully control the California red scale in coastal and intermediate climatic areas, but that the more severe climate of interior citrus areas so reduced the efficiency of the parasite that satisfactory biological control generally was not achieved.

It was found that locally extreme, low winter temperatures, as well as high summer temperatures and low humidities, caused high mortality of various stages of *A. lingnanensis,* greatly reduced the reproductive potential of adult females, adversely affected sex-ratios and even caused very localized extermination at times. However, weather did not limit the range of the parasite, merely its efficiency, and it was shown that what appeared to be rather small differences between weather in different areas (or even in the same area) such as Riverside and Escondido, resulted in major differences in the degree of success in biological control (DeBach, Fisher and Landi, 1955; DeBach, 1965).

These studies indicated that *A. lingnanensis* was not eradicated in extreme climatic areas because protective shelters and genetically resistant individuals permitted limited survival of a population under otherwise period-ically intolerable conditions. Evidently the number of protected micro-habitats, along with a degree of intrinsic genetic variation, set an average minimum population density below which *A. lingnanensis* did not fall. This minimum may vary from year to year. Following an adverse period, the remainder of the season would be spent in rebuilding the population toward the previous higher density. However, if adverse periods occur at regular intervals, as they do in the more extreme climatic areas, it is evident that weather acts in the regulation of *A. lingnanensis* populations.

We hasten to point out that this does not necessarily mean that *A. lingnanensis* would be rare in an unfavorable or suboptimal climatic regime. Our results suggest some interesting anomalies are involved with respect to parasite density (i.e., commonness) and host density. For instance, abundance of a parasite in a habitat does not necessarily indicate efficiency in host population regulation, nor does scarcity indicate inefficiency. A parasite may be rare because it is intrinsically ineffective or because of adverse environmental conditions, or because it is highly effective in host population regulation.

The latter instance may be considered as the axiom that "effective parasites regulate their own average population densities at low levels by regulating host population densities at low levels. Thus we may find a host insect species and its parasite rare in a climatic zone optimal for both if the parasite is intrinsically *effective,* whereas if the parasite species is an intrinsically *ineffective* one, then the host may be common to abundant and the parasite more or less common, depending upon its degree of ineffectiveness. On the other hand, in a climatic zone suboptimal for the host but even more so for an intrinsically effective parasite, the host may be unregulated and become abundant, while the parasite remains less common but perhaps occasionally abundant. Thus, both would be more common on the average than they would in a climatic zone optimal for both. Here we have the anomaly of scarcity being associated with ideal weather conditions." (DeBach, 1965)

Because of the climate-related restrictions to the effectiveness of *A. lingnanensis,* emphasis was placed on the discovery and importation of better adapted species. This resulted in *Aphytis melinus* DeBach, among others, being introduced from India and Pakistan in 1956-1957. About 2.5 million mated females were cultured and released in California during 1957-1959, and the new species became well established and dispersed rapidly. It soon proved to be better adapted to extreme climatic conditions than *A. lingnanensis,* and the further development of selected strains of the latter species became unnecessary. The pattern of competitive displacement soon repeated itself (Fig. 5), and by 1965 (Fig. 6) *A. lingnanensis* was entirely displaced by *A. melinus* in interior areas, was greatly decimated in intermediate areas, and was still dominant only in some coastal areas (DeBach and Sundby, 1963; DeBach, 1966).

Aphytis fisheri DeBach, considered a sibling species of *A. melinus,* was simultaneously introduced from Burma and released in large numbers in California during 1957-1958. Laboratory tests indicated it to be a potentially effective parasite, but competition by *A. melinus* and *A. lingnanensis* appar-

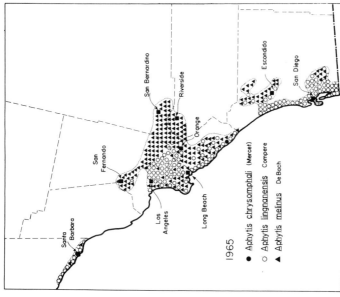

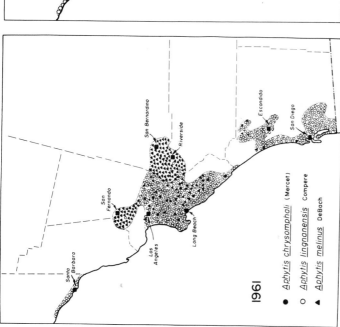

Figs. 5 and 6. Distribution and relative abundance of *Aphytis chrysomphali* (Mercet), *Aphytis linganensis* Compere and *Aphytis melinus* DeBach in California red scale populations on citrus in Southern California. Semi-schematic. Fig. 5 (from DeBach and Sundby, 1963) illustrates the virtual extinction of *A. chrysomphali* and the early progress of *A. melinus* in interior areas, in 1961; Fig. 6 shows further spread of *A. melinus* and gradual displacement of *A. linganensis* by 1965 in interior areas.

ently precluded it from becoming established.

Today, *A lingnanensis* remains dominantly responsible for control of red scale in coastal areas, where *Prospaltella perniciosi* acts as a complementary factor, and *A. melinus* is the dominant parasite in interior areas, with *Comperiella bifasciata* acting to supplement it somewhat (see Table 5).

Table 5.   Composition of populations of California red scale parasites on citrus in coastal vs. interior regions of California, 1968.   (Percentages)

Species	Coastal	Interior
Aphytis melinus	5.7	76.6
Aphytis lingnanensis	68.2	0.0
Aphytis chrysomphali	0.3	0.0
Prospaltella perniciosi	25.4	0.1
Comperiella bifasciata	0.4	23.3

These striking results with *Aphytis* clearly illustrate the need for intensive foreign exploration and introduction of natural enemies, as well as the advantages of multiple species introductions, and show the superiority of the importation method over various methods of augmentation.

By 1962, a general decline of red scale was evident in California, and the pest reached its lowest level in many years. It has been kept under satisfactory control by the introduced parasites in several untreated groves or districts in various areas (see Fig. 7), demonstrating the potential general effectiveness of biological control of the scale in the state, although the final solution in interior and desert areas may be difficult.

Groves in which the California red scale is actually held at low levels by parasites usually show an average year-round active parasitization of about 15-20 per cent. This percentage does not appear impressive, and obviously does not present the entire picture; numerous additional scales are killed by direct host-feeding by the adult parasite females. However, from accumulated empirical knowledge, we know that such per cent parasitization figures may be used as an approximate index of the progress or success of the parasite in host population regulation. When the parasite is just getting started in a grove, parasitization is low and the proportion of live scales is high; as the per cent parasitization approaches the "equilibrium" average of about 15-20 per cent,

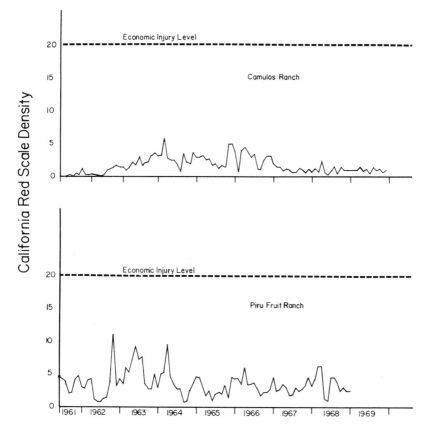

Fig. 7. Population density of California red scale in two untreated citrus groves in Southern California. The effectiveness of natural enemies, mainly parasites, has kept the scale populations in these and other groves at subeconomic levels.

the proportion of live scales becomes low. This relationship is seen in Fig. 8, which emphasizes how relatively small initial increases in parasitization are reflected by relatively great increases in red scale mortality (DeBach, 1969a).

The effectiveness of the introduced parasites has been repeatedly demonstrated by the rapid increase of the scale populations upon parasite elimination from selected trees or branches in untreated plots by various experimental methods (Fig. 9) (see also Chapter 5). In the Fillmore district alone, of some 8,500 acres of citrus, the successful biological control of

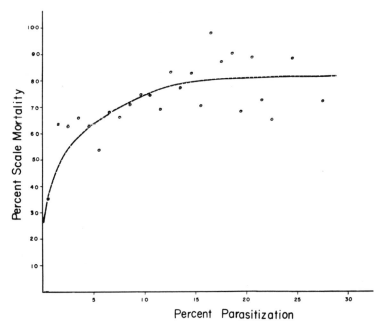

Fig. 8. Relation between per cent parasitization by *Aphytis melinus* DeBach and total field mortality of California red scale on citrus (from DeBach, 1969*a*).

California red scale is estimated to save the growers about one-half million dollars annually in reduced chemical applications. Nevertheless, most of the citrus acreage of California still receives chemical control regardless of whether or not the practice is ecologically sound or even necessary. This situation illustrates one of the most important problems of biological control. The natural enemies are, of course, largely precluded from groves receiving frequent treatments with non-selective pesticides, and the transition to biological control over large areas practically devoid of natural enemies is very difficult even with full cooperation from the growers. Figure 10 illustrates the length of time required for California red scale populations to return to normal after the cessation of non-selective treatments. Such a transition from chemical control procedures to biological control of spider mites on grapes in California required four years before final success (Flaherty and Huffaker, 1970).

Although the biological control program for California red scale has not yet achieved complete success, it has been very useful in elucidating and demonstrating some of the more important principles and practices of biological control. Most of the experimental methods for the evaluation of the

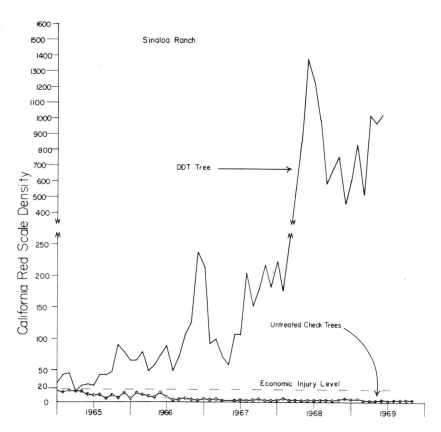

Fig. 9. Population density of California red scale on a citrus tree regularly treated with DDT, as compared to adjacent untreated trees. The scale outbreak on the DDT tree, caused by the elimination of natural enemies, demonstrates the effectiveness of these enemies on the untreated trees.

effectiveness of natural enemies, discussed in Chapter 5, were developed and demonstrated during the course of research on the biological control of red scale. The need for basic ecological studies and the adverse effects of ants, dust, insecticides, and climatic factors on the natural enemies of scale insects were likewise demonstrated by this project. Competition between natural enemies for a common prey or host was found to have no adverse effect on the degree of biological control attained—in fact, complete displacement of

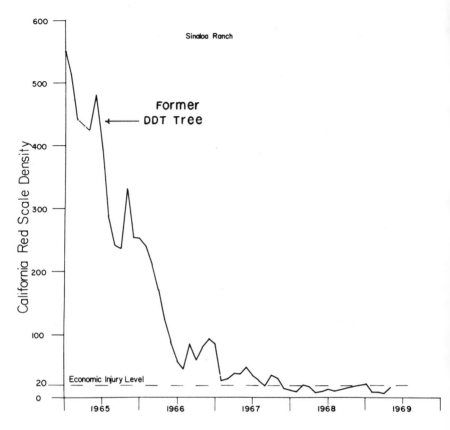

Fig. 10. Lengthy time required to reattain biological control of California red scale on a citrus tree after a DDT-induced outbreak.

one species by another newly introduced one led to improvement in biological control in two successive cases of displacement. Some direct benefits have also accrued in other countries. For example, *Aphytis melinus* was recently introduced into Greece, and has been responsible for substantial gains in the biological control of California red scale as well as of two other scale insect species in that country (DeBach and Argyriou, 1967).

SUMMARY AND CONCLUSIONS

Biological control projects have been directed against some 223 insect pest species and have resulted in some appreciable degree of success with over 50 per cent of these pests. Coccids account for more than half of all these projects against insects showing some degree of success, and about two-thirds of all complete successes. Of about 64 species of coccids subjected to biological control attempts, 50 species, or 78 per cent, have shown at least partial control. With all other insect pests (159 species), some success was obtained with 70 species, or 44 per cent.

The higher proportion of success with coccids appears due in part to greater research emphasis on this group; in part to the frequency with which coccids invade new territories without accompanying natural enemies and seriously attack high-priced crops while being difficult to control with chemicals; and possibly in part to various biological and ecological characteristics of coccids.

This is not meant to imply that the degree of success with other pest insect groups ultimately cannot be as high as that obtained with coccids or that the current extent of success observed with coccid species represents an expected ceiling. Indeed, we expect it to increase. It is emphasized that short-term lack of success with any given species does not mean that the problem is insoluble, as was shown by the ultimate degree of success obtained with the California red scale and olive scale projects.

Financial savings to the agricultural industry from successful biological control have been great. It is conservatively estimated that from all biological control projects in California alone some $200,000,000 savings have resulted from reduced need for chemical control and decreased pest damage during the past 45 years. The environmental and sociological benefits from diminished chemical pollution are obvious.

Four successful biological control projects—the cottony cushion scale, the Florida red scale, the olive scale, and the California red scale—were selected as illustrative models and are discussed in some detail. The results from these, as well as from other projects with coccids, lead to the following conclusions:

1. Importation of new natural enemies from abroad is the single best approach to biological control and needs increased emphasis. Search for natural enemies should extend throughout the entire range of distribution of a pest.

2. Accurate taxonomy, especially biosystematics, is necessary and basic to success in applied biological control. Current knowledge is woefully inadequate.

3. Basic ecological studies of the pest-natural enemy complex at home and abroad may furnish the ultimate key to success but need not necessarily precede, and certainly should not retard, the importation of new natural enemies.

4. From the successful projects studied, the most effective natural enemies characteristically show very high, if not complete, host or prey specificity (preference). They are often multivoltine with respect to the prey. They are well adapted to physical conditions of the pest habitat or their potential will not be realized. They are obviously good searchers, although this attribute is difficult to measure, *a priori,* without detailed field studies abroad.

5. There is no single "best" natural enemy for a given pest species as has sometimes been postulated.

6. In most cases of complete success, one or perhaps two species of natural enemies are responsible for control in a given habitat or ecologically uniform area. A large combination or sequence of species, as also has been postulated, is not necessary.

7. When the pest ranges over wide areas, different species of enemies may be responsible for control in different habitats and these may be mutually exclusive. The interspecific competition between natural enemies occurring in the areas of overlap, or when they are imported and released in the same habitats, is not detrimental to regulation and control of host populations. In all cases studied, the addition of a second or third "competing" species of enemy only led to better pest population regulation.

8. Multiple importation of natural enemies, either simultaneously or chronologically, is the only means of assuring that the best single species or combination of species is established in each habitat and throughout the pest's range. Addition of a second enemy species, which by itself might be inconsequential, can make the difference between partial and complete success in biological control.

9. It has been suggested that so-called "direct" pests (i.e., ones that cause direct damage to such marketable items as fruit at low pest densities) are not suitable subjects for biological control. This is refuted by the successful control of the olive scale, which is a direct pest. However, *probability* of success with direct pests may be lower than with indirect pests.

ACKNOWLEDGMENTS

This chapter is a contribution of the IBP Project on Biological Control of Scale Insects. It was supported in part by NSF grants GB-6776 and GB-14489.

LITERATURE CITED

Bedford, E. C. G. 1968. An integrated spray programme. S. Afr. Citrus J., No. 417, pp. 9-28.

Broodryk, S. W., and R. L. Doutt. 1966. The biology of *Coccophagoides utilis* Doutt (Hymenoptera: Aphelinidae). Hilgardia 37: 233-254.

Clausen, C. P. (Editor) In press. Introduced parasites and predators of arthropod pests and weeds. A world review (1974). USDA.

Compere, H. 1955. A systematic study of the genus *Aphytis* Howard (Hymenoptera, Aphelinidae) with descriptions of new species. Univ. Calif. Publ. Entomol. 10: 271-319.

Compere, H. 1961. The red scale and its insect enemies. Hilgardia 31: 173-378.

DeBach, P. 1960. The importance of taxonomy to biological control as illustrated by the cryptic history of *Aphytis holoxanthus* n. sp. (Hymenoptera: Aphelinidae), a parasite of *Chrysomphalus aonidum,* and *Aphytis coheni* n. sp., a parasite of *Aonidiella aurantii.* Ann. Entomol. Soc. Amer. 53: 701-705.

DeBach, P. (Editor) 1964. Biological Control of Insect Pests and Weeds. Reinhold Publ. Co., N. Y. 844 pp.

DeBach, P. 1965. Weather and the success of parasites in population regulation. Can. Entomol. 97: 848-863.

DeBach, P. 1966. The competitive displacement and coexistence principles. Ann. Rev. Entomol. 11: 183-212.

DeBach, P. 1969a. Biological control of diaspine scale insects on citrus in California. Proc. 1st Int. Citrus Symp. Riverside, Calif. (1968) 2: 801-816.

DeBach, P. 1969b. Uniparental, sibling and semi-species in relation to taxonomy and biological control. Israel J. Entomol. 4: 11-28.

DeBach, P., and L. C. Argyriou. 1967. The colonization and success in Greece of some imported *Aphytis* spp. (Hym. Aphelinidae) parasitic on citrus scale insects (Hom. Diaspididae). Entomophaga 12: 325-342.

DeBach, P., and B. R. Bartlett. 1951. Effects of insecticides on biological control of insect pests of citrus. J. Econ. Entomol. 44: 372-383.

DeBach, P., E. J. Dietrick, C. A. Fleschner, and T. W. Fisher. 1950. Periodic colonization of *Aphytis* for control of the California red scale. Preliminary tests, 1949. J. Econ. Entomol. 43: 783-802.

DeBach, P., T. W. Fisher, and J. Landi. 1955. Some effects of meteorological factors on all stages of *Aphytis lingnanensis,* a parasite of the California red scale. Ecology 36: 743-753.

DeBach, P., and R. A. Sundby. 1963. Competitive displacement between ecological homologues. Hilgardia 34: 105-166.

Doutt, R. L. 1954. An evaluation of some natural enemies of the olive scale. J. Econ. Entomol. 47: 39-43.

Doutt, R. L. 1958. Vice, virtue and the vedalia. Bull. Entomol. Soc. Amer. 4: 119-123.

Flaherty, D. L., and C. B. Huffaker. 1970. Biological control of Pacific mites and Willamette mites in San Joaquin Valley vineyards. Hilgardia 40(10): 267-308.

Flanders, S. W., J. L. Gressitt, and T. W. Fisher. 1958. *Casca chinensis,* an internal parasite of California red scale. Hilgardia 28: 65-91.

Hafez, M., and R. L. Doutt. 1954. Biological evidence of sibling species in *Aphytis maculicornis* (Masi) (Hymenoptera, Aphelinidae). Can. Entomol. 86: 90-96.

Huffaker, C. B., and C. E. Kennett. 1966. Biological control of *Parlatoria oleae* (Colvee) through the compensatory action of two introduced parasites. Hilgardia 37: 283-335.

Huffaker, C. B., C. E. Kennett, and G. L. Finney. 1962. Biological control of the olive scale, *Parlatoria oleae* (Colvee), in California by imported *Aphytis maculicornis* (Masi) (Hymenoptera: Aphelinidae). Hilgardia 32: 541-636.

Kennett, C. E., C. B. Huffaker, and G. L. Finney. 1966. The role of an autoparasitic aphelinid, *Coccophagoides utilis* Doutt, in the control of *Parlatoria oleae* (Colvee). Hilgardia 37: 255-282.

Kennett, C. E., C. B. Huffaker, and K. W. Opitz. 1965. Biological control of olive scale. Calif. Agr. 19(2): 12-15.

Maltby, H. L., E. Jiminez-Jiminez, and P. DeBach. 1968. Biological control of armored scale insects in Mexico. J. Econ. Entomol. 61: 1086-1088.

Muma, M. H. 1969. Biological control of various insects and mites on Florida citrus. Proc. 1st Int. Citrus Symp., Riverside, Calif. (1968) 2: 863-870.

Rosen, D. 1965. The hymenopterous parasites of citrus armored scales in Israel (Hymenoptera: Chalcidoidea). Ann. Entomol. Soc. Amer. 58: 388-396.

Rosen, D. 1967. Biological and integrated control of citrus pests in Israel. J. Econ. Entomol. 60: 1422-1427.

Taylor, T. H. C. 1935. The campaign against *Aspidiotus destructor* Sign., in Fiji. Bull. Entomol. Res. 26: 1-102.

White, E. B., P. DeBach, and M. J. Garber. 1970. Artificial selection for genetic adaptation to temperature extremes in *Aphytis lingnanensis* Compere (Hymenoptera: Aphelinidae). Hilgardia 40: 161-192.

Chapter 8

CONTROL OF PESTS IN GLASSHOUSE CULTURE BY THE INTRODUCTION OF NATURAL ENEMIES

N. W. Hussey
Glasshouse Crops Research Institute
Littlehampton, England
and L. Bravenboer
Proefstation voor de Groenten- en Fruitteelt
onder Glas
Naaldwijk, The Netherlands

INTRODUCTION

While the predominately warm and sheltered glasshouse environment favors the rapid growth of pest populations, it also provides favorable conditions for their natural enemies. Thus, any interaction between a pest and its primary natural enemy can, in the absence of hyperparasites or secondary predators, proceed at a predictable rate. It is therefore surprising that, in the 30,000 acres of glasshouses in the world, biological control has been exploited to such a limited extent since the pioneering work by Speyer (1927) on the control of whitefly *Trialeurodes vaporariorum* Westwood by the parasite *Encarsia formosa* Gahan. In 1930 the Cheshunt Experimental Research Station (since transferred to the Glasshouse Crops Research Institute) was annually supplying 1½ million of these parasites to about 800 nurseries in Britain. At about the same time, 18 million *Encarsia* were shipped to Canada, and later this parasite was introduced into Australia and New Zealand and subsequently became widely dispersed in those countries. After World War II, distribution was discontinued in Britain as the newly-introduced insecticide DDT provided convenient and efficient control on most glasshouse crops. Later, Doutt (1951, 1952) demonstrated that the mealybug *Planococcus citri* (Risso) could be successfully controlled on gardenias by two encyrtid parasites (*Leptomastix dactylopii* Howard and *Leptomastidea abnormis* Girault) and the ladybird *Cryptolaemus montrouzieri* Mulsant. Earlier, McLeod (1937) had defined the conditions in which *Aphidius phorodontis* Ashmead could control *Myzus persicae* (Sulz) in glasshouses and more recently Richardson and Westdal (1965) controlled this aphid with *Aphelinus semiflavus* Howard.

Interest in biological control in glasshouses was stimulated by the introduction of the phytoseiid predator *Phytoseiulus persimilis* (Athias-Henriot) into Europe by Dosse (1959). Within a few years, several workers (Chant, 1961; Bravenboer, 1963; and Bravenboer and Dosse, 1962) demonstrated the efficiency of this predator as a control of *Tetranychus urticae* Koch on both runner beans and cucumbers. While European workers were convinced of the efficiency of this natural enemy under experimental conditions, they were not impressed with its performance under commercial conditions. They argued that its effective use demanded too much skill on the part of the grower.

In Britain, however, the importation of *Phytoseiulus* coincided with the completion of studies on the effect of leaf-damage by *T. urticae* on crop production of cucumbers. These studies had been started after other work on acaricide resistance in this mite (Hussey, 1966) had shown that the problem was so serious and complicated by cross-resistance patterns in different strains that the acaricide selection pressure had to be reduced.

Hussey and Parr (1963) were able to show that an efficiently applied and effective acaricide could prevent the damage caused by red spider mites from exceeding the economic threshold with less than half the applications used by the average commercial grower in the early 1960's (Gould and Kingham, 1964). This damage assessment work also showed that a mature cucumber plant could withstand quite severe mite injury (equivalent to a loss of about 30 per cent of the photosynthetic area) before the yield was reduced. Furthermore, this plant has remarkable powers of recovery from such damage and we have several documented records where the "recovery" yield over-compensated the original loss, so that the ultimate yield of damaged plants was superior to that of the undamaged controls. It should not be overlooked, however, that the crop value is not a function of yield alone but is much influenced by variations in price as the season progresses. Late increases in yield are not necessarily profitable to the grower.

These studies on leaf-damage showed that the damage caused by uncontrolled mite populations increased at predictable rates, so it was decided to test whether the predator could influence leaf-damage in a predictable manner. This proved to be the case (Hussey *et al.*, 1965) and it naturally follows that we should start red spider mite/*Phytoseiulus* interactions by introducing the pest uniformly through the crop. Experience in the Netherlands suggests that there may be some unpredictability in the pattern of predatory control by *Phytoseiulus* but this conclusion was based on actual numbers of mites. Leaf-damage is a better measure of control. While numbers may vary widely, leaf-damage does not. Using this criterion, the margin of safety ensures effective control consistently below the economic threshold. The success of this technique (Gould *et al.*, 1969) in large-scale commercial experiments increased the demand for the predator, but despite the good control it was soon apparent that, in the absence of routine acaricide applications, other pests became a problem. In the ensuing sections of this chapter techniques for the biological control of some of these important glasshouse pests will be elaborated and some pest-management systems suitable for the complex of pests on chrysanthemums and cucumbers are discussed.

SOME BIOLOGICAL CONTROL TECHNIQUES

Control of Red Spider Mite

In temperate regions the key to successful control of spider mites under glass is the prevention of dispause. Elimination of the overwintering population throughout a glasshouse nursery can lead, at least temporarily, to the extinction of the pest. Predators are therefore used to contain, and finally eliminate, mites which would otherwise hibernate in the autumn within cracks and crevices and so remain protected from standard winter hygiene treatments such as washing-down and sulphur fumigation.

Normally, spider mites begin to emerge from hibernation in the glasshouse structure soon after the houses are heated in January and continue to do so for the next three or four months. These mites descend onto the young plants indiscriminately and, if uncontrolled, develop into "patchy" infestations of varying severity. To contain these infestations, spider mites are deliberately introduced onto every cucumber plant within a few days of planting by placing a small section of bean leaf bearing 10-20 female mites on each plant. These prey mites are allowed to increase until the mean leaf-damage index is 0.4 (Hussey and Parr, 1963) when two predators are introduced on every second plant (Fig. 1). The predator quickly moves to the adjacent plants and within two or three weeks will be widely distributed, for Hussey and Parr (unpublished data) showed that every colony of mites was accompanied by a predator 18 days after introducing *Phytoseiulus* onto one plant in ten. Gould *et al.* (1969) have shown that the continual invasion of the plants by post-diapause mites keeps the interaction going throughout the summer without further manipulation. Establishment of a uniform prey population to support the increase of a well distributed population of a predator was also used successfully by Huffaker and Kennett (1956) when they made introductions into newly planted fields of strawberries.

If no spider mites remain on the plants in early September to be triggered into diapause by the shortening daylength (Parr and Hussey, 1966) a very different situation occurs the following year. No mites are present to infest the plants in January so that, if the spider mite/predator interaction is set-up as before, the introduced mite population will be eliminated within 4 to 5 weeks and, although the predators may survive for a further 2 to 3 weeks, they then die out. There is then no continuing interaction as described earlier. We therefore advocate delaying the introduction of spider mites until late April or early May. It is also necessary to prolong the mite/predator interaction as long as possible. Since it is unlikely that mites will be brought in from outside in any numbers, there is less need, than in the primary year, to establish large numbers of predators as rapidly as possible. We therefore introduce the predator onto every fifth (rather than every second) plant, which delays control for eight weeks (Parr and Hussey, 1967).

Despite this delay in introducing the predator it will, in the absence of fresh invasions of red spider mite, die out. We therefore advocate that the red spider mite should be re-introduced three weeks after control is achieved and

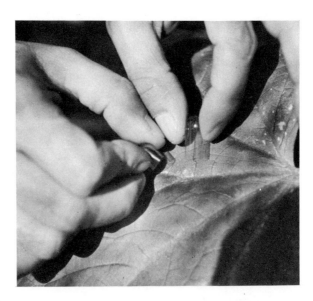

Fig. 1. Releasing *Phytoseiulus* from gelatin capsule on cucumber leaf.

that further introductions are made at four week intervals. This program maintains a low population of predators uniformly distributed over the plants which can deal with any invasions by the pest that may occur from outside sources.

A feature in the predatory efficiency of *Phytoseiulus* is the deleterious effect of high temperatures. Force (1967) claimed from laboratory experiments that its efficiency declines above the optimum temperature of 20°C (68°F). In the traditional European cucumber house with a low pitched roof the plants are trained up to the ridge, immediately below the glass. It has been commonly observed that, despite successful predatory control up to the end of June, severe spider mite damage may develop at the tops of the plants immediately below the ridge. By shading portions of the ridge-glass in trial houses, this damage, which seems to be due to decreased efficiency of *Phytoseiulus* or merely its avoidance of excessive heat at or about the summer equinox, was eliminated. However, growers dislike pest control practices which affect cultural techniques and so the difficulty can best be avoided by timing the original predator introduction so as to achieve almost complete control of the spider mites by late June. There are then no mites to concentrate on the upper foliage. The question may well be asked—why do these mites move to the tops of the plants at this time? The key to this question possibly lies in the more obvious behaviour of *T. urticae* on year-round chrysanthemums. Here the mites swarm to the upper leaflets when the flower buds are formed. This is presumably a response to senility (or maturity) in the plant and possibly occurs in the long-growing cucumber plant after the longest day. Another method of preventing "top" damage is by spraying below the ridge with tetradifon which has only a slight deleterious effect on the predator.

All these techniques are designed to avoid the indifferent control which follows predator introduction on to natural mite infestations. The predators tend to stay at the primary infestation site on which they have been placed until the prey is almost eliminated, when starvation and infrequent contact with food cause them to wander on to other plants. In the meantime, however, other infestations may have developed. These can increase for some days before predators find them. In this way severe patches of damage appear in different parts of the glasshouse. One of us (L.B.) has shown that if the predator is present in large numbers only at one point in the center of a large glasshouse, and spider mites are introduced on all other plants, economic damage will be prevented only up to 10 meters from the original concentration of predators. While certain small growers have successfully utilized the predator in this way, almost constant surveillance is required if damage is to be avoided (Gould, 1968).

In Holland, biological control of mites has been successfully achieved on about 50 nurseries during 1969. A single grower has reared the predator and sold it to others. *Phytoseiulus* is introduced on to every second plant when spider mite infestation is first observed. The supplier remains responsible for control over the first three weeks and he charges 20-28 cents/m² treated. Obviously the costs of rearing *Tetranychus* to obtain uniform infestations are avoided but, in Britain, growers could probably achieve this very cheaply themselves, using a few adequately illuminated bean plants.

Control of Whitefly

Although *Encarsia* has been used to control whitefly in glasshouses for almost 40 years there are no precise recommendations for the numbers to be used or the techniques of introduction. At first, the parasite was introduced merely by hanging up plant material bearing black parasitized scales. This practice inevitably fell into disrepute due to the frequency with which plant diseases were spread from the distribution center.

The first attempt at precision in the introduction technique was due to McLeod (1938) who claimed that one parasite should be used for each sq.ft. of glasshouse area. While practical biological control workers were using vague "hit or miss" methods, Burnett (1960a, 1960b and 1964) analyzed a series of carefully conducted experiments. He showed that the parasite killed the whitefly in two ways. Characteristically, whitefly scales are killed by the development of *Encarsia* larvae after the adult parasite has laid its egg in the sedentary scale stage. The eggs, active first-instar larvae and the pupae are never attacked. Parasitized scales turn black 9 to 10 days after being attacked. The parasites may also attack second-instar larvae just after they have settled into permanent feeding positions. These attacks normally occur where there is severe competition for the older, preferred stages. These young scales may be probed many times and, although no eggs are laid, the scales are killed.

It is also known that the adult parasites may probe pupae and host-feed, so obtaining sufficient protein to prolong survival whether or not scales are available for parasitism (Gerling, 1966).

Burnett also concluded that *Encarsia* would not control *Trialeurodes* at temperatures below 24°C. The overriding effect of temperature is well-known and may be summarized as follows: At 18°C the fecundity of the whitefly is 10 times that of *Encarsia* though the rate of development is equal, while at 26°C the fecundity is equal and the rate of development of the parasite is twice that of the host.

Speyer's (1927) work suggested that small numbers of tomato shoots bearing parasitized scales should be hung in the glasshouse for three weeks and that black, parasitized scales should appear on the plants 2 to 4 weeks later. In severe whitefly infestations he suggested that the houses first be fumigated with hydrogen cyanide. It was further recognized that the night temperature should not fall below 13°C if *Encarsia* was to establish itself.

It is important to remember that in a glasshouse crop which normally lasts from six to nine months, only a single oscillation of the host/parasite interaction is involved, unlike the many successive oscillations which Burnett arranged in his artificial situation. Parr (1968) obtained complete control of whitefly in 19 weeks when parasites were released at 1 per sq.ft. on to four plots of tomato plants infested with different numbers of whiteflies. A most significant conclusion from this experiment was the larger numbers of whiteflies on the plots when the infestations had been started either on 4 or 12 of the 50 plants in each plot, in contrast to the much smaller numbers where the whiteflies had been released on 20 of the 50 plants. This again suggested the need to initiate all artificial biological control programs with a uniform infestation of the pest. In a large-scale trial of this technique on cucumbers (Hussey, 1969), the pest was established by liberating eight whiteflies between each pair of cucumber plants. *Encarsia* was released at the rate of eight per plant 14 days after releasing the whiteflies (Fig. 2). Within twelve weeks, control had been achieved on 75 per cent of the plants. On a few plants the pest continued to increase despite the presence of large numbers of parasites; indeed the proportion of parasitism tended to decrease. This partial failure was apparently due to the large amount of honeydew then present which interfered with parasite oviposition and emergence of the adults from parasitized scales. This honeydew can, however, be easily sprayed off with water.

More recently (Gould, pers. comm.), has completed an exactly comparable experiment in which, although the introduction rate of whiteflies was the same, establishment of the pest was lower. In this case, complete control was achieved on all the plants.

The most important prerequisite for such biological control is establishment of the economic threshold of damage by the pest. In the case of whitefly, this is the population producing the largest quantity of honeydew that is insufficient for damaging sooty mold (*Cladosporium sphaerospermum*) establishment and spread at optimal humidity (Hussey *et al.,* 1957). On tomatoes, a whitefly population of 10 adults per leaf on the apical foliage produced only 3 per cent "dirty" fruits when larger populations were causing more severe damage. On cucumbers, little or no honeydew was evident on the leaves where less than 40 whiteflies were present on each upper leaf. Copious quantitites of honeydew accumulated where 135 flies were present on the upper leaves, though no sooty mold developed. A maximum of 50 to 60 whiteflies can, therefore, probably be safely tolerated without economic loss.

It would appear that unless the temperature is too low, *Encarsia* can be safely used to control whitefly on all except the most highly pubescent hosts (Milliron, 1940). A minimal night temperature of 15°C should be adequate so long as there is a reasonable number of sunny days to raise the mean temperature to around 18°C.

Fig. 2. Black, parasitized whitefly scales in an open tube, supported on a cane, for emergence of *Encarsia* adults.

Control of the Cotton Aphid

The cotton aphid, *Aphis gossypii* Glover, is not normally a problem in glasshouse culture in Europe, but when the acaricide spray program is relaxed or eliminated it may become a problem on cucumbers or even chrysanthemums.

There is doubt as to how this aphid survives the winter outdoors in temperate climates. Kring (1959) showed that the primary hosts, on which over-wintering eggs are laid, include members of the Malvaceae, Bignoniaceae and Rhamnaceae. Possibly the most important and widespread host in Europe is the Virginia Creeper, *Parthenocissus* sp. However, wild, outdoor hosts may not be necessary as the source of new infestations in glasshouses each spring, for the aphid is very common on *Convolvulus* and can persist on isolated specimens of this plant that may grow below the glasshouse benching and so re-infest the new crops.

Once established, the cotton aphid can increase at a phenomenal rate. Studies by Passlow and Roubicek (1967) and Khalifa and Sharaf El-Din (1965) showed that the developmental period at $27°C$ was about seven days, the longevity of viviparous females was 16 days, and about 40 nymphs were produced in this time. In commercial cucumber houses the first immigrant females breed even faster in the absence of competition from their fellows and produce about 40 progeny in seven days. In this time, the progeny are themselves reproducing so that the total population increases about 10-fold in each subsequent week until the whole plant is overwhelmed in 4 to 5 weeks. Large concentrations of aphids cause collapse of the older leaves (as if by wilting) while the youngest leaves become very dark-green and are severely distorted (rolled under at the edges). Associated with this direct damage, large quantities of honeydew are deposited on the upper surfaces of the lower leaves and a dense growth of sooty molds develops. Where plants were artificially infested at the 8-leaf stage and aphid increase was permitted until leaf symptoms showed, yield was reduced by 90 per cent over the first two weeks after the aphids were controlled, and by 50 per cent over the next six weeks; later, yield became normal. The economic threshold, i.e., above which losses could be detected, was 7 aphids/cm^2. In view of its rapid reproductive rate, control must be achieved very soon after the aphid is first detected.

It is also essential that control measures applied become effective very rapidly. Of the parasites of *A. gossypii* which were available at other research centers, *Aphelinus flavipes* Kurdjumov, obtained from India through the Commonwealth Institute of Biological Control, was successfully established and reared at the Glasshouse Crops Research Institute and has been used in all our experiments.

Even when *A. flavipes* was introduced within one week of artificially infesting the plants with aphids it was unable to overtake the pest population because the reproductive increase of the pest was virtually unaffected by the presence of the parasite. Only when overcrowding occurred so that the aphid's rate of increase was self-limited, did the rate of parasite increase (6-fold per week) exceed that of the aphid. Two leads emerged from these studies which provided clues to effective control techniques. Firstly, reducing the glasshouse temperature from 23°C to 19°C resulted in a slower rate of aphid increase, thus permitting the parasites to contain the pest before severe leaf-distortion occurred. Secondly, and more important, where parasites were inadvertently present at the time the aphids were introduced, aphid reproduction was immediately suppressed and effective control ensued.

In practice, growers would probably not detect the presence of the first few aphids as quickly as would an entomologist and so it would seem essential in commercial practice either to reduce the aphid numbers before introducing a partially parasitized aphid population or to make a series of introductions of the latter as an insurance against invasions of *A. gossypii* from outside.

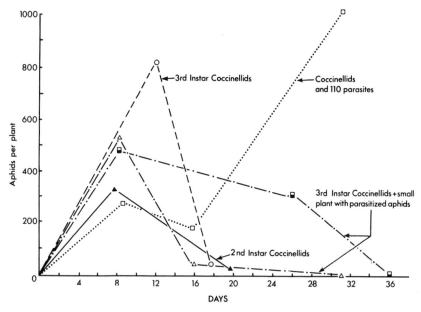

Fig. 3. Control of *Aphis gossypii* Glover on cucumbers by coccinellid larvae and the parasite *Aphelinus flavipes* Kurdjumov. Third instar and second instar larvae introduced at a ratio of 20 aphids: 1 predator almost eliminated the aphids after 16 and 18 days, respectively. (Where a parasitized aphid population was introduced at the same time there was no resurgence of aphids but adult parasites, artifically released, failed to sustain control).

It has been shown (B. Gurney and N. W. Hussey, *unpublished data*) that the coccinellid *Cycloneda sanguinea* L. is capable of rapidly effecting control of this aphid on cucumbers. In the course of its 15 to 17 day development a single larva consumes about 270 aphids. A long series of experiments on both young and mature plants showed that almost complete control could be achieved by liberating third-instar larvae at the rate of one larva to 20 aphids. Second-instar larvae also effected control, although only after a slightly longer period of time (Fig. 3) for the aphid continues to reproduce as the coccinellid develops through its second instar during which it consumes relatively few aphids.

Where parasites were liberated soon after the coccinellid larvae were introduced, the aphids were checked at first, but the parasite failed to establish and continue the control. We have had great difficulty in establishing the parasite either by introducing "mummies" or liberating adults, but if the parasite is allowed to spread from an infection already established on other plants, it establishes itself readily on the main crop. We have, therefore, had reproducible successes by following the releases of ladybird larvae with the introduction of cotton plants, dwarfed with B-9, growing in small peatblocks, and each carrying up to 50 aphids, half of which are parasitized (Fig. 4). As the parasite spreads only slowly from plant to plant, one of these infested

Fig. 4. Dwarfed cotton plants with parasitized population of *Aphis gossypii* Glover for introduction of *Aphelinus flavipes* Kurdjumov on cucumber.

cotton plants is introduced for every 2 or 3 cucumber plants.

It is possible that more effective parasites of this aphid exist and a screening program is in progress at Bangalore, Pakistan, under the aegis of the Commonwealth Institute of Biological Control. It is, of course, vital to ensure that aphids introduced in this way are not infested with viruses (e.g., cucumber virus 1) so that tested stock must be mass-reared under controlled conditions.

Control of the Aphid *Myzus persicae* (Sulz.)

Myzus persicae is a widespread and important pest in glasshouses on tomatoes and a wide range of ornamental plants. The most important problems at present concern year-round chrysanthemum culture.

This crop is grown from cuttings raised by a few specialist producers—a practice which encourages the spread of pesticide resistance in the aphid through the few survivors which are distributed on the cuttings. The continuous culture ensures that, in every glasshouse, all growth-stages of the plant are present simultaneously to facilitate pest increase.

In the early 1960's, organo-phosphate resistance appeared within a few weeks in flower-growing nurseries throughout the country. Unexpectedly, a crisis was averted because demeton-methyl continued to be effective when applied as a systemic. However, the problems posed by this concentrated monocultural practice had been exposed and interest in biological control was stimulated.

A most important feature of the biology of this aphid on chrysanthemum is its change of behavior when the initiation of flower buds stimulates movement to the tops of the plants (Wyatt, 1969). Further, there are marked varietal differences in susceptibility to aphids.

These factors have important repercussions on attempts at biological control for it is essential that control be effected before the buds form and this stage of growth will depend on the daylength response of the cultivar concerned. To induce flowering of this short-day plant throughout the year, cultivars have been selected which can be brought into flower by exposure to short-day photoperiods for between 9 to 14 weeks. Any biological control program must, therefore, be varied according to the response group of the particular cultivar.

Wyatt (1969) showed that the preference of *M. persicae* extends over a range of cultivars from "Mefo" and "Tuneful" to "Portrait" which supported populations only 1/40 that of the preferred cultivars. He also showed that, in a bed of mixed cultivars, the rate of increase of the *M. persicae* population was governed by the rate at which it could increase on the most susceptible cultivar present. This effect arises from the restless, migratory behavior of the population which ensures that a significant proportion of the feeding occurs on the preferred variety.

These cultivar preferences have important implications for biological control, for on the aphid-prone cultivars "Tuneful" and "Mefo," the parasite *Aphidius matricariae* Haliday, which has been successfully used to control *M. persicae* in many experiments, is able to overtake the aphid only when its reproductive rate has been checked by overcrowding. The interaction between aphid and parasite in this situation is largely influenced by their relative multiplication rates and, on most cultivars, the parasite increases as, or more, rapidly than its host.

Wyatt (in press) showed that small numbers of *Aphidius* can check the aphid's increase in eight weeks and almost exterminate it within 15 weeks. Distinct parasite generations are produced every 17 days with a seven-fold increase between peaks. This rate of increase is equivalent to a weekly multiplication of only 2.2-fold, whereas in the absence of parasitism the aphid commonly increases 3 to 4-fold weekly. The balance between aphid and parasite required for quick control, i.e., before too many aphids have developed, can only be achieved by suppressing the aphid's reproductive potential below that of the parasite. Bombosch (1963) and van Emden (1966) showed, theoretically, that small differences in time of introduction of a given number of a natural enemy species could markedly influence results. Furthermore, an outstanding feature of the parasite's increase in Wyatt's studies was the marked synchronization of generations over the first eight weeks of interaction. This presents disadvantages for biological control. The delay between parasite introduction and the first aphid mortality allowed a period of free expansion of the pest population. At the time of peak emergence the adult parasites competed for hosts and wasteful superparasitism resulted—yet, a few days later new aphid nymphs could complete their development safely. Thus, the pest population is alternately suppressed by, and released from, parasite attack. Thus, for effective control, the pressure in this situation should be exerted continuously so that the aphids have no such periods of release from attack. It is, therefore, necessary to introduce a heavily parasitized aphid population which has already become stabilized so that many parasites emerge every day. Control is then inevitable if the parasite population is introduced early enough in the life of the crop to check aphid increase before damage occurs. In this case damage is a relative term, for distortion of growth occurs only when enormous numbers of aphids develop on the plants. A smaller, yet still considerable number of aphids produces unsightly deposits of honeydew and cast-skins on the upper leaf surfaces. Populations of less than two hundred aphids per plant cause no growth distortion or other undesirable effects on   young plants. At flowering, however, control must be absolute, as even dead aphids impair appearance and prevent top-quality blooms. For six months during 1969, control of *M. persicae* on ½ acre of chrysanthemums has been successfully achieved by regular introductions of *Aphidius* on the newly planted beds (N. Scopes, *unpublished data*).

Another interesting possibility for controlling this aphid on year-round chrysanthemums is the use of entomogenous fungi. *M. persicae* is infected by a number of these fungi, but only two, *Entomophthora coronata* Kevorkian and *Cephalosporium aphidicola* Petch, are common in glasshouses. The former is not regarded as suitable for artificial distribution as a biotic pesticide for it has been shown to be pathogenic to mammals under certain circumstances (Hurpin, 1967). However, *C. aphidicola* presents attractive possibilities as it has water-dispersed spores, as distinct from the "self-propelled" spores shot from the basidia of *Entomophthora* species. This fungus has been successfully mass-cultured on sterilized grain media (Muller-Kogler, 1967). We are currently studying the dispersal of these spores in the mist which is applied for 16 hours daily during the ten-day rooting program for the cuttings.

Control of the Leaf-Miner *Phytomyza syngenesiae* Hardy on Chrysanthemums

No careful population studies have been made on this leaf-miner of chrysanthemums but several experiments have been made in commercial glasshouses on this crop. *Phytomyza syngenesiae* (= *atricornis*), like *M. persicae,* shows marked cultivar preferences (Hussey and Gurney, 1962) and in a bed of several cultivars some may be completely free of mines. The adults, which feed at slits cut by the ovipositor, feed on almost all cultivars but they are much more selective when ovipositing. While the growth of the plant is unaffected by the leaf-mines, the foliage may be severely disfigured and the sale value of the crop depreciated.

In British glasshouses leaf-miner resistance to the standard chemical treatments, based on control of the larvae by BHC or diazinon (Hussey, 1969*b*), has become widespread and has increased interest in biological control.

In this case, the full economic effect of the pest is felt as soon as an egg hatches and the larva begins to tunnel. The control agent should, therefore, attack the pest in the earliest larval stage. At the present time, insufficient is known about the parasites of this leaf-miner to choose the ideal biotic agent. However, despite this ignorance, we have succeeded in controlling *P. syngenesiae* with parasites collected in the field.

Leaves of *Sonchus* sp. mined by *Phytomyza* were collected and placed in emergence cages. As the parasites emerged, those of the genera *Rhizarcha* and *Chrysocharis* were collected and released into a commercial glasshouse. In one case, the leaf-miner was already established but the introduced parasites effected control within five weeks and almost no further mines developed. In another trial on a commercial nursery, we introduced leaf-miners on to newly planted cuttings every week over a six-month period. Ten *Phytomyza* were released on to every 300 cuttings and ten days later ten *Chrysocharis melaenis* were released in the same area. This pre-establishment technique prevented further infestation and, as this parasite attacks very young larvae, no mines were found throughout the experiment. In other houses on this nursery, regular insecticidal sprays were needed to control the pest.

Although these successful trials have encouraged us to investigate parasitic control of this pest more closely, there is some doubt as to whether parasites could be economically produced on a large scale. The miners can only colonize their host plants at a rate of up to 50 mines per leaf so that the production of thousands of parasites would require a very large number of plants. We are, therefore, interested in the possibility of utilizing the sterile-male release technique which would only demand mass-production of the host.

A Biological Control Program for the Pest Complex on Cucumbers

The red spider mite is a universal pest of cucumbers but if the routine use of acaricides is discontinued, aphids and whitefly rapidly become established. Another common pest is thrips, which damage the leaves rather like spider mites; they are especially troublesome in mid-summer. Frequently, a tyrophagid mite is also troublesome (*vide infra*).

Our preliminary trials showed that it was not possible to integrate biological and chemical techniques for these pests without upsetting the predictability of predator control of spider mites for which we were striving. Despite extensive laboratory studies which suggested that there were significantly wide differences between the L.D. 50's of the predator *Phytoseiulus* used for control of spider mites and the aphid *A. gossypii,* the use of discriminating dosages of chemicals on a commercial scale failed to control the aphid effectively. As this aphid reproduces so rapidly, higher dosages, toxic to the predator, had to be applied. This showed that, if the advantage of using the predator was to be exploited, biological control of all the principal pests must be used simultaneously.

The advantages of biological control on glasshouse cucumbers are considerable. Addington (1966) reported yield increases of at least 20 per cent after using this technique on a crop worth £15,000 ($36,000) per acre. These increases in yields are due partly to superior mite control by *Phytoseiulus* and partly to the lack of phytotoxicity associated with the alternative, frequent applications of petroleum oils and acaricides. Moreover, benefits to the grower are not merely financial for, in many cases, nurseries are situated near large towns, thus creating serious competition for labor. Since horticultural wages are hardly competitive with those paid by light industry, the glasshouse grower must utilize the smallest labor force possible. Each high-volume spray application requires about 20 man-hours per acre which, in an average season, would mean 200 man-hours of spraying. This must be compared with a maximum number of 60 man-hours needed to distribute all the necessary organisms to operate the biological management programs for aphids, white-flies and mites.

The complete absence of toxic residues on the harvest crops may be regarded by many as an additional advantage of non-chemical control though this should not be a problem where chemicals are used correctly.

We believe that if any system of biological control of glasshouse pests is to be widely adopted commercially, the appropriate natural enemies must be made available from some centralized mass-production center. It would not be economically feasible for such a center to fully service the scheme for hundreds of growers. Thus, the biotic agents must be packaged and distributed in such a way that the pest/natural enemy interactions can be set up in the glasshouses with a minimum of skill. In general, we are packing the organisms in gelatin capsules of the type commonly used for the administration of medicines. These have the advantage that they disintegrate in the presence of moisture without requiring further attention.

For each pest or its enemy, we package the necessary numbers and indicate to the grower when, and how, they must be distributed in the glasshouse.

There are two other pests which we have not adequately considered, namely, the so-called "French fly"—a mite *Tyrophagus longior* (Gerv.) and *Thrips tabaci* Lindeman. The former swarms out, soon after planting, from the straw bales or raw horse-manure used to make the growing beds. Usually, they eat only minute pits in the foliage which, as the leaves grow, become ragged holes but, more seriously, they may attack the growing points, retarding and distorting plant growth. Soil drenches of 0.01 per cent parathion effectively prevent such damage.

It is fortunate that the species of thrips attacking cucumbers is one that pupates in the soil for, otherwise, we would be forced to find some suitable natural control agent which preliminary investigations by the Commonwealth Institute of Biological Control suggest would be difficult. A very convenient control of *T. tabaci,* which can be successfully integrated with the management systems for other pests at any stage in crop growth, is to drench the beds with a 0.02 per cent wettable powder formulation of BHC at a rate of one gallon per yard run of bed.

The complete procedure for pest control on cucumbers is therefore as follows:

Days relative to planting	*Procedure*
-7	Drench plants in propagating house with demeton-methyl (5cc per 5 gals.) to keep plants clean until artificial infestation with mites.
0	Planting
+20	Infest each plant with 10-20 spider mites.
+30	Leaf-damage should have reached 0.4. Introduce two predators on to each alternate plant.
+80) +101) +122) +143) +164) +185)	Re-seed plants with spider mites if they do not infest plants from hibernation sites or elsewhere.
April 1st (or earlier if whitefly is present elsewhere on nursery)	Introduce whiteflies at 4 per plant.
April 14th	Introduce *Encarsia* at rate of 8 per plant.
May 1st	Introduce cotton plants bearing *A. gossypii* heavily parasitized with *Aphelinus flavipes.*

Throughout the program, cucumber mildew must be controlled with dimethirimol applied every six weeks at the rate of 20cc per plant.

Biological Control System for Pest Complex on Year-Round Chrysanthemums

In the absence of routine pesticide applications, aphids, leaf-miners and red spider mites become serious problems in glasshouses throughout Britain.

As stressed earlier, control of aphids and mites on chrysanthemums must be complete by the time the flower-buds become visible. As the light response-groups of the different cultivars vary, the program of natural enemy introduction will itself depend on the response-group. Chrysanthemums are normally grown at 62°F (16°C) so that the respective pest/natural enemy interaction proceeds much more slowly than on cucumbers.

Upon receipt from the specialist producers, the young cuttings are usually contaminated by low populations of aphids and parasites so that the first parasites and predators must be introduced within a few days of planting. The beds are usually covered, every 12 feet, with metal hoops which support both the black polythene used for shading during "long" days and the electrical wiring for the lights used to increase the day-length in autumn and winter. About 360 cuttings (planted 5" x 5") occupy each bed-section between successive hoops. These artificial divisions provide a convenient pattern for introducing natural enemies evenly.

As control of red spider mites takes about six weeks from introduction of the predator, the cuttings should be infested with mites at four points in each bay followed, three weeks later, by the introduction of 10 predators at each of the same points. Control should then be achieved before the budding stage on a nine-week response cultivar. The timing for longer response cultivars should be delayed accordingly. It is essential to achieve control by budding time as, at this time, the mites swarm to the tops of the plants and disfigure the flowers.

Aphids respond in a similar manner but, as the control by parasites seldom proceeds to complete elimination, a standard program can be used irrespective of response group. In the unlikely event that the cuttings are completely free of aphids, the pest must be introduced at two points in each bay within a day or so of planting. Seventeen, twenty-four and thirty-one days later parasites are introduced at these same points at the rate of 10 parasites per bay. Normally, these are introduced as "mummies" which can easily be shaken out of containers and distributed over the young plants. Serial introductions are made to avoid the synchrony of parasite development, and so intermittent adult activity, which follows a single introduction when parasites lay eggs only in the youngest nymphs.

Experiments currently in progress suggest that these procedures can be further short-circuited by applying both pests and natural enemies directly to the boxes used to distribute the cuttings. This is a simple and economical technique which, in a series of small trials, has consistently produced completely clean flowers at harvest.

Leaf-miners and their parasites are liberated as outlined in an earlier section when the full pest complex is being controlled.

The Commercial Future of Biological Control in Glasshouses

As stressed earlier, the advantages of biological control are not restricted to the economy of the techniques to the grower. The most important advantages are: (1) reduced man-power requirements in crops where most insecticide treatments must be made in the evenings, at overtime rates, to avoid phytotoxicity due to sun-scorch, (2) absence of direct phytotoxic effects (the spray-damage to cucumbers has already been mentioned) for, in chrysanthemums, intense spraying programs are made more complicated by the extreme sensitivity of many cultivars and, in some cases, the rooting capacity of treated cuttings may be impaired, and (3) more effective control—the dense planting makes adequate coverage of high-volume sprays difficult. Coupled with the contribution which biological control could make towards the prevention of pest resistance to chemicals, one might expect that its development would follow readily and naturally. However, there are serious difficulties. Firstly, all cucumber growers in a climatic regime would require their natural enemies at approximately the same time so that a centralized rearing unit would require other markets geared to different seasonal production so as to operate throughout the year. Possibly, the year-round chrysanthemum crop will conveniently provide this extended market. Secondly, while the pest complex on a crop may be controlled biologically, any fungal diseases would still require chemical treatment. Development of the selective, systemic fungicidal carbamate, dimethirimol, makes possible an ideal integrated program but the recent development of tolerance or resistance in cucumber mildew seriously threatens this program. Such sudden changes in the circumstances within which biological control must be used introduces a very uncertain element in its commercial use. Remembering that the systems advocated demand precise production techniques and packaging as well as carefully appraised instructions to grower operatives, an efficient, centralized production unit seems necessary.

The production systems developed by Scopes (1968, 1969) would demand facilities costing at least £150,000 ($360,000) even if only half the British glasshouse acreage were to be served. The uncertainties mentioned would probably deter private investment unless the project were financed as a diversification operation of a large horticultural enterprise. Some might argue that the State should play a role in such developments, but as pest control costs constitute only about 1 per cent of the value of protected crops, it is unlikely that any reduction in production costs would be passed on to the taxpayer. As public money should generally be spent to the "common good," public support of a rearing unit would amount to a direct subsidy.

Casting a further shadow over ideas for commercial biological control is the difficulty of patenting the organisms or, at least, their production techniques.

It seems unlikely that any large-scale scheme will be implemented unless the difficulties caused by resistance to pesticides become almost insurmountable. This, however, might occur earlier than expected as the increasing costs of developing new pesticides reduce the choice of alternative materials.

LITERATURE CITED

Addington, J. 1966. Satisfactory control of red spider mites on cucumbers. Grower 66: 726-727.

Bombosch, S. 1963. Untersuchungen sur Vermehrung von *Aphis fabae* Scop. in Samenrubenbestanden unter besonderer Berucksichtigung der Schwelfliegen. Z. angew. Entomol. 52: 105.

Bravenboer, L. 1963. Experiments with the predator *Phytoseiulus riegeli* Dosse on glasshouse cucumbers. Mitt. Schweiz. Entomol. Ges. 36: 53.

Bravenboer, L., and G. Dosse. 1962. *Phytoseiulus riegeli* Dosse als Predator einiger Schadmilben aus der *Tetranychus urticae*-Gruppe. Entomol. Exp. Appl. 5: 291-304.

Burnett, T. 1960a. An insect host-parasite population. Can. J. Zool. 38: 57-75.

Burnett, T. 1960b. Effects of initial densities and periods of infestation on the growth-forms of a host and parasite population. Can. J. Zool. 38: 1063-1077.

Burnett, T. 1964. Host larval mortality in an experimental host-parasite population. Can. J. Zool. 42: 745-765.

Chant, D. A. 1961. An experiment in biological control of *Tetranychus telarius* (L.) in a greenhouse using the predacious mite *Phytoseiulus persimilis* A.-H. Can. Entomol. 93: 437-443.

Dosse, G. 1959. Uber einige neue Raubmilbenarten (Phytoseiidae). Pflanzenschutzberichte 21: 44-61.

Doutt, R. L. 1951. Biological control of mealybugs infesting commercial greenhouse gardenias. J. Econ. Entomol. 44: 37-40.

Doutt, R. L. 1952. Biological control of *Planococcus citri* on commercial greenhouse *Stephanotis*. J. Econ. Entomol. 45: 343-344.

Force, D. C. 1967. Effect of temperature on biological control of two-spotted spider mite by *Phytoseiulus persimilis*. J. Econ. Entomol. 60: 1308-1311.

Gerling, D. 1966. Biological studies on *Encarsia formosa*. Ann. Entomol. Soc. Amer. 59: 142-143.

Gould, H. J. 1968. Observations on the use of a predator to control red spider mite on commercial cucumber nurseries. Plant Pathol. 17: 108-112.

Gould, H. J., and H. G. Kingham. 1964. The efficiency of high-volume spraying with acaricides on cucumbers under glass. Plant Pathol. 13: 60-64.

Gould, H. J., N. W. Hussey, and W. J. Parr. 1969. Large scale commercial control of *T. urticae* on cucumbers by the predator *Phytoseiulus riegeli*. Proc. 2nd Intern. Congr. Acarol. (1967). Pp. 383-388.

Huffaker, C. B., and C. E. Kennett. 1956. Experimental studies on predation: Predation and cyclamen mite populations on strawberries in California. Hilgardia 26: 191-222.

Hurpin, B. 1967. Symposium sur l'action des mins et champignons entomopathogenes surbs vertebres. Entomophaga 12: 321-323.

Hussey, N. W. 1966. Aspects of the development of resistance to chemicals in British insect and acarine pests. Proc. 3rd Br. Insectic. and Fungic. Conf. Pp. 28-37.

Hussey, N. W. 1969a. Greenhouse whitefly. Rep. Glasshouse Crops Res. Inst. (1968).

Hussey, N. W. 1969b. Differences in susceptibility of different strains of chrysanthemum leaf-miner (P. syngenesiae) to BHC and diazinon. Proc. 5th Br. Insectic. and Fungic. Conf. Pp. 28-37.

Hussey, N. W., and B. Gurney. 1962. Host selection by the polyphagous species Phytomyza atricornis Mg. (Dipt. Agromyzidae). Entomol. Monthly Mag. 98: 42-47.

Hussey, N. W., W. J. Parr, and H. J. Gould. 1965. Observations on the control of T. urticae Koch on cucumbers by the predatory mite Phytoseiulus riegeli Dosse. Entomol. Exp. Appl. 8: 271-281.

Hussey, N. W., W. J. Parr, and B. Gurney. 1957. The effect of whitefly populations on the cropping of tomatoes. Rep. Glasshouse Crops Res. Inst. (1958). Pp. 79-86.

Hussey, N. W., and W. J. Parr. 1963. The effect of glasshouse red spider mite (Tetranychus urticae Koch) on the yield of cucumbers. J. Hort. Sci. 38: 255-263.

Khalifa, A., and N. Sharaf El-Din. 1965. Biological and ecological study of Aphis gossypii. Soc. Entomol. Egypt Bull. 48: 131-153.

Kring, J. B. 1959. The life-cycle of the melon aphid, Aphis gossypii Glover, an example of facultative migration. Ann. Ent. Soc. Amer. 52: 284-286.

McLeod, J. H. 1937. Some factors in the control of the common greenhouse aphid Myzus persicae Sulz. by the parasite Aphidius phorodontis Ashm. Ann. Rep. Entomol. Soc. Ont. 67: 63-64.

Milliron, H. E. 1940. A study of some factors affecting the efficiency of Encarsia formosa, an aphelinid parasite of the greenhouse whitefly, Trialeurodes vaporariorum. Tech. Bull. Mich. Agr. Expt. Sta. 173: 1-23.

Muller-Kogler, E. 1967. On mass cultivation, determination of effectiveness and standardization of insect pathogenic fungi. Pp. 339-353. In Insect Pathology, P. A. van der Laan (ed.). Proc. Intern. Colloq. Wageningen (1966).

Parr, W. J. 1968. Biological control of greenhouse whitefly (Trialeurodes vaporariorum by the parasite Encarsia formosa on tomatoes. Rep. Glasshouse Crops Res. Inst. (1967). Pp. 137-141.

Parr, W. J., and N. W. Hussey. 1966. Diapause in the glasshouse red spider mite (T. urticae): A synthesis of present knowledge. Hort. Res. 6: 1-21.

Parr, W. J., and N. W. Hussey. 1967. Biological control of red spider mite on cucumbers: Effects of different predator densities at introduction. Rep. Glasshouse Crops Res. Inst. (1966). Pp. 135-139.

Passlow, T., and M. S. Roubicek. 1967. Life-history of the cucurbit aphid (A. gossypii). Queensland J. Agr. Anim. Sci. 24: 101-102.

Richardson, H. P., and P. H. Westdal. 1965. Use of Aphidius semiflavus Howard for control of aphids in a greenhouse. Can. Entomol. 97: 110-111.

Scopes, N. 1968. Mass-rearing of Phytoseiulus riegeli Dosse in commercial horticulture. Plant Pathol. 17: 123-126.

Scopes, N. 1969. The economics of mass-rearing Encarsia formosa, a parasite of the whitefly Trialeurodes vaporariorum, for use in commercial horticulture. Plant Pathol. 18: 130-132.

Speyer, E. R. 1927. An important parasite of the greenhouse whitefly (Trialeurodes vaporariorum Westwood). Bull. Entomol. Res. 17: 301-308.

van Emden, H. G. 1966. The effectiveness of aphidophagous insects in reducing aphid
    populations. *In* Ecology of Aphidophagous Insects, I. Hodek (ed.). W. Junk, The
    Hague. 227 pp.
Wyatt, I. J. 1969. Factors affecting aphid infestation of chrysanthemums. Ann. Appl.
    Biol. 63: 331-337.
Wyatt, I. J. In press. The distribution of *Myzus persicae* (Sulz.) on year-round chrysanthe-
    mums. II. Effect of parasitism by *Aphidius matricariae* Hd. Ann. Appl. Biol.
    (1970).

Chapter 9

THE BIOLOGICAL CONTROL OF THE WINTER MOTH IN EASTERN CANADA BY INTRODUCED PARASITES

D. G. Embree

Canadian Forestry Service
Fredericton, New Brunswick

The basic fact of this account of biological control is that an insect pest, the winter moth (*Operophtera brumata* L.) was accidentally introduced, presumably from Europe, into the eastern seaboard of North America. A serious defoliator of hardwoods, and having a natural range that includes most of Europe and parts of Asia, the insect is capable of spreading across North America. But it has been effectively controlled at its point of entry by the deliberate introduction of two parasites, *Cyzenis albicans* (Fall.), a tachinid fly, and *Agrypon flaveolatum* (Grav.), an ichneumonid wasp, both introduced from a generally endemic environment in Europe. While any control short of eradication is not likely to prevent the eventual spread of the winter moth to other areas favorable to it, the actions of these parasites may reduce the rate of spread. Moreover, based on their demonstrated ability to disperse in their new environment, they are most likely to spread with their host.

The point of entry was Nova Scotia where, in the span of 10 years, in just two counties, the damage by the winter moth to oak forests amounted to about 26,000 cords per year. The total amount of wood lost would be valued today at close to $2,000,000. Other forest species, shade trees, and orchards were also damaged and the insect spread over most of the Province.

Had its depredations continued, the more than 1½ million cords of oak now standing in the Province would in all probability have been killed, and the potential loss to the forest economy of the region may have been far greater than the $12,000,000 roadside value of the wood. The cost of research related to the actual liberation was just over $160,000 at today's values so

217

that the potential return per dollar invested was relatively high. No birds were killed, no streams polluted, and no sinister residual chemicals now lurk in the ecosystem to endanger ourselves or our descendants.

This is a real testimonial for biological control but the dollars and cents values that are quoted are potential values of a resource that is just now beginning to be utilized. In 1954 when the study began, the winter moth was considered to be just a nuisance pest of shade trees. It dispersed slowly and was considered a threat only because it might eventually reach more valuable trees elsewhere. It also posed a minor problem in apple orchards.

Today the situation is different. Hardwoods are being increasingly utilized in the Province and a heavy investment has been made in new paper- and hardboard-producing mills designed to use hardwoods. Had these mills existed in 1954, we might have attempted to control the winter moth with insecticides because it would have been impossible to predict, then as now, how successful any biological program might be.

The general research policy was directed toward understanding the population dynamics of forest insects and was particularly conducive to biological control efforts and research evaluation. As a result, most of the effort (at a cost of approximately $500,000) was concentrated on studies on the population dynamics of the host and later the parasites, once they became established. The results of these studies have been reported in detail (Embree, 1965; Embree and Sisojevic, 1965; Embree, 1966). It is impossible to determine the return per dollar invested on this aspect of the problem. Such research efforts yield results well beyond the practical benefits achieved by the biological control effort itself. Although the benefits achieved once the parasites were introduced would have been virtually the same whether or not this research effort had been made, one of the values of such efforts is to gain insights and to stimulate other such efforts. We have managed to document a successful biological control experiment, and the sort of basic information that has been collected will eventually guide us in embarking on those biological control programs that offer highly predictable probabilities of success.

Six species of parasites were introduced from Europe during the period 1955 to 1960 (Graham, 1958). Four of these species were never recovered but *C. albicans* and *A. flaveolatum* became established within three years. This was established by rearing large numbers of host larvae collected in the immediate areas of the liberations and obtaining the parasites issuing from this material.

Before the parasites were introduced, winter moth populations fluctuated in an erratic fashion at high density levels. It was found that the eggs, which are laid on the tree trunks and limbs in the late fall, hatch early in the spring and the newly hatched larvae are carried to the foliage by air currents.

If this occurred after the buds had opened, larval survival was relatively high; often as high as 60 per cent. But if the buds were still closed, as was usually the case, most of the larvae that successfully reached the branch tips starved. This larval loss was the key factor in the population dynamics of the winter moth before the parasites became established. It was predominantly responsible for the fluctuations in numbers, although at times starvation of larger larvae also exerted a partial effect at high densities (Embree, 1965).

Most of these observations were based on intensive studies on a 2-square-mile area in the center of the infestation; supplementary studies were also made in outlying areas. Initial parasite releases were made at one point on the edge of the main study area, beginning in 1955, and thus it was possible to record dispersal rates as well as the increasing effectiveness of the parasites in a succession of nine plots distributed throughout the area (Embree, 1960). After 1958, additional parasite liberations were made at other locations in the Province but large gaps were left between release points so that the spread of the parasite into new areas could be documented.

Table 1 shows two simplified life tables for one of the plots in the study area. One life table was compiled in 1958 before apparent parasitism by the introduced species had reached 10 per cent and the other was compiled in 1960, one year before the collapse of the host population (e.g., see Fig. 3, p. 221). These life tables are shown here because they were the only ones from a total of 37 tables in which both egg density and dispersal loss of first instar larvae were essentially identical. For a stable winter moth population, the

Table 1. Abbreviated life tables for winter moth, Plot 1, Oak Hill, N.S.

Age interval	1958		1960	
	Number/tree	% Mortality between age intervals	Number/tree	% Mortality between age intervals
Eggs	24,695		24,174	
		78		77
First instars	5,477		5,500	
		74		73
Prepupae	1,392		1,469	
		37[†]		94[‡]
Adults	872		82	
Generation mortality		96.5%		99.7%

[†]Mortality due to parasitism – 8%; other causes – 29%.
[‡]Mortality due to parasitism – 72%; other causes – 22%.

generation mortality would have to be 99.04 per cent, a figure calculated
from the average sex ratio and fecundity of the winter moth. Thus the 96.5
per cent mortality in the 1958 table indicates an increasing population; while
99.7 per cent in the 1960 table indicates a decreasing population. In fact,
population density showed a two-fold increase in 1959 and a ten-fold decrease
in 1961. The reason for the difference in generation mortalities between the
life tables is of course related to mortalities within age intervals. These
mortalities are all similar, except at the prepupal stage where prepupal (and
pupal) mortality was much higher in 1960. Mortality due to other causes was
roughly equal for both years. However, parasitism increased from 8 per cent

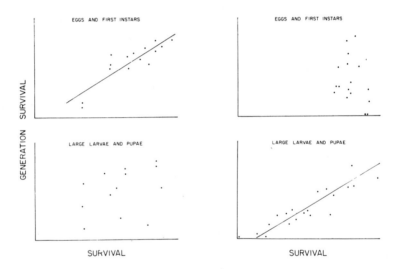

Fig. 1. Relationship between winter moth generation survival and age-
interval survival before parasitism had reached 10%, based on life table data.
Oak Hill, N.S. 1954-1962 (Embree, 1965).

Fig. 2. Relationship between winter moth generation survival and age-
interval survival after parasitism had reached 10%, based on life table data.
Oak Hill, N.S. 1954-1962 (Embree, 1965).

in 1968 to 72 per cent in 1960 and was alone responsible for the pronounced decrease in the winter moth population in 1961.

From an analysis of the 37 life tables which were compiled during the study, a population model was developed which showed that parasitism had indeed become the key factor which controlled the winter moth (Embree, 1965). Supporting data are shown in Fig. 1. Generation survival was correlated with egg and early instar survival before parasitism was effective but was not correlated with large larval and pupal mortality. However, when parasitism increased beyond 10 per cent, this relationship was completely reversed (Fig. 2) and generation survival was correlated with late larval and pupal survival and not with early instar survival. Finally, pupal survival was correlated with pupal parasitism (Fig. 3).

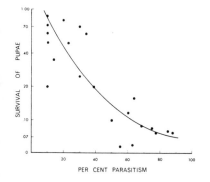

Fig. 3 (Left). Relationship between survival of winter moth pupae and per cent parasitism based on life table data. Oak Hill, N.S. 1954-1962 (Embree, 1965).

Fig. 4 (Below). Typical history of a winter moth infestation and parasitism by *Cyzenis albicans* (Fall.) and *Agrypon flaveolatum* (Grav.). Time refers to the number of years the outbreak persisted (Embree, 1966).

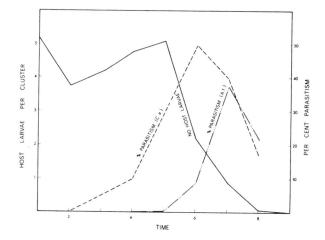

All this has been based on the analyses of life tables from one area. When data from other areas are included we obtain a picture of the outbreak as shown in Fig. 4. Three years after parasitism by *C. albicans* reached 10 per cent, the population had collapsed to an endemic level. The increase in *A. flaveolatum* has so far always occurred toward the end of an outbreak.

Finally, from the life tables we obtained functional response curves for two densities of *C. albicans* (Fig. 5) and the shape of these curves show one reason why *C. albicans* was so successful. The curves were S-shaped which meant that over a certain range of host density, *C. albicans* had regulatory properties through its functional response alone. The reasons for this appear to be complex and were based on the oviposition behavior of *C. albicans* when other defoliators are present (Embree, 1966), and on differences in synchronization of the parasites with well-fed hosts or those that were underfed. Normally, for most parasites, the rate of attack decreases with increasing host density, as is shown in Fig. 6 for one density of *A. flaveolatum.* Numerical response curves were not compiled, but the numerical response of each species was high, as can be interpreted from Fig. 4.

There are 63 known parasites of the winter moth (Wylie, 1960); fortunately the two which were established were compatible and were strongly supplementary. At high host densities *C. albicans,* which oviposits near feeding

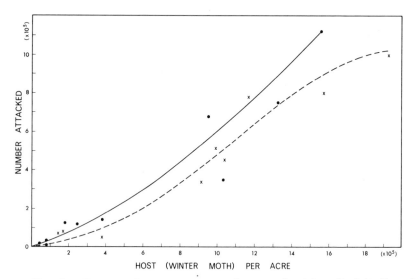

Fig. 5.   Functional response curves for two densities of adult *Cyzenis albicans* (Fall.) per acre (Embree, 1966).

damage of the host, is the most efficient parasite, and parasitism can reach as high as 80 per cent. Moreover, the nature of its functional response enables it to react immediately to increases in host densities. However, at low host densities its efficiency decreases because under these conditions, the proportions of other defoliators, such as the fall cankerworm, increase and the parasite wastes its eggs on non-susceptible species. *A. flaveolatum* is more efficient at low host densities because it oviposits directly on its host.

When multiparasitism occurs it does not jeopardize *C. albicans,* presumably because *A. flaveolatum* develops more slowly than *C. albicans* in the host, and *C. albicans* rather than *A. flaveolatum* survives. Consequently, both species can survive at low host densities, *A. flaveolatum* because of its more

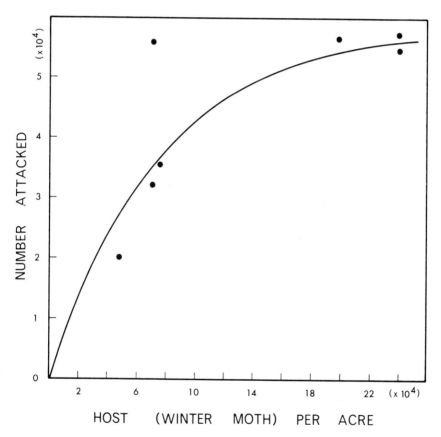

Fig. 6. Functional response curve for *Agrypon flaveolatum* (Grav.) at a parasite density of 5×10^4 adults per acre (Embree, 1966).

selective oviposition behavior and *C. albicans* because of its better capacity to survive instances of multiparasitism (Embree, 1965). No data are available on actual survival of *C. albicans* in host larvae also parasitized by *A. flaveolatum.* However, the relationship between host density and the percentage parasitism by the two species, as demonstrated by the rates of attack in one area (Fig. 7), suggests that *A. flaveolatum* is least able to survive. Once the host density increased, the proportion of hosts attacked, based on number of parasites within host pupae, by *A. flaveolatum* decreased and those attacked by *C. albicans* increased.

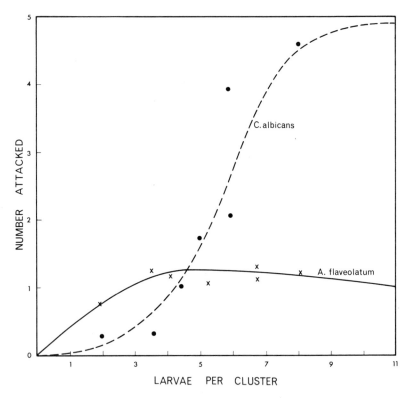

Fig. 7. Attack curves for *Cyzenis albicans* (Fall.) and *Agrypon flaveolatum* (Grav.) at Waverley, N. S., 1964. C. *albicans* density was 4 x 10^4 adults per acre. *A flaveolatum* density was m 1 x 10^4 adults per acre. Number attacked refers to number of winter moth larvae successfully attacked (Embree, 1966).

How long will the control last? Studies in England (Varley and Gradwell, 1968; Hassell, 1969) based on mathematical models of host/parasites interactions have forecast that in Canada strong oscillations in interacting populations of the winter moth and its parasites will cause damaging outbreaks of the winter moth at 9 or 10 year intervals. These studies are based solely on the predicted behavior of *C. albicans,* while synchronism between the winter moth and its host tree is essentially treated as a constant. Our studies suggest that other factors besides the parasites will determine the cause of future outbreaks. These are factors such as the virus disease which appeared after the collapse of the major outbreak, and the low probability of the occurrence of a series of years of favorable synchronism between winter moth hatching and bud-break, such that the increase in host numbers might be too great for the parasites to contain for a time. An unpredictable aspect of the problem is that as densities of hosts and parasites decrease subtle changes in their genetic make-up may also occur. We do not know what happens to the heterozygosity of a population when overdispersion results in substantial inbreeding. For example, selection might favor winter moths that hatch late and are therefore synchronized with bud-break of oak.

If winter moth outbreaks do develop again in forest areas, they will probably result from invasions of winter moths from shade and orchard trees which are better synchronized with winter moth hatching.

In retrospect, the success of this experiment can be described as mere good fortune. But now there is a good chance of producing the kind of information needed for the more sophisticated methods which hopefully will be employed in biological control programs in the future. We are in the process of explaining the various reasons why the combination of *C. albicans* and *A. flaveolatum* was successful in Canada. Analysis of the behavior of the parasites under a variety of simulated conditions, based on mathematical models now being constructed, may define criteria for the selection of parasites and the planning of biological control operations against pests similar to the winter moth.

If such investigations continue to be pursued as a part of biological control research, ultimately the selection of biological control agents and the manner in which they are to be manipulated will be based on the results of simulation studies using standard mathematical models, encompassing the prospective controlling agent and the ecosystem in which it is expected to act.

LITERATURE CITED

Embree, D. G. 1960. Observations on the spread of *Cyzenis albicans* (Fall.) (Tachinidae: Diptera), an introduced parasite of the winter moth, *Operophtera brumata* (L.) (Geometridae: Lepidoptera) in Nova Scotia. Can. Entomol. 92: 862-864.

Embree, D. G. 1965. The population dynamics of the winter moth in Nova Scotia, 1954-1962. Mem. Entomol. Soc. Can. 46. 57 pp.

Embree, D. G. 1966. The role of introduced parasites in the control of the winter moth in Nova Scotia. Can. Entomol. 98: 1159-1168.

Embree, D. G. and P. Sisojevic. 1965. The bionomics and population density of *Cyzenis albicans* (Fall.) (Tachinidae: Diptera) in Nova Scotia. Can. Entomol. 97: 631-639.

Graham, A. R. 1958. Recoveries of introduced species of the winter moth, *Operophtera brumata* (L.) (Lepidoptera: Geometridae), in Nova Scotia. Can. Entomol. 90: 595-596.

Hassell, M. P. 1969. A population model for the interaction between *Cyzenis albicans* (Fall.) (Tachinidae) and *Operophtera brumata* (L.) (Geometridae) at Wytham, Berkshire. J. Animal Ecol. 38: 567-576.

Varley, G. C., and G. R. Gradwell. 1968. Population models for the winter moth. Pp. 132-142. *In* Insect Abundance, T. R. E. Southwood (ed.). Symposia Roy. Entomol. Soc. London, No. 4.

Wylie, H. G. 1960. Insect parasites of the winter moth, *Operophtera brumata* (L.) (Lepidoptera: Geometridae) in Western Europe. Entomophaga 5: 111-129.

Chapter 10

BIOLOGICAL CONTROL OF RHODESGRASS SCALE BY AIRPLANE RELEASES OF AN INTRODUCED PARASITE OF LIMITED DISPERSING ABILITY

Michael F. Schuster

Texas A&M University Agricultural Research and Extension Center at Weslaco

J. C. Boling

Currently, Department of Entomology, Kansas State University, Manhattan

and J. J. Marony, Jr.

Currently, Museum of Zoology Louisiana State University, Baton Rouge

INTRODUCTION

The rhodesgrass scale, *Antonina graminis* (Maskell), was first found in Texas in 1942 attacking rhodesgrass, *Chloris gayana* Kunth (Chada and Wood, 1960). Since that time this scale insect has been found infesting 94 species of grasses in North America (Chada and Wood, 1960; Schuster, 1967), some of which it affects severely. Introduction of the scale parasite, *Neodusmetia sangwani* (Rao), from India was made in 1959 (Dean *et al.,* 1961). Evaluation of this parasite in Texas as a control agent for rhodesgrass scale was begun in 1961.

Rhodesgrass scale was first described from grass at Hong Kong, China, as *Sphaerococcus graminis* (Maskell, 1897). It was later described from India by Green (1908) as *Antonina indica.* The scale has been recorded from 26 countries, generally located between the $30°$ north and south latitudes (Schuster and Boling, in press). The distribution within the United States, embracing the southern states from California to Florida, was given by Chada and Wood (1960). These authors reported 62 counties infested by the scale in Texas.

The number of recorded hosts is increasing each year. Chada and Wood (1960) recorded 69 host species in the United States; Brimblecrombe (1966), 14 in Australia; Guagliumi (1963), 22 in Venezuela; and Williams and Schuster (in press), 86 in Brazil. The large host range, wide distribution and inherent effect on its hosts, indicate that forage losses on ranges could well be great under climatic conditions favorable for the scale. Chada and Wood (1960) reported that bermudagrass, *Cynodon dactylon* (L.) Pers., and rhodesgrass, *C. gayana,* were the only forage grasses affected by the scale in Texas. Schuster (1967) found that in greenhouse tests, it developed on most range grasses common to southern Texas.

Interviews with ranchers in Brooks, Kenedy, Willacy, Kleberg and Duval Counties indicated that the grazing or carrying capacity of native ranges had been reduced approximately 30 per cent since the introduction of rhodesgrass scale in Texas in the early 1940's. This heavy loss has not been regained, presumably due to scale infestation of native grasses.

The life history of rhodesgrass scale was described by Chada and Wood (1960). The adult scale is parthenogenetic and reproduces ovoviviparously. The crawlers are positively thigmotropic and attach themselves to the plant nodes under the leaf sheath. The legs are lost at molting and the second and third instar larvae are saclike and resemble the adults. There are five generations annually in southern Texas. About 85.4 per cent of the scales are found below the first plant node or on the crown node. Adults live up to six weeks without food. Dissemination of individuals is mainly by crawling or by wind.

Control measures were soon attempted using chemicals (Wene and Riherd, 1950; Richardson, 1953; Chada and Wood, 1960). However, Chada and Wood (1960) pointed out that insecticides were too costly and are useless under range conditions.

The areas of Texas to benefit most from the establishment of effective biological control would be the area of some 45,000 square miles south of the Balcones Escarpment, east to San Antonio and then south to Corpus Christi. Effective agents should also be employed along the Coastal Bend to the Louisiana border in an area of about 15,000 square miles. The large area over which natural enemies would have to be distributed requires careful consideration of methods of distribution (Schuster and Boling, in press).

Attempts to control the scale biologically were begun in 1949 when the parasite, *Anagyrus antoninae* Timberlake, was introduced from Hawaii (Riherd, 1950). In 1954 and 1955, several parasites, *Xanthoencyrtus phragmitis* Ferr., *Boucekiella antoninae* (Ferr.), *Timberlakia europaea* (Mercet), and *Anagyrus diversicornis* Mercet., were introduced from France (Dean and Schuster, 1958), but no establishment was obtained. Dean and Schuster (1958) concluded that the effect of *A. antoninae* on rhodesgrass scale populations was of little value. The parasite could not withstand the high temperatures and low humidities of southern Texas, except in certain ecologically modified areas around lakes and canals in the Lower Rio Grande Valley. The lowest parasite activity occurred during periods of highest scale population.

The parasite, *Neodusmetia sangwani* (Rao)[1] was first found attacking rhodesgrass scale near Delhi and Bangalore, India, by Narayanan *et al.* (1957) and was originally described as *Dusmetia sangwani* by Rao (1957). Introduction and establishment of *N. sangwani* into Texas was in 1959 (Dean *et al.,* 1961; Rao, 1965). Dean *et al.* (1961) described the method of rearing the parasite for releases on range sites. Colonies established by Dean *et al.* (1961) were the source of parasites introduced into Arizona, California, the Bermuda Islands, Mexico and Brazil. Methods of laboratory rearing and distribution of individuals are also described by Machado da Costa *et al.* (in press).

Schuster (1965) studied the biology of the parasite under laboratory conditions. At 30° C, 17-20 days were required to complete the life cycle; at 20°C, 53-56 days were required. Later studies indicated that some individuals complete the life cycle in 33 to 35 days at 20°C, indicating that 20°C is near the threshold temperature of arrested development. The pupal period in the laboratory was extended by low temperatures. Apparently, the parasite overwinters as pupae near Delhi, India, in this manner (Narayanan *et al.,* 1957). Pre-imaginal stages include a caudate first instar, a hymenopteriform second instar, a similar but nonfeeding prepupa, and a pupa. Reproduction by unfertilized females resulted in male progeny. When exposed singly to 20 to 30 scales, the short-lived brachypterous females laid an average of 6.2 eggs in each of 5.7 scales and produced an average of 35.3 offspring. The sex ratio ranged from 6.7 - 7.8 females to 1 male for field-collected scales. The male has functional wings.

No comprehensive research has been done to determine effective and practical methods for mass distribution of entomophogous parasites on the scale required to infest the 60,000 square miles of southern Texas which require control. Even DeBach (1964), in his monumental work, offers no solution for such a great task. The problem is further complicated by the fact that the female of the parasite to be distributed is functionally wingless, and

---

[1]The genus *Neodusmetia* was established with *N. sangwani* as the type by Kerrick in 1964. *Dusmetia indica* Burks (1957) is a synonym.

the life span of the adult female is very short, about 24 hours; therefore, rapid contact with hosts is essential. Hence, human assisted distribution is suggested (Schuster, 1965).

When it was apparent in 1962 that *N. sangwani* was established and was resulting in apparent scale control, investigations were begun to establish the degree of effectiveness and to develop colonization techniques. The brachypterous condition of the female necessitated considerable effort for colony distribution. In this chapter we describe (1) effectiveness of *N. sangwani* as a population regulating agent, (2) control of scale damage due to this regulating effect in both laboratory and in range sites, (3) colonization techniques, (4) colony spread under range conditions, and (5) methods of distribution.

POPULATION REGULATION OF RHODESGRASS SCALE
BY *N. SANGWANI*

That the parasite *N. sangwani* is inherently capable of regulating the population of its host, rhodesgrass scale, is evidenced by results of a number of related lines of investigation. This capability was exhibited by model-type experiments using four grass species grown in containers and many plots on natural range comprising a variety of grass species. The statistics on which the conclusion is based include, most importantly, crop yield, and less so, plant survival (longevity), scale population densities, percentage of nodes infested, and percentage of scales parasitized, contrasting parasite-present and parasite-absent plots.

The population regulation of rhodesgrass scale has been described by Schuster and Boling (in press). The percentage parasitization of the third instar and adult rhodesgrass scales varied considerably throughout the year. Yearly parasitism from 1961 to 1965 varied from 28.1 to 34.6 per cent. Monthly means varied from 23.3 to 37.8 per cent. Parasitization of scales increased after April and remained at a high level (45 to 85 per cent) throughout the summer and fall. The percentage parasitization was lower during periods of rapid scale increase, due to the apparent reluctance by the parasite to oviposit in the first two instars. At these seasonal times the scale population consisted almost exclusively of the first two instars and there naturally was a lag in parasitization until the scales became suitable for oviposition by the parasite.

Peaks in scale numbers occurred in June and November in the normal population (Fig. 1). Reduction of rhodesgrass scales in parasite regulated areas was greatest during periods of peak scale numbers but less in periods of fewer scales, i.e., the relation was density-dependent. During the summer, high temperatures killed large numbers of scales. In the parasite regulated popula-

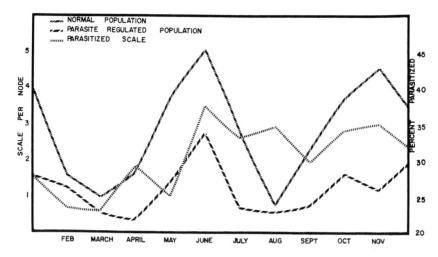

Fig. 1. Average rhodesgrass scale numbers in a normal and parasite-regulated population and average percentage of third-instar and adult scales parasitized during 1961 to 1965 in south Texas at nine locations, on seven grass species.

tion, the June peak of scales occurred normally but the November peak was delayed. However, about 50.0 per cent fewer scales were present at both peaks in parasite-regulated populations as contrasted to the check populations. The rate of increase of scales in parasite-regulated populations was less than in the check populations where parasites were absent.

As scale numbers decreased, the survivors were found mostly in crevices under leaf sheaths or on underground culms. The searching ability of the female parasite may be limited in these areas. Therefore, some nodes were always infested although the general population was low. This feature could well be an important refuge damping aspect against too severe exploitation of hosts by the parasites (see Chapter 2).

It was evident from the data that a percentage parasitization figure was of little value in evaluating the population-regulating ability of *N. sangwani* (Schuster and Boling, in press). The number of scales per node is a better indicator of the regulating effect than percentage nodes infested or percentage scales parasitized (see Chapter 5). The best measure is of course yield.

The scale population in parasite-regulated populations increased at a slower rate during both the April and September periods of scale increase and reached lower levels at the peaks. The reduced effective fecundity by parasitized scales probably contributed greatly to this lower population performance (Table 1). Parasitized scales produced 93.7 and 90.1 per cent fewer crawlers on rhodesgrass and paragrass, respectively, than did non-parasitized scales (Schuster and Boling, in press). The full production of crawlers probably was not realized, since scales were removed from the plant; although, Chada and Wood (1960) found that the scales would live six weeks without food. Greater numbers of crawlers were produced by scales collected from rhodesgrass than from paragrass. This indicates that rhodesgrass was a more suitable host than paragrass and may further indicate that other species might have different reproductive potentials for the scale.

Table 1.   Average number of rhodesgrass scale crawlers produced by parasitized and non-parasitized scales.

Year	Host grass	Average numbers of crawlers	
		Parasitized	Not parasitized
1963	rhodesgrass	7.7	107.5
1964	rhodesgrass	3.3	43.9
	paragrass	0.0	26.7
1965	rhodesgrass	1.8	89.4
	paragrass	6.7	56.8
1967	rhodesgrass	6.4	63.6
	paragrass	5.6	41.0
Average:	rhodesgrass	4.8	76.1
	paragrass	4.1	41.5

Scale Control and Yield Response
in Model-Type Experiments

Model experiments were conducted by Schuster and Boling (in press) to obtain detailed information on scale control and subsequent yield increase of four grass species. Each grass [rhodesgrass, fringed signal grass (*Brachiaria ciliatissima* (Buckl.)), cane sourgrass (*Bothriochloa barbinodis* Lag.), and Texas grass (*Vaseyochloa multinervosa* (Vasey) Hitchc.)] was planted in 5-gallon clay pots and after seedling emergence the pots were divided into 6 to 9 replicates of each treatment. The test treatments were: (1) scale-free, (2) scales present (adult rhodesgrass scales placed among the seedlings) and (3) bio-control (same as (2) above but after the scales became full grown, 10 mated female *N.*

sangwani per replicate were introduced). Scales on Texasgrass in the number (2) treatment were later exposed to parasites and the recovery of the grass from the damage inflicted was evaluated. Grass yields were taken by clipping all of the plants in each plot at flowering. Plants were clipped five inches above the soil and yield was recorded as grams of oven-dry hay. The average number of plants alive at the end of the experiment was recorded for rhodesgrass. Cane sourgrass did not have a number (1) treatment.

Rhodesgrass data are shown in Model Experiment 1, Table 2. It is noteworthy that there was no significant difference in yield or plant stand for the bio-control, or parasite plot, and the scale-free plot. The scales significantly reduced yield and plant stand in the absence of the parasites.

Table 2. Yield response and survival of rhodesgrass as affected by control of rhodesgrass scale by *Neodusmetia sangwani* (Rao), Model Experiment 1[†].

Treatment	Yield in g of oven-dry hay	Plants alive at end of test
bio-control	560.5 a[‡]	5.14 a
scale	454.2 b	1.57 b
scale free	632.5 a	5.14 a

[†]Fifteen clippings over a two-year period.
[‡]Means bounded by the same letter do not differ at the .05 level according to Duncan's Multiple Range Test.

Fringed signal grass responded to the above treatments in the same manner as rhodesgrass (Table 3). There was no difference between the biological control and scale-free treatments, and both had greater yields than the scale treatment.

Texasgrass treated as above reacted like the two previously mentioned plant species (Table 4). After parasites were introduced into the scale treatment plots, there was no difference between the treatments, indicating that the plants recovered quickly once the scale population was reduced. There appeared to be an increase in yield in the bio-control, and in the scale-delayed-parasite release plots, compared to scale-free plots. The yield was significantly greater than the scale-free plot during period 2. Since the plots were fertilized equally, disregarding treatment, it is probable that as the parasite achieved control the residual fertilizer in the pots was utilized by the plants and greater yield resulted.

Model Experiment 4 shows that cane sourgrass recovered rapidly after

Table 3. Yield response of fringed signalgrass as affected by control of rhodesgrass scale by *Neodusmetia sangwani* (Rao), Model Experiment 2 (4 clippings).

Treatment	Yield in g of oven-dry hay
bio-control	55.85 a[†]
scale	30.85 b
scale-free	58.03 a

[†]Means bounded by the same letter do not differ according to Duncan's Multiple Range Test at the .05 level.

Table 4. Yield response of Texasgrass as affected by control of rhodesgrass scale by *Neodusmetia sangwani* (Rao) and recovery of plants following release of parasites into scale plots, Model Experiment 3.

Treatment	Yield in g of oven-dry hay[†]	
	Period 1	Period 2
bio-control	39.77 a[‡]	48.21 a
scale	27.65 b	47.91 a
scale-free	44.61 a	43.68 a

[†]Period 1 harvested four times; period 2-parasites introduced into scale plots, harvested three times.
[‡]Means bounded by the same letter do not differ according to Duncan's Multiple Range Test at the .05 level.

the scales were controlled by *N. sangwani* (Table 5). The scale treatment plots recovered rapidly after scales were controlled and yielded a slightly but not significantly greater yield than the scale-free treatment plots.

Scale Control, Population Data, and Yield Response on Natural Range Sites

The results from the above general data and experiments show that, contrary to a common conception, the percentage of parasitization by a

Table 5. Yield response of cane sourgrass as affected by control of rhodesgrass scale with *Neodusmetia sangwani* (Rao); recovery of plants following release of parasites into scale plots, Model Experiment 4.

Treatment	Yield in g of oven-dry hay[†]	
	Period 1	Period 2
scale	62.17	129.29
scale-free	80.93	126.26
t,p=.95	2.74*	1.05 ns

[†]Period 1 - clipped two times after scale infestation in February; period 2 - parasites introduced into scale plots in May and clipped four times, July through December.

parasite is often not a very meaningful indicator of its population regulation efficiency—i.e., whether or not it can control its host population at a low, uneconomic density. On the contrary, the only real measure of this is afforded by more direct experimental means, i.e., by comparing crop yields over a period of time or longevity of the stand between parasite-present and parasite-absent plots. Also, the comparative scale population densities per grass node is also a better measure of control effect than is percentage of parasitization.

The effectiveness of *N. sangwani* was tested in range experiments conducted in a transition paspalum-fringed signalgrass site previously described by Nord (1956) and Marrow (1959). *N. sangwani* adults were introduced into cages and six months later, grass from six subsamples of 1 m^2 were clipped from within each cage and weights were recorded as grams of oven-dry hay (Schuster and Boling, in press).

A second experiment was conducted in the same area, but the parasite-released sample area was separated from a check area by a distance of 1.5 miles. Ten random 1 m^2 samples were clipped at flowering in each area from 1963 through 1965. In 1965 the check area (no parasite releases made) was infested with *N. sangwani* and the areas were clipped during 1966 through 1968. Each month, data were collected on scales per node, percentage of nodes infested and percentage parasitization.

A single clipping six months later showed that parasite-infested cages yielded 511.3 pounds air-dry hay while the control cages (no parasites) yielded 145.9 pounds air-dry hay. Scales had been reduced 50 per cent by the parasites. However, grasses died in the cages shortly thereafter due to reduced sunlight penetration as the cages darkened with age.

Comparison of biological control areas with check areas (without parasites) proved to be the best method of demonstrating scale control (see also Chapter 5). Near Encino the first three years of clipping showed a 44.2 per cent greater yield from the bio-control area than from the check area (Table 6). Population data showed that the scale had been reduced 52.6 per cent and there was an average parasitization of 73.9 per cent of the third instar and adult scales (Table 6). Following release of *N. sangwani* into the check area in 1965, only a slight difference in yield and scale populations were found between the bio-control and check plots in that area for the next three years. These data also indicate that a reduction of 52 per cent in total scale numbers reduced the damage below economic loss.

Table 6. Comparison of rhodesgrass scale regulation and forage yield of a *Paspalum*-fringed signalgrass range site near Encino before and after the check area was infested with *Neodusmetia sangwani* (Rao).

Period of study	Scale/node		Nodes infested, %		Parasitism, %		Yields (lbs/acre)		t,p=.95
	Bio-control	Check	Bio-control	Check	Bio-control	Check	Bio-control	Check	
	Prior to check being infested with *N. sangwani*								
1963–65	.91	1.91	26.6	42.1	30.9	0.0	1850.7	1021.2	10.39**
	After check was infested with *N. sangwani*								
1966–68	.59	.66	27.0	26.9	24.1	14.0	2117.0	2262.0	.93 ns

Rhodesgrass Pasture Longevity with Bio-control

Chada and Wood (1960) stated that when rhodesgrass plants infested with rhodesgrass scales were grazed, they died within three years. Similar observations were recorded in Queensland, Australia, and fertilization and controlled grazing were recommended to restore vigor (Anon., 1940).

In our results, pure grass stands were conducive to wide fluctuations in scale populations. As scale populations increased, severely damaged culms died and new culms were produced. Finally the entire plants died. At certain times greater scale numbers were found on grass in the biological control area as contrasted to the check area. This resulted indirectly, since scale control by the parasite resulted in more robust plants and greater areas on the plants to support scale development. But parasite activity soon reduced such high scale densities. Schuster and Boling (in press) compared scale-infested rhodesgrass longevity at Kingsville, Weslaco, and Armstrong with paired fields in which *N. sangwani* was released. In all cases plants in the check areas which did not have the parasite began to show considerable damage by the end of the second year and were dead at the end of the third year. Plants in the bio-control areas were still alive and produced good yields at the end of five years when the observations were stopped.

THE DEVELOPMENT OF MASS RELEASE PROCEDURES AND THE RESULTS OBTAINED

A number of factors of importance in utilizing the inherent capacity of this internal parasite in the control of natural infestations of its host species were investigated in order to show that it can be used economically, and extensively. Thus, it was shown that since the adult female is brachypterous and may live only for about 24 to 48 hours on the range, it needs human assistance in the form of rapid and adequate distribution over the range if it is to overtake and control its host before great losses ensue. Through the use of either truck or airplane in distributing the parasite at frequent intervals (not closer than 1/2 x 1/2 mile intervals), it was found that the parasites could then spread adequately within a few years, coalesce and effect substantial control of the scale. The experiments conducted to ascertain the best methods for distributing the parasites are treated in the next subsections.

A number of factors required close study before a suitable program of distribution could be developed. Such factors include the inherent mobility of the brachypterous females, agencies of natural dispersal, female longevity, whether or not ants or other forms of life would likely interfere with the parasite, the best stages of the insect for use in distribution, the manner of distribution, the frequency of distribution points, and the resultant rate of coalescing and adequate population increase over the whole area. The latter would depend upon and measure the value of different combinations of the various patterns of distribution used.

Female *N. sangwani* Longevity

The longevity of the females is important in making contact with the host, especially if searching power is greatly restricted. In these studies, we were also concerned with the impact longevity would have on the mode of delivery. Schuster and Boling (in press) found that at 20° and 30°C only about 40 per cent of the females lived 24 hours. Food supplements did not increase longevity. These data indicated that in a rearing program the female would have to be transported to the point of release within a short period, and this did not appear to be practical in our case.

Mobility of *N. sangwani* Females

Mobility of parasites indicates the relative searching ability of the female, as well as the ability of the species to invade new contiguous areas inhabited by its host. The wingless female of *N. sangwani* presented obvious questions to be answered before mass distribution could be made; e.g., how far and at what speed can a population expand the area occupied?; and what is the mode of movement of the individual which contributes most to this expansion (hopping, wind carried, et cetera)?

To obtain pertinent answers to the questions we (Schuster and Boling, in press) investigated the means and speed of dispersal by use of sticky traps, flourescent tagging of individuals, and by determining (by sampling) the scale population perimeter expansion following their release in southern Texas.

Weekly inspection of sticky traps placed at different altitudes showed that females were caught at three and four feet above the ground during 1965. During 1966 females were caught six feet above the ground. Parasites were caught in greatest numbers during July and August, with a lesser peak in October. These peaks coincide with periods of greatest parasitism of scales by *N. sangwani.*

Females coated with fluorescent powder and released in a lawn were found trapped two days later on the sticky traps placed flat on the ground, at distances up to six feet from point of release. The grass was two inches in height and was almost free of rhodesgrass scale. The test was repeated in a rhodesgrass pasture which averaged six scales per node and 65 per cent parasitization. Not one of the 1500 females marked and released was recovered although several hundred unmarked parasites were captured. Apparently, females on the rhodesgrass did not have to search as actively for hosts because of the existing scale population, and thus failed to reach the traps.

Activity of marked females was closely observed with the aid of an ultraviolet light. Females released at the grass crown moved rapidly upward in search of suitable scales. Females were found at the apex (1 m high) of rhodesgrass plants within a few minutes of release; they then reversed their pattern and proceeded down the plant. Active hopping by the female was infrequent and appeared to be an avoidance reaction. Wind dislodged some of them from the plants.

These data suggest that distant dispersion of the brachypterous female by wind may be greater than by crawling and hopping. Windborne movement was greatest in grasslands with large scale populations and in the periods following the reduction of the June and September-October scale peaks. This indicates that the movement was density induced. As long as there were scales available for oviposition there was little need to increase their searching. This passive windborne spread would be most important in rapid distant distribution.

At Armstrong, *N. sangwani* colony spread as a whole, in a pasture of rhodesgrass and sandburgrass, was at the rate of about 1/2 mile per year. However, the spread through a saline area of predominately *Spartina spp.* (non-host) was less than 1/2 mile in four years and this was in the direction with the prevailing wind. At Encino in a *Paspalum*-fringed signalgrass site, the parasite spread 0.4 mile each year for three years. Near Three Rivers, colony spread in bermudagrass and sandburgrass along a highway was about 1/2 mile each year for three years; however, the parasites were also found 21 miles downstream on the Neuces River in the same period of time. Apparently, flooding washed grass clones containing scales and parasites down the river channel. Thus, parasite spread downstream may be an important means of more rapid distant distribution. In a rhodesgrass pasture near Kingsville the rate of spread was 1/3 mile per year for 2 1/2 years. These observations indicate that colony spread is greatest in grasslands which provide the most contiguous scale populations.

Colony Establishment Influenced by Life Stages Used, Number of Females Released and Timing of Releases

The ease by which parasites become established on natural populations is governed by the ability to find their hosts, environmental conditions at the time of release, and their behavioral responses due to manipulations during transportation. The ability of *N. sangwani* to find hosts is considerably limited by its short life span as an adult and its lack of an active long-range mobility. Thus, establishment on range sites was investigated, comparing dates of release, numbers of females released and the life stages at time of release.

Colony establishment on range lands, as influenced by the number of female parasites and the time of year of release, was investigated during 1961-1963 (Schuster and Boling, in press). Parasites were reared in the insectary (Dean *et al.*, 1961) and released in multiples of 100 at points two or more miles apart. Colony establishment was determined one year later.

The data in Table 7 show that as the numbers of females released per site increased, the percentage of colony establishment increased; however, 100 per cent establishment was never attained (Schuster and Boling, 1970). The release of 100 to 200 females established colonies 64 per cent of the time.

The best month for release was August, when 76.9 per cent of the lots released became established. The poorest months were November and December when only 35.7 per cent of releases resulted in established colonies. An average of 53.4 per cent of the total lots released became established.

Table 7. Colony establishment as influenced by numbers of female *Neodusmetia sangwani* (Rao) and the period of release on range sites in South Texas.

Numbers of females	
Numbers of females released per release point, range	Colony establishment, %
0 - 100	41.8
100 - 200	64.0
200 +	90.0
Over-all average (133 points)	53.4

Period of release	
Month of release	Colony establishment, %
August	76.9
September	66.7
October	53.1
November–December	35.7
Over-all average (126 points)	51.6

Range sites for release of parasites were chosen without regard to vegetation, scale population, temperature and precipitation. These factors obviously contributed to low colony establishment in some of the above tests. Parasites were collected each morning between 7:00 and 9:00 a.m. and delivery time to range sites varied from 1 to 8 hours. Many females were thus probably unable to oviposit before death. These factors would be encountered in any mass release program. Thus, dispersal of adults does not appear practical unless individuals can be collected periodically during the day and released quickly.

Another phase of this study was a comparison of colony establishment by the release of adult female parasites versus the distribution of grass sprigs with parasitized scales attached. For the latter, an estimate of the natural parasite population was made by dissections. An area near Encino was infested by each method, with approximately 150 female parasites per site during August, 1966. The release sites were alternated at 0.5 mile intervals until releases had been made at 50 sites by each technique. Observations on colony detection were made every three to four months following release. The results indicated the investigators' abilities to detect colonies with passage of time, as well as comparing the two techniques of distribution.

The data in Table 8 show that there were only slight differences in the number of established colonies resulting from the releases of adults or pre-imaginal stages while still within rhodesgrass scales. There was a definite increase in the number of colonies detected as more time elapsed for population increase.

Table 8. Releases of 150 adult females or pre-imaginal stages of *Neodusmetia sangwani* (Rao) on range sites, and frequency of colony establishment following release in August, 1966 at Encino.

Sample date (interval)	Established colonies detected, %	
	Pre-imaginal	Adult
Nov. 2, 1966 (3 months)	21.7	18.2
Jan. 12, 1967 (5 months)	39.1	31.8
May 19, 1967 (9 months)	47.8	50.0

The results indicate that the distribution of pre-imaginal parasites still within the scale boxes would be as effective as the release of adults. Furthermore, the expense would be far less than if maintenance of a laboratory culture was required to produce the adults for distribution. We used natural parasite populations. Also, the problem of the short life of the female parasite would be minimized as the parasite would be transported in the pre-imaginal stages and the adults would emerge in the field under natural conditions. They would have their entire life span as adults available to search for hosts.

Results from Large Area Distribution Studies

The best procedures for mass distribution of the parasite on range sites was determined by a study conducted on a 43 square-mile pasture near Encino. Rhodesgrass sprigs with a natural infestation of scales and parasites were collected from a pasture and distributed at one-mile intervals from a pickup truck in October, 1962. The numbers of parasites per release point were estimated beforehand by determining the average number of parasites per scale and the average number of third-instar and adult scales per culm. We dropped 10 culms which contained approximately 100 female parasites at each point. Due to the rough terrain only 40 points were used. It required 16 hours of labor and 125 miles of driving with a 3/4 ton truck to infest the 43 square mile pasture. The cost was about $1.28 per square mile. Grass stems ready for distribution are shown in Fig. 2.

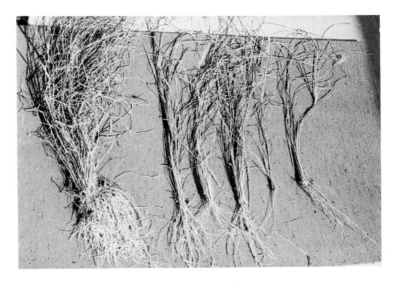

Fig. 2. Grass stems separated and prepared for distribution to release points as described in the text. Parasites will emerge from scales on the grass and oviposit in scales at the release site.

The time involved and the inability of driving into certain areas suggested that an airplane could be used to distribute the grass stems provided they could be cut into short pieces and placed in small boxes for delivery.

Airplane distribution of the grass sprigs was evaluated in 1964, 1965 and

1966. In 1964 the infested grass culms were collected near Weslaco. The grass was cut below the ground-line with a hoe and the tops cut off just above the third node. The clones were separated by hand into smaller pieces and placed in boxes (4 1/2" x 5 1/2" x 2") of the type used for distribution of adult screw worms. The boxes and contents were stored at 45°F for two days before delivery to the range site. An area of 161 square miles was infested by dropping 1 box per 1/2 square mile from an airplane at 1000 ft. altitude.

In 1965 and 1966, grass sprigs with scales and parasites were collected near Kingsville, Texas by use of a sod-lifting device (Table 9). Grass sprigs were packed in one pound shrimp food boxes. The plane was flown at 500 feet in 1965. One box was dropped at 1 x 1 mile-intervals at Kingsville, and at 1 x 2 mile intervals at Armstrong, while two boxes were dropped at 2 x 2 mile intervals at Kingsville. In 1966 we restocked the original 1964 site. All distributions were made in August. Data were kept on the man hours employed, flying time, area of grassland required to fill each box, area covered, and number of females per drop (Table 10). Cost per square mile was calculated from these data.

The Moony airplane was rented from the firm which was dispersing sterile screw worms for the U.S.D.A. and the dispensing equipment on the plane was made available to us for trial by the Mission Center. Essentially, the

Table 9. Operational data on collection of grass sprigs with scales and parasites and the delivery and distribution by airplane.

Man hours	Flying time for area[†]	Area grassland required per box[‡]	Drop interval, miles	Area covered	Parasites per site
48	6 hr 58 min	—	1/2 x 1/2	161 sq mi	226
66[§]	3 hr 0 min	3.06 sq ft	1 x 1	370 sq mi	212
39	1 hr 45 min	2.5 sq ft	1 x 2	318 sq mi	212
39	2 hr 15 min	2.5 sq ft	2 x 2[¶]	373 sq mi	424

[†]Piper Tri-pacer used for area 1; Moony single engine for other areas.
[‡]Rhodesgrass pasture at Kingsville used to collect scales and parasites.
[§]Sod-lifter used two hours.
[¶]Dropped two boxes at each site.

Table 10.   Frequency of *Neodusmetia sangwani* (Rao) detection as determined
by line transect sampling of areas where parasites were released at various densities and comparing releases from ground vehicle vs. airplane at 500 feet.

Area 1. Released from ground vehicle at 1 x 1-mile intervals, 1962						
Time from release, months	12	23	36	43	47[†]	56
Frequency, per cent	4	32	62	48	76	83

Area 2. Released from air at 1/2 x 1/2-mile intervals, 1966	
Time from release, months	12	16
Frequency, per cent	18	24

Area 3. Released from air at 1 x 1-mile intervals, 1965				
Time from release, months	11	14	23	28	34
Frequency, per cent	16	32	40	‡	64

Area 4. Released from air at 1 x 2-mile intervals, 1969				
Time from release, months	11	14	23	28	34
Frequency, per cent	22	42	68	‡	83

Area 5. Release from air at 2 x 2-mile intervals, 1965				
Time from release, months	11	14	23	28[†]	34
Frequency, per cent	4	6	14	14	27

[†]Based on incomplete sample.
[‡]Not sampled.

device was geared to the air speed of the plane so that at predetermined intervals a box was dispersed through the bottom of the plane. A stop watch was used to time the drop of sprigs when using the Piper Tripacer. The sprigs were merely dropped out of the door. The slower speed of the Tripacer was desirable because the area was being dropped at 1/2 mile intervals.

Vegetative descriptions of the areas studied are as follows: the major portion of the 2 x 2 mile areas near Kingsville were of the Kleberg clay soil type and the major grasses did not support large scale populations. In the sandy loam sites near Kingsville, of the 1 x 2 mile area, the major grasses were all attacked by the scale and as a result had a greater scale population. The deep sandy sites around Armstrong and Encino were primarily populated with grasses that contained as many scales as those around Kingsville. Saline sites are found throughout each of the areas and the vegetation (*Spartina* spp.) of these sites does not support rhodesgrass scale populations. Major portions of the area infested at the 1/2 x 1/2 mile interval near Encino are composed of this type of vegetation and about 30 per cent of the area near Armstrong was of this type. Thus, in these areas the expansion of colonies would be reduced. The greatest colony expansion would be expected in the sandy loam and sandy areas as they would have a more contiguous scale population.

Colony establishment and colony coalescence was determined at six-month intervals, beginning 12 months after release date. Transect samples were made at 0.5 mile intervals along predetermined lines. Parasites were detected by the "thumb nail" test described by Schuster and Boling (1970). The same sample points were examined on subsequent dates to estimate colony movement.

In the preliminary test in 1964, the boxes used were too small so that the sprigs of grass were compressed to the extent that they failed to be scattered from the boxes on hitting the airstream. As a consequence, the box and entire contents would fall to the ground intact and thereby reduce the chance of the emerging parasites in finding scales. Also, some speedier method of digging the grass culms was indicated from this test.

The essential data for the airplane experiments are shown in Table 9. The technique for collecting parasites was refined further in 1965. The rhodesgrass area to be harvested was mowed 1 to 2 inches high with a lawn mower. Two types of sod-lifters were then used to lift the sod. One type was a blade mounted to a draw-bar which cut under the sod 3 to 4 inches deep. In coarse-textured soils this was satisfactory but not in fine-textured soils. The second type of sod-lifter was a commercial lifter used in both Florida and Texas in carpetgrass and St. Augustine grass nurseries. This machine is adjustable for depth of cut and cuts the sod slice into 1 sq. ft. blocks.

The clones of grass were broken up by hand into as many individual stems as possible and placed into the box. Greater scattering of scales resulted by the breaking up of the clones. One crew made up boxes while another closed them and placed them in larger boxes for movement to the plane. The plane crew consisted of a loader and pilot. Boxes filled with grass sprigs are shown in Fig. 3.

Costs for collecting and delivery with a single-engine Mooney aircraft were calculated as $0.419, $0.392 and $0.367 per sq. mi. where distributed at 1 x 1 mile, 1 x 2 mile and 2 x 2 mile intervals, respectively. Plane cost was $25.00 per hour.

Colony coalescence on range sites following truck and airplane distribution are shown as the frequency of parasite detection along the transect lines following distribution as given in Table 10. Comparison can be made between ground delivery (Area 1) and air delivery. There was no real difference in the number of colonies detected at 11 and 12 months following release at the rate of 1/2 x 1/2 mile, 1 x 1 mile and 1 x 2 mile intervals. There was no difference for the 1 x 1 mile intervals when releases were made by ground or airplane up to the 36 month sample date. The great number of saline areas mentioned previously for area 3 resulted in less complete parasite coverage of

Fig. 3. A one-pound shrimp box, filled with grass sprigs, upon which the parasitized rhodesgrass scales were attached, was dropped from an airplane to distribute the parasite.

the area. The difference in coverage of the area in area 3, when compared with area 4, also was influenced by the saline areas in area 3.

Area 5, where parasites were distributed at 2 x 2 mile intervals, was in the poorest condition for the release of parasites at the time of the experiment. The area was in a drought stricken one and much of the ground was bare of grass hosts of the scale. In addition, most of the principal grasses grown in this area are not infested by this scale. Large areas have been planted to non-host grasses. Thus, frequency of colony establishment, although low, was greater than expected.

The data collected from areas 1, 2, 3 and 4 indicate that four or more years would be required before a range site could be considered totally infested by *N. sangwani* with the densities of distribution used in these tests. Presence of greater scale numbers and a more even scale population throughout the year would probably decrease the time required to complete coverage or coalescence, and thus biological control.

DISCUSSION

The low value of grasslands favors the utilization of biological control agents. Rhodesgrass scale fitted the category of a pest which required large numbers over a protracted period of time to cause damage. Grasslands provide a relatively stable environment for the operation of biological control agents. Chemical control was economically feasible only on lawns and golf greens. Clearly, it lent itself well for a biological effort.

The rapid control experienced with *Neodusmetia sangwani,* while several other parasites failed to establish themselves even with repeated introductions, supports Clausen's (1951) view that a fully effective parasite or predator is normally easily and quickly established and thus control is usual in three host generations, or three years. We always got substantial host reductions in the first year.

N. sangwani reduced its host primarily by preventing its reproduction. The percentage reduction of total scale populations was always greater than the percentage of parasitism. For this reason, percentage parasitism as an index of scale control, is misleading. Yield comparisons between treatments was the most indicative, and laboratory data gave results similar to, and as good as, field data.

Airplane distribution proved a practical and inexpensive method for infesting large areas with parasites. The ability to mechanically harvest the grass host of the scale and thus utilize natural infestations, was by far the major factor contributing to the success of the mass distribution program. Utilization of this technique by other researchers should be considered in the light of the following conditions: (1) the parasite to be distributed over a large area has limited dispersing ability, (2) where laboratory rearing problems are commonly encountered, (3) if suitable and adequate quantities of host plant species are available to rear the parasites in the field, and (4) the host species are rugged enough to withstand the mechanical manipulations necessary for gathering and transporting to sites of release. Laboratory reared individuals could also be distributed in this manner, but the utilization of field reared individuals for distribution would result in considerable savings in cost.

SUMMARY

Investigations during 1961 to 1968 demonstrated the effectiveness of the parasite, *Neodusmetia sangwani* (Rao), as a controlling agent of rhodes-grass scale, *Antonina graminis* (Maskell). This internal parasite reduced scale populations 68.6 per cent during the year. The two normal yearly scale population peaks were reduced by 50 per cent. The yearly mean parasitism varied from 28.1 to 34.6 per cent of third instar and adult scales. Parasitized scales produced 93.7 per cent fewer crawlers on rhodesgrass and 90.1 per cent fewer crawlers on paragrass.

N. sangwani was successful in eliminating losses in yield due to scale damage on a *Paspalum*-fringed signalgrass site. This range control was sub-stantiated with model experiments in greenhouses with rhodesgrass, fringed signal grass, Texasgrass and cane sourgrass. Longevity studies at four biological control locations showed that rhodesgrass was not killed by scales even under severe grazing conditions.

Female *N. sangwani* were found to have a short life span even under the most favorable conditions. Colony establishment experiments showed that release of adults or distribution of parasitized scales (allowing adults to emerge) gave about equal establishment of colonies. The best month for release was August when 76.9 per cent of the colonies became established. From 100 to 200 females were required per release site to insure 64 per cent establishment.

Colony spread was found to occur at the rate of 1/2 mile per year in grasslands with normal scale populations. The female moved six feet or less by means of hopping and crawling in her life-time; however, wind transport was found to be important during periods of peak parasite activity in July-August and October. Females were caught at heights of six feet with sticky traps.

Techniques developed for mass distribution of *N. sangwani* from natu-rally infested rhodesgrass scale were as follows: the grass (infested with scales which were in turn infested by *N. sangwani*) was cut off just below the ground with sod-lifting machines and the ground node plus one or two other nodes were placed in boxes. The grass plant was then dropped at the desired density from an airplane at points along predetermined lines. The study showed that under the conditions of this test, four years would be required for the parasites to invade 80 per cent of the range. The expansion was restricted in saline areas which supported only *Spartina* sp., a plant that is not a host of rhodesgrass scale. Cost of distribution by airplane was about 39 cents per sq. mi.

ACKNOWLEDGMENTS

Support for this project came from a Research Marketing Act grant through the Forage Insects Research Branch and grants from the King Ranch, Inc.

LITERATURE CITED

Anonymous. 1940. The felted grass-coccid. Queensland Agr. J. 54: 398.

Brimblecrombe, A. R. 1966. The occurrence of the genus Antoninae (Homoptera: Coccoidae) in Queensland. J. Entomol. Soc. Queensland 5: 5-6.

Burks, B. D. 1957. A new parasite of the rhodes-grass scale (Hymenoptera, Encyrtidae). Bull. Brooklyn Entomol. Soc. 52: 124-127.

Chada, H. L., and E. A. Wood, Jr. 1960. Biology and control of the rhodesgrass scale. U.S. Dept. Agr. Tech. Bull. 1221.

Clausen, C. P. 1951. The time factor in biological control. J. Econ. Entomol. 44: 1-9.

DeBach, P. (Editor). 1964. Biological Control of Insect Pests and Weeds. Reinhold Publ. Co., New York. 844 pp.

Dean, H. A., and M. F. Schuster. 1958. Biological control of rhodes-grass scale in Texas. J. Econ. Entomol. 51: 363-366.

Dean, H. A., M. F. Schuster, and J. C. Bailey. 1961. The introduction and establishment of *Dusmetia sangwani* on *Antoninae graminis* in south Texas. J. Econ. Entomol. 54: 925-954.

Green, E. E. 1908. Remarks on Indian scale insects (Coccidae). Pt. 3. Dept. Agr. Mem., Indian Entomol. Ser. 2. Pp. 15-46.

Guagliumi, P. 1963. Insectos y Arachnidos de las plantas communes de Venezuela. DIPUVEN.

Kerrich, G. J. 1964. On the European species of *Dusmetia* Mercet, and a new oriental genus (Hym., Chalcidoidea, Encyrtidae). Entomophaga 9: 75-79.

Machado da Costa, J., R. N. Williams, and M. F. Schuster. In press. Cochonilha dos Capins, *"Antonina graminis"* no Brasil-Parte II-Introducae de *Neodusmetia sangwani,* inimigo natural da cochoxilha. O. Biologico. (1969)

Marrow, J. 1959. The relationship of soils, precipitation, phosphorous fertilization and live stock grazing to vegetational composition and forage production of native vegetation on the Encino Division of the King Ranch. Ph.D. Dissertation. Texas A&M University.

Maskell, W. M. 1897. On a collection of coccidae principally from China and Japan. Entomol. Monthly Mag. 33: 239-244.

Narayanan, E. S., B. R. S. Rao, and H. S. Sangwan. 1957. New species of the parasites of the rhodes-grass scale from the Indian Union. Indian J. Entomol. 19: 65-66.

Nord, E. C. 1956. The influence of drought, fertilization and clipping on native range vegetation in south Texas. Ph.D. Dissertation. Texas A&M University.

Rao, B. R. S. 1957. Some new species of Indian Hymenoptera. Proc. Indian Acad. Sci. 46: 385-390.

Rao, V. P. 1965. Shipment of *Dusmetia sangwani* Subba Rao, a parasite of the rhodes grass scale to the U.S.A. Commonwealth Inst. of Biol. Control. Bull. 5.

Richardson, B. H. 1953. Insecticidal control of rhodesgrass scale on St. Augustine grass lawns. J. Econ. Entomol. 46: 426-430.

Riherd, P. T. 1950. Biological notes on *Anagyrus antoninae* Timberlake (Hymenoptera-Encyrtidae) and its host *Antonina graminis* (Maskell) (Homoptera-Coccidae).

Florida Entomol. 33: 18-22.

Schuster, M. F. 1965. Studies on the biology of *Dusmetia sangwani*. (Hymenoptera-Encyrtidae). Ann. Entomol. Soc. Amer. 58: 272-275.

Schuster, M. F. 1967. Response of forage grasses to rhodesgrass scale. J. Range Management 20: 307-309.

Schuster, M. F., and J. C. Boling. In press. Biological control of rhodesgrass scale in Texas by *Neodusmetia sangwani* (Rao); effectiveness and colonization studies. Texas A&M Univ. (1970)

Wene, G. P., and P. T. Riherd. 1950. Oil emulsion to control rhodes grass scale. J. Econ. Entomol. 43: 386.

Williams, R. N., and M. F. Schuster. In press. Cochonilha dos capin (*Antonina graminis*) no Brasil. Parte I. Distribuicao E. Plantas Hospldeiras. Pesquisa Agropecuaria Brasilerira. (1969)

SECTION III

THE UNHERALDED NATURALLY-OCCURRING
BIOLOGICAL CONTROL

Chapter 11

THE IMPORTANCE OF NATURALLY-OCCURRING BIOLOGICAL CONTROL IN THE WESTERN UNITED STATES

K. S. Hagen, R. van den Bosch, and D. L. Dahlsten

Division of Biological Control
University of California, Berkeley

INTRODUCTION

Sweet are the fruits of adversity—for from a patch of forest defoliated perhaps once in a century, to the yearly attacks of bolls by borers in manicured cotton fields, we find tongues in trees, books in running brooks, sermons in stones and good in everything, for the defoliated forests and rotten bolls tell a story of disrupted nature. To paraphrase Shakespeare here would be appropriate, for it was in "As You Like It" that Shakespeare reflected on nature and saw beauty and reason in what appeared ugly. Indeed, defoliated trees and rods of rotten bolls are unpleasant sights, but such products of insect outbreaks tell a story of disturbance, be it natural or man made. We also are made much aware of the potential destructiveness of unleashed phytophagous insect populations.

In the biomes of the western United States of America, as in other temperate as well as tropical regions, there are countless species of phytophagous insects that rarely, if ever, manifest severe population outbreaks. In our coniferous forest and chapparral biomes there are *occasional* population explosions of insects that defoliate and at times kill patches of plants. It would be an exaggeration to suggest that the entire homeostatic nature of these biomes results from natural biological control, but we suggest that natural enemies are normally responsible for regulating and keeping under control many phytophagous species which occasionally attain outbreak proportions in wooded areas. But as we turn our attention to the various agro-ecosystems, we see a mosaic of pest "fires" being fought in fields and orchards. The frequency and degree of these outbreaks are related directly to man's modern agricultural practices. The greater the artificiality and exoticness of the agricultural components the greater are the chances of crop destruction by insects.

Outbreaks of pests are readily induced by inhibiting the effectiveness of their natural enemies. Acres of succulent salad growing profusely in a bleak faunal community of a few scattered phytophagous insects presents a most favorable environment for them. The pests meet very little resistance here because the entomophagous insects are not present—they have been inadvertently discriminated against in the monoculture environment where no refuges, no alternate hosts or prey, no pollen, and no honeydew are present on which to survive. The pests reproduce quickly and geometrically; only the depletion of their plant food can stop the increase. From one gravid female moth, 1,000 progeny, 500 daughters are easily produced. Virtually unrepressed, these daughters are capable of depositing a half million eggs in these expansive lush lettuce fields. By chance, perhaps an occasional natural enemy has "zeroed" in on a few of the pest larvae—too few and too late, however, to prevent widespread damage.

In this situation, a broad spectrum insecticide is usually applied. Damage, however, is only temporarily diminished; for now, even if 99 per cent of the pests are killed, pest foci are generally scattered throughout the field. From these strategic foci the pests resurge, exploding into a general outbreak. Again, because there are too few natural enemies, another application of insecticide is then applied. Fortunately, not all crops are annuals or so completely divorced from a wild setting, or treated so indiscriminately with broad spectrum pesticides. In the biomes and other relatively persisting natural communities, in certain perennial crops or in annual crops surrounded with plant and animal diversity, natural biological control plays its greatest role, saving farmer and consumer millions of dollars by preventing economic damage and preserving the beauty of the landscape which has no dollar measure.

We view insect numbers as being regulated by density-dependent action, operating as a negative feedback system. Indeed, the classical example describing a negative feedback loop in ecology is the interaction of populations of a predator and its prey (Margalef, 1968).

Negative feedback is evoked and intensified as the population's density increases. The usual factors that react with sensitivity to density changes within a low density range are natural enemies (predators, parasites or pathogens), while intraspecific competition normally comes into play as densities go higher. The theories of population regulation and control were dealt with in detail in Chapter 2, but we wish to add to their criticism of the contention that genetic feedback constitutes a mechanism of regulation of organisms' abundance. We certainly agree that the structure and composition of the community, and the tolerance of organisms to the physical environment are products of evolution through genetic feedback or co-evolution, but the very nature of a mortality factor that relaxes in intensity as the population decreases, minimizes the selection pressure relative to that factor. This preserves and makes available in the community valuable co-evolved predator-prey (or host-parasite) links of a density-dependent, mutually regulating nature. Such mechanisms promote long-term community stability. Haldane (1953, 1956) and Nicholson (1954) discussed the interrelationship between density-dependent factors and genetics at the population level.

One of the best understood undisturbed natural forest communities is a small oak wood near Oxford, England. G. C. Varley and G. R. Gradwell have monitored the animal populations living in this woodland for over 15 years. Of the many species of Lepidoptera that are part of this oak fauna, populations of nine moth species have been closely censused annually. Only one or two of these species attained densities that caused defoliation some years. Some were characteristically rare, others of intermediate abundance. The winter moth, *Operophtera brumata* (L.), is generally the most abundant species. Varley and Gradwell (1963) noted that, generally, while the different species' average densities were quite different, each of them although they varied stayed at about the same population level. Varley and Gradwell developed techniques and sorted out the key factor responsible for changes in abundance of the winter moth.

The analytical determination of the complex population relationships in this long-term study of Varley and Gradwell's, together with the many successful reductions of average densities of agricultural pests following introductions of natural enemies, and the induction of pest population increases by natural enemy exclusions give us confidence to extrapolate from casual observations that natural enemies are important in regulating the abundance of many phytophagous insects in nature.

Beginning with the most diverse communities which are usually the most stable, i.e., forests, and concluding with the simplest and most unstable type of ecosystem, an annual agricultural crop, we here present examples which we believe represent important action of naturally-occurring natural enemies in the western United States.

NATURAL ECOSYSTEMS

Coniferous Forests

The coniferous forest is probably the oldest natural community to persist in the West. Coniferous forests were flourishing in the north temperate region by the Cretaceous and already harbored a complex fauna. The major superfamilies of parasitic Hymenoptera were represented in the Alaskan and Canadian forests of 100 million years ago (McAlpine and Martin, 1969). It is probable that the Mymaridae and Scelionidae had already evolved the ability to internally parasitize insect eggs. It is little wonder that rather homeostatic conditions prevail in many of our undisturbed coniferous forests today, for the components of such have been co-evolving for over 100 million years. Widespread devastating insect outbreaks are relatively rare in such forests (Balch, 1960).

The dynamics of insect populations in western forests are poorly understood. Only recently has there been a concerted effort to probe deeply into the dynamics of bark beetle populations (Stark, 1966; Stark et al., 1968; Stark and Dahlsten, in press, and Chapter 14). In scanning the most extensive ecological studies on major forest caterpillar pests of the world, one finds no general agreement on what factors regulate their abundance, and the factors which account for cyclic outbreaks of some of the species are poorly understood. Researchers agree, however, that intensive investigations should be made of forest pest insects when they are in an endemic (non-outbreak) phase and not just when they are epidemic, as has been common in most studies. Many authors agree that when forest pest populations are in the crash phase of an epidemic (outbreak), natural enemies cause heavy mortality, and may be responsible for the crash, but not all these authors agree that natural enemies are responsible for pest population regulation at endemic levels.

Lepidoptera. Today, in the Northwestern states and in the Sierra Nevada, the pine butterfly, *Neophasia menapia* Feld., is a relatively rare insect in ponderosa pines. It is difficult to believe that it is capable of becoming so abundant as to cause the loss of a billion board feet of timber during a single outbreak. But historically this has been the case. Over 150,000 acres came under its devastating attack in Washington in the late 1800's (Keen, 1952). Another large outbreak appeared in Idaho in 1922 and 1923. The ichneumonid *Theronia (Theronia) atlantae fulvescens* Cresson was considered the most important natural enemy (Evenden, 1926). Over 90 per cent of the butterfly pupae were found parasitized during collapse of an outbreak (Aldrich, 1912).

It is incredible that so little is known about the ecology of such a potentially destructive pest as *N. menapia*. Only one parasite (the species mentioned above) is listed as attacking this butterfly in Thompson's parasite catalogue. The fact that this ichneumonid is so widespread and not host-specific, suggests the existence of other more host-specific parasites that are possibly influential at low (endemic) densities, but do not compete well with *T. atlantae fulvescens* at outbreak densities of this butterfly.

The larvae of the pandora moth, *Colradia pandora* Blake, attacks pines and the larvae of the hemlock looper, *Lambdina fiscellaria lugubrosa* Hulst, are capable of defoliating spruce-hemlock forests. Both moth species have manifested huge population outbreaks every 15 to 30 years. These epidemics appear to be brought under control by natural enemies (Patterson, 1929; Keen, 1952). The most important natural enemies in the crash phases of both species appear to be viruses (Wygant, 1941; Keen, 1952). However, as with the pine butterfly, the factors in the environment that prevent *C. pandora* and *L. f. lugubrosa* from reaching outbreak status for periods of 15 to 30 years are largely unknown.

The Douglas-fir tussock moth, *Hemerocampa pseudotsugata* McD., has been more closely followed in its population ecology (both outbreak and crash phases) than any of the three Lepidoptera discussed above. An outbreak during the late 1920's in Washington destroyed at least 300,000,000 board feet of Douglas-fir and true fir. In the West there appear to have been scattered outbreaks about every 10 years. The most recent outbreaks in California occurred from 1962 to 1967. The outbreak phase and collapse of the moth population was closely followed in northeastern California (Luck and Dahlsten, 1967; Dahlsten and Thomas, 1969; Dahlsten *et al.*, 1970). The decline of the tussock moth population was so complete that no living larvae could be found in 1966, and only 25 to 30 egg masses were found in a study area in a white fir forest. In addition to a nuclearpolyhedrosis virus, three predaceous insects and nine parasites were found associated with the immature stages. Parasitization was an important factor in the collapse of the population in the area studied by Dahlsten *et al.* (1970). Furthermore, the females that emerged did not oviposit. The lack of oviposition was believed to be due to sublethal virus infections (Dahlsten and Thomas, 1969). Here again the population ecology during the endemic infestation phase has not been studied because the extremely low densities pose many difficulties.

The lodgepole needle miner, *Caleotachnites (Recurvaria) milleri* (Busck), is restricted to the Sierra Nevada of California. The larvae of this moth, which take two years to develop, can attain population levels that weaken or kill lodgepole pine. Currently, the damage is greatest in forests managed for recreation (Struble, 1967). Unlike the previous Lepidoptera discussed, the population dynamics of this species has been studied in some detail in both its endemic and epidemic phases. In recent years, it has become an almost chronic pest in the Yosemite National Park. In spite of the effects of twenty or more species of primary parasites and a granulosis virus disease, it sometimes reaches outbreak proportions (Patterson, 1921; Struble and Bedard, 1958; Struble and Martignoni, 1959; Telford, 1961*a, b*; Struble, 1967).

A steady, general decline of this needle miner has occurred since 1965, and the population is approaching endemic levels (Struble, 1967). Parasites were not regarded by Struble as responsible for this decline; however, in previous years he believed that parasites were important in causing cessation of outbreaks. Telford (1961*b*) critically analyzed the interrelationships between seven primary, five secondary, and two tertiary parasites associated with the needle miner and alternate hosts. He postulated that outbreaks could be induced by action of hyperparasites. Struble (1967) concurred that the effectiveness of the needle miner parasites may be impaired by a concurrent complex of hyperparasites.

Scale Insects. A striking contribution to the meager experimental evidence on the importance of natural enemies in controlling forest insects in the western United States has come, not from designed field experiments, but from inadvertent "experimentation," i.e., by following man's use of insecticides (Dahlsten *et al.*, 1969). This concerns an outbreak of several scale insects on pines at South Lake Tahoe, California. The perimeter of the "experimental area" was distinctly outlined by the presence of extremely high densities of these scale insects, with *Phenacaspis pinifoliae* (Fitch) predominating. The foliage of lodgepole and Jeffrey pines in the growing new city literally became so heavily infested that the trees appeared white. The area of infestation coincided precisely with an area that had been under aerosol ("fogging") treatments with the organophosphate Malathion in the attempt to control *Culex tarsalis* Coquillett. For five successive summers the fogging was aimed at killing the adult mosquitoes about the dwellings under the pines. But besides killing mosquitoes, the insecticide evidently also destroyed the inconspicuous tiny parasites and predators that keep the scale insects, *Phenacaspis pinifoliae, Nuculaspis californica* (Colemen), *Pineus* sp., *Physokermes* sp., and *Matsucoccus* sp. at such low levels of abundance that they are usually difficult to find in the forest. The insecticide fogging was stopped in July, 1969 after it was shown to be relatively ineffective against the mosquitoes, anyway (Dahlsten *et al.*, 1969), and already parasitization of the scales is increasing in some areas. The populations were apparently entering the crash phase in some areas during the spring of 1970.

Additional evidence of the importance of the parasites on *Phenacaspis pinifoliae* comes from outbreaks on pines bordering dusty roads. The dust evidently interferes with or kills the parasites (DeBach, ed., 1964), which only a few yards away from the roads—on dust-free trees—effectively control the scale.

Bark Beetles. Bark beetle attacks can commonly be detected by the presence of dying or dead trees. In the undisturbed forest ecosystem, the bark beetles are "nature's loggers" attacking mainly weakened or older trees in the community or trees otherwise predisposed to attack (can be disease, lightning, etc.). While in unlogged forests or ones not suffering blow-down or fire damage, et cetera, the populations of bark beetles thus tend only to "log" the over-mature timber. When such disturbing factors are pronounced, bark beetle populations may reach such abundance that quite vigorous, undamaged trees of any age-class are attacked, and a devastating epidemic may be initiated. The role of pheromones in the initiation of attacks is currently under intensive study.

The beetles themselves and their young are usually under heavy attack by parasites and predators. If it were not for these natural enemies bark beetles would presumably become far more destructive in the forest community. In the Western States in recent years the natural enemies of a few species of scolytids have been studied. To what degree natural enemies may keep bark beetle populations, in general, from becoming epidemic is an unanswered question. However, there is mounting evidence which indicates that natural enemies may be a key factor in regulating the population densities of some scolytid species.

For example, the most important scolytid in the West is the western pine beetle, *Dendroctonus brevicomis* LeConte. This beetle is a major factor in the ecology of ponderosa pine stands for it has the capacity of destroying large quantities of this valuable timber species. In the past extensive stands nearing maturity were thinned by *D. brevicomis* epidemics, and new trees grew up to replace those killed (Miller and Keen, 1960). Recently, the population dynamics of this species has been intensively studied by a team of scientists (Stark and Dahlsten, in press; and Chapter 14).

The diversity of natural enemies attacking *D. brevicomis* is shown by surveys and rearings made by Bushing (1965), Dahlsten (in press), Dahlsten and Bushing (in press), and Miller and Keen (1960). Among the predators, the clerid beetles are most common (Berryman, 1966; Dahlsten, (in press); Person, 1940). The predators (mainly clerids) were observed to attack first those beetle colonies located about one-third the way up the tree. There is then a gradual movement toward the base of the tree (De Mars *et al.,* in press *a*). Berryman (1966) found that in each larval instar, *Enoclerus lecontei* (Wolcott) consumed at least one mature prey larva, accounting for up to seven prey during its larval development. The adult consumed 44 to 158 adult bark beetles. Person's (1940) feeding studies on the same insect indicated that more than 25 beetles are eaten by an adult and at least an equal number by a larva.

Person found that western pine beetles were nearly three times as abundant in clerid-excluded logs as they were in logs where clerids were allowed access.

Natural enemies of *D. brevicomis* and the hyperparasites associated with them appear to follow a regular sequence in their attacks on bark beetles in *D. brevicomis*-infested trees. The first to appear are predators (*Enoclerus, Temnochila* and *Medetera*). A group of small mycetophagous flies (Sciaridae and Cecidomyiidae) arrives next, followed shortly by their natural enemies. Parasites of the western pine beetle do not appear in numbers until three to five weeks after the first bark beetle attack (Dahlsten, in press *a*). Further detailed studies of patterns of distribution of natural enemies, including woodpeckers, were made during all seasons by De Mars *et al.* (in press). During winter, for example, the woodpeckers fed especially on bark beetles but turned to alternative foods during warmer months.

Mortality of western pine beetle caused by parasites (the complex) was found to be extremely variable, but usually they cause only a small fraction of total beetle mortality. They may be, however, a key factor in population regulation of the pest (Dahlsten and Bushing, in press). Their rapid responses to host density may result in control of a population even though they cause only a small mean percentage of mortality.

In Dahlsten and Bushing's study, predators, on the other hand, had the greatest effect on brood reduction, but they were present at relatively constant densities over much of an infested tree and therefore had less effect on inter-sample variability in mortality. Woodpeckers were sporadic in their activity. When they were active in a tree or sample they caused heavy local mortality—thus much inter-sample variability. The interaction between parasites and predators was complex indeed (Berryman *et al.*, in press).

Analyses of survivorship from eight generations of western pine beetle in California and a preliminary life table show two conspicuous periods of loss, (1) between egg deposition and the second larval instar, and (2) between pupation and just prior to adult eclosion, but the *key factors* for either trend changes or regulation of population size have not yet been determined (De Mars *et al.*, in press *b*). In the meantime, intensive analysis of the ponderosa pine ecosystem is being continued, and this program could serve as an example of the proper team approach to study of insect infestations in a natural ecosystem.

In unpublished U.S. Forest Service reports of the 1930's W. D. Bedard and D. DeLeon recommended that parasite samples be taken prior to any control work. They also recommended leaving high stumps when cutting timber to protect and foster bark beetle predator populations. These predators usually concentrate in the stumps and outnumber the scolytids. A resolution was passed at the California Forest Pest Control Action Council meeting in 1966 restating the old overlooked recommendations of Bedard and DeLeon.

Berryman (1967) confirmed this earlier work. He showed that the emergence of *Dendroctonus brevicomis* from the basal five to ten feet of the infested pine is considerably lower than that from upper positions, and he suggested avoiding treating the butt section during spray operations in order to protect the predator population.

Dendroctonus pseudotsugae (Hopkins) is an important pest of Douglas-fir, *Pseudotsuga menziesii* (Mirb.) Franco. Predators play a part in the natural control of this pest (Hopping, 1947; Kline and Rudinsky, 1964), but a parasite seems to be of more importance. One method of control of this pest is the rapid cutting, transportation and milling of infested trees so that beetles will be destroyed in the slabs before they emerge to infest other trees. Ryan and Rudinsky (1962) discovered that in the course of this harvesting practice, certain natural enemies were being inadvertently destroyed along with the beetles. Because of this, it was thought that serious beetle infestations were being unnecessarily prolonged. They found that the most abundant and effective insect parasite of the bark beetle was the braconid wasp *Coeloides brunneri* Viereck. Ryan and Rudinsky thought that the value of *C. brunneri* in Douglas-fir beetle population dynamics could be enhanced by manipulating trees of certain trunk diameter. Smaller trees with thinner bark permitted much higher parasitization by *C. brunneri* because the parasite could reach the scolytid larvae (with its ovipositor). In larger trees with thicker bark the larvae were out of range of attack by *Coeloides*. Thus, it was recommended to leave the smaller infested trees. Ryan (1962, 1965) described the biology of *C. brunneri*.

The fir engraver, *Scolytus ventralis* Leconte, is another species in which mortality has been investigated. Natural enemies take a varying toll of engravers from tree to tree, and the type of natural enemy varies considerably between trees (Stark and Borden, 1965). Berryman (1968) improved sampling techniques and constructed life tables for this insect. The greatest mortality was attributed to intraspecific competition and predation.

Before leaving the subject of natural enemies and bark beetles we should mention that beetle predators, mainly clerids and an ostomid, and a predaceous fly, are considered important in controlling mountain pine beetle, *Dendroctonus ponderosae* Hopkins (DeLeon, 1935a; Struble, 1943), and a braconid parasite *Coeloides dendroctoni* Cushman is considered to cause considerable mortality of this species (DeLeon, 1935b, Reid, 1963).

Sawflies. The potential destructiveness of sawflies is revealed when species are accidentally introduced into a new area without their natural enemies. Fortunately, such accidental importations have occurred only in eastern United States and Canada. In western forests the pine sawflies, which mainly belong to the genus *Neodiprion* (Fam. Diprionidae), rarely reach epidemic levels. There is evidence that natural enemies are mainly responsible for the repression of these sawflies (Keen, 1952).

Preliminary life table data for pine sawflies during high density endemic phase constructed by Dahlsten (1967) showed over 99 per cent total brood mortality in *Neodiprion fulviceps* attacking young *Pinus ponderosa* in three study areas in California (Table 1, Fig. 1). Predators and parasites caused much of the mortality, but there was also a large percentage of unexplained deaths. Predators implicated in this mortality were small mammals which attacked the cocoons, two pentatomid species, a reduviid, a nabid, eight species of spiders and undesignated ants which were observed feeding on *N. fulviceps* larvae. The parasites were: one egg parasite (*Derostenus*), two

Table 1. Modified life table for the 1960-1961 generation of *Neodiprion fulviceps* complex in Area III, Mt. Shasta, California. (From Dahlsten, 1967)

x	1_x	d_xF		d_x	$100q_x$
Eggs	5740	Hymenopterous parasite		39	0.68
		Competition		356	6.20
		Unknown		1560	27.18
		Total		1955	34.06
Early first instar larvae	3785	Ants and unknown		1394	36.83
Feeding larvae	2391	Grosbeak predation		0	0
		Various predators and			
		unknown		724	30.28
		Total		724	30.28
Prespinning larvae	1667	Unknown		1124	67.43
Cocoons	543	Hymenopterous parasites		154	28.36
		Dipterous parasites			
		Villa s. jaennickeana		26	4.79
		tachinids		21	3.87
		Predation		179	32.96
		Fungus disease		18	3.31
		Unknown		57	10.50
		Total		455	83.79
Adults	88	Sex (SR 1.0:0.4)		36	40.8
Females x2	52				
Generation				5688	99.09

Mean number eggs per female − 40.2
Expected eggs   − 1045
Index of population trend   − 18% (1045/5740 x 100)

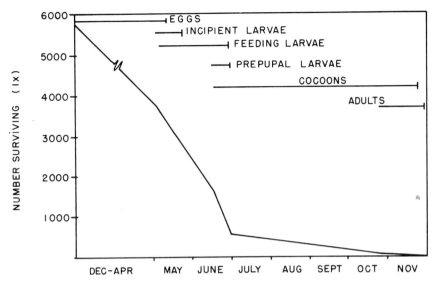

Fig. 1. Survivorship curve of *Neodiprion fulviceps* complex, 1960-1961 in Area III, Mt. Shasta, California. (From Dahlsten, 1967)

tachinids (*Spathimeigenia, Phorocera*), three ichneumonids (*Exenterus, Lamachus* and *Agrothereutes*), a pteromalid (*Tritneptis*), a cleptid (*Cleptes*), and a bombyliid (*Villa*).

Oak Woodland

The black and white oaks of the western United States support a large complex of insect species. Only recently have the microlepidoptera of these western *Quercus* been surveyed. These studies have revealed that over 35 species occur on the California live oak (*Q. agrifolia*) alone in central California. Numerous hymenopterous parasites have been associated with most of these (Opler, 1970). Rarely are outbreaks of these moths observed. However, the macrolepidoptern, *Phryganidia californica* Packard, the California oak moth, commonly and cyclically defoliates *Q. agrifolia* in northern California, but rarely do these trees die, even when the rare, successive defoliations occur. In southern California, populations of this moth appear to be more stable, non-cyclic in nature. Harville (1955) regarded the northern California area as a relatively new area of invasion by a species originating from a southern tropical moth group. The several· parasite species and the virus disease, though at times even abundant, are evidently not able to prevent the frequent outbreaks of *P. californica* in northern California.

The tent caterpillars,[1] *Malacosoma* spp., are indeed conspicuous at times, but normally are apparently under fairly good biological control. Their parasites are not too host-specific. Langston (1957) records 22 known hymenopterous parasites occurring on western North American species of tent caterpillars. Fifteen of these attack the larval and pupal stages, and seven parasitize the egg stage.

There appear to be no host preferences within the genus *Malacosoma* for most of these parasites occur on several species. Stehr and Cook (1968) found the diversity of the parasite fauna to be great and virus diseases also common in high natural populations of *Malacosoma* larvae. General predation on young larvae by ants can be important if spring temperatures are warm (Ayre and Hitchon, 1968). Root (1966) judged from the behavior exhibited by birds while handling *Malacosoma* larvae, that the caterpillar hairs appear to serve as a protective device against avian predators.

Outbreaks of these moths are not usually cyclic. When monitored carefully, epidemics were found to arise in different local areas, which gives the impression of an area having cyclic outbreaks. Since there is spatial variation in the area of outbreaks, Stehr and Cook (1968) pointed to varying natural enemy activity and not macro-environmental weather influences. Pure stands of trees are more apt to generate *Malacosoma* outbreaks than mixed stands (Tothill, 1958).

Chaparral Biome

Chaparral floral communities cover 11 million acres in the West. These communities are not as old as the coniferous forest communities. They arose during the Tertiary, moving north and eventually invading the western United States from the south (Axelrod, 1958). Chaparral, dense stands of evergreen, shrubby vegetation, commonly covers steep slopes; therefore, many chaparral areas persist today for these sites are not suitable for agriculture. However, belts of chaparral once adorning many gradual slopes have been replaced by grasslands, with the help of man.

The insect fauna appears sparse at first glance on various species of shrubs that make up the various chaparral stands. But diligent collecting

---

[1]Tent caterpillars are included under "Oak Woodland" because they attack mainly oaks; however, they are also pests of many different kinds of trees.

produces a great diversity of insects. There is at least one species of mealybug (Pseudococcidae) on each species of shrub (McKenzie, 1967). These mealybugs all have at least one species of parasite and share coccinellid predators. Rarely are outbreaks of pseudococcids observed in chaparral stands. Occasionally, along dusty roads in the Sierra Nevada, *Arctostaphylos* sp. will be seen harboring a conspicuous outbreak of mealybugs. Here again we suggest that the dust has interfered with the natural enemies.

Collecting with a beating sheet at night during the spring reveals at least one species of scarab and weevil to each species of shrub. These beetles, again, do not seem to reach destructive levels on their host shrubs. Boring beetles also occur in great diversity on chaparral. For example, during one summer in a small area where Manzana Creek meets the Sisquoc River in Sunset Valley (Santa Barbara County, California), 23 species of the buprestid genus *Acmaeodera* were collected from seven or eight species of chaparral shrubs. None of the beetles were common nor was there any conspicuous damage to the plants. We suggest again that these insects are under natural biological control as it is unlikely that other environmental factors could be so restrictive for this varied assortment of species as to prevent any of them from commonly attaining damaging densities. One shrub species of California chaparral, coyote brush (*Baccharis pilularis* DeCandolle) has a well known insect fauna. Tilden (1951) found 144 species of permanent insect associates on *B. pilularis*. Of these, there were 54 species of herbivores, 23 predators, 62 parasites and 5 honeydew feeders. One native moth species and one chrysomelid occasionally defoliate *Baccharis*, but Tilden felt that insects were not in any sense determining factors in the abundance of *B. pilularis*. However, black scale, *Saissetia oleae* (Olivier) an introduced pest, may weaken or kill the plant. This is the most serious "pest" of *B. pilularis* and it is particularly abundant when attended by the Argentine ant which "protects" it against parasitization by the introduced parasite *Metaphycus helvolus* (Compere).

The distribution of a native, a gall-forming midge *Rhopalomyia californica* Felt, was studied in relation to this host plant *Baccharis pilularis* by Doutt (1961). Twelve species of parasitic wasps were found to attack *R. californica*. From 1,000 galls, 95.6 per cent produced parasites. The gall midge did not extend over the entire range of its host and certain parasites of the midge showed a narrower range than their host, a phenomenon that has been noted in many instances (e.g., Wilson, 1943).

Sagebrush, Grasslands and Range

The vast area in western United States considered rangeland and pastures constitutes about one billion acres. Much of this still supports its native brush species, but the native perennial bunchgrasses are largely replaced by Mediterranean annual grasses. Presently, there is a move toward removing the brush and seeding imported grasses to increase grazing productivity (Bentley, 1967; Love, 1970).

We know amazingly little about the insects of our western rangelands. We have some idea of the diversity of the Insecta associated with some of the plants from general collecting and taxonomic studies. A serious study has been made of the insects on one or two common species of brush or grass. Also, a few studies have focused on a particular insect living in the desert or semidesert.

Grasshoppers have been collected extensively; hence, we know considerable about the kinds of rangeland grasshoppers. Over 100 species occur in Nevada (La Rivers, 1948), over 104 in Idaho (Hewitt and Barr, 1967), and over 200 in California where a greater habitat diversity is found (Strohecker *et al.*, 1968). Only 12 of these are of economic importance in California (Harper, 1952). Yet, very little is known as to what factors govern the abundance of any species. A greater degree of food specificity has been found among the grasshoppers than what was thought earlier, and this may account for the limited distributions of some species. The few authors that have touched upon the natural enemies view them (scelionid egg parasites, bombyliid, nemestrinid and sarcophagid flies and at times fungus diseases) as being of only local importance in regulating grasshopper populations (Greathead, 1963; Middlekauf, 1958, 1959; Parker and Wakeland, 1957; Pickford and Reigert, 1964; Smith, 1958; York and Prescott, 1952).

La Rivers (1945) found considerable mortality of the Mormon cricket, *Anabris simplex* Haldman, from an egg parasite *Sparasion pilosum* Ashmead. He also estimated that in a rectangular area one mile long by one-half mile wide, a total of over 500,000 crickets had been paralyzed and buried by approximately 30,000 black cricket wasps, *Chlorion laeviventris* (Cresson). Mormon cricket outbreaks are not frequent, but since the time of the early settlers in the Great Basin, crops there have occasionally been attacked by the flightless crickets moving out of the desert hills.

In southern Idaho, sagebrush (*Artemisia* spp.) occupies about 17 million acres of rangeland. In addition to the high grazing values of some species for deer and antelope, it is important as a ground cover, reducing soil erosion by wind and water. Most of the *Artemisia* insects are rarely abundant but the moth *Aroga websteri* Clarke occasionally becomes numerous enough to defoliate extensive stands. In a survey, Fillmore (1965) determined the parasites and the percentage parasitism of this moth over a two year period. He observed fluctuations in moth populations from area to area, and found the density of local populations to be directly associated with this parasitism. However, in some areas he also noted fluctuations in densities in sites of apparently low parasitism. Predation was not considered in Fillmore's study, but among the 28 parasite species Fillmore reared, 20 were primary parasites. Of these, 14 emerged from larvae and 6 from pupae. The commonest parasites were *Copidosoma bakeri* (Howard), *Phaeogenes* sp. and *Spilochalcis leptis* Burks. In some local areas where the highest host populations occurred, parasites were felt to exercise control. In one study site, the hyperparasite *Catolacus aeneoviridis* (Girault) killed 75 per cent of the parasites. A kill of the moth's primary parasites would result in less sagebrush (Ritcher, 1966*b*).

Sagebrush supports several species of coccids, but they seem rarely to be abundant. Some that attack the roots, however, do become abundant enough to kill plants. Barr (1953) observed the scale insect *Orthezia annae* Cockerell infesting roots of *Atriplex confertifolia* (Torr.) and *A. nuttalli* S. Wats. The ant *Campanotus vicinus* Mayr was associated with this scale and the coccinellid *Brumoides parvicollis* Casey was commonly encountered on saltsage, *A. nutalli,* feeding on *O. annae.*

On the mountain slopes ringing the Great Basin, common bitter brush, *Purshia tridentata* (Pursh) D.C. is periodically defoliated by the tent caterpillar *Malacosoma californicum fragile* (Stretch). Since bitter brush is an important browse plant, some research has been directed at determining the factor involved in the inception and termination of the moth outbreaks. Clark (1955) found a virus disease to be correlated with each sudden decrease (crash) in *Malacosoma* population density. However, he gave no clues to the factors that regulate the populations between epidemics. Some authors believe that shifts in genetic composition are involved and account for population fluctuations (e.g., Wellington, 1965).

The insect component of grassland ecosystems is even more poorly known than for the brush ecosystems. Natural biological control of aphids infesting grasses is conspicuous by the sheer number of predators one commonly sees. The coccinellids, particularly *Hippodamia convergens* Guerin, *H. quinquesignata* Kirby and other *Hippodamia* species, with their facultative adult diapause, usually repress the aphids each spring (Hagen, 1962; Smith and Hagen, 1966). Larvae of the syrphids *Metasyrphus, Syrphus, Scaeva, Allograpta* and *Sphaerophoria* species are often seen in grasses infested with aphids, especially if the coccinellids are not abundant.

Since most of the grass cover of rangelands in western United States is represented by introduced species, the three introduced aphid species *Sitobion fragariae* (Walk.), *Metopolophium dirhodum* (Walk.), and *Rhopalosiphum padi* L. that attack these grasses find nearly equivalent conditions as in their Mediterranean native home area except that they are minus their parasites.

The native *Monoctonus paulensis* Ashmead, however, has adapted well to parasitizing the grass aphid *Sitobion fragariae*. In an average square foot of coast grassland Calvert (1970) counted as many as 330 aphids. This amounted to about three aphids per grass stem. Over half of them were parasitized by *M. paulensis*. The interior California valley grass aphids mentioned above are less parasitized, but the predaceous beetles and flies usually keep the aphids from doing much damage.

Only recently has there been some illumination as to what foods the great diversity of our western scarabs eat. Ritcher (1966a) found a variety of food substance to be utilized by the adults and larvae. Some larvae do feed on grass roots. However, we do not see much apparent damage. Little is known about what prevents the densities of scarabs, or the wireworms from being perpetual pests in our western rangelands.

AGRO-ECOSYSTEMS

The best evidence indicating importance of natural biological control comes, paradoxically, from observations and experiments with insects and spider mites occurring in agricultural crops. The relative simplicity and the lesser number of species living in crop monocultures, permits easier detection and more fruitful analysis of interrelationships between pest and natural enemy than is true in natural communities, with their more entangling biotic interrelationships. Among the agro-ecosystems there are various degrees of ecological diversity and stability, i.e., from crops that rarely exhibit pest outbreaks to ones that may present frequent epidemics. This variation in degree of ecological stability is often correlated with the longevity of the crop, and with the degree of "exoticness" and the ecological "relatedness" of the biotic components. Thus, an exotic species in a region is more likely to be attacked effectively by endemic natural enemies if it has close relatives that are endemic in the area than if it does not.

We shall discuss crop ecosystems in the general order of their stability in regards to the frequency of pest outbreaks as apparently correlated with activities of endemic natural enemies. This sequence of crops begins with the long-lived tree fruit and nut crops, followed by vineyards and perennial field crops, and concluding with the least stable, the annual vegetable crops.

Fruit and Nut Trees

Nearly all of our fruit and nut trees are introduced into the United

States as are their pest insects and spider mites. However, some of our native natural enemies have adapted to preying upon or parasitizing the new food source so effectively that a number of exotic pest species are prevented from reaching damaging levels (Fleschner, 1958, 1959; Pickett, 1959).

Avocados, *Persea americana* Mill., are native to tropical America but are extensively cultivated in southern California. Some of their insect pests have come along with the trees and fortunately some of these were accompanied by certain of their natural enemies. A striking example of parasite and predator effectiveness was dramatically demonstrated by Fleschner, Hall and Ricker (1955). They hand-removed each predator or parasite they saw from branches of a tree during the day, for a period of 84 days. The foliage in the area where they had removed the natural enemies showed distinct increases of omnivorous looper, *Sabulodes cabernata* Guenee, six spotted mite, *Eotetranychus sexmaculatus* (Riley), long-tailed mealybug, *Pseudococcus adonidum* (Linnaeus), latania scale, *Hemiberlesia lataniae* (Signoreti) and avocado brown mite, *Oligonychus punicae* (Hirst) (Fig. 2). The increase of latania scale and mealybug populations was due to the removal of imported parasites, not native ones, but that of the other species involved action of native natural enemies. McMurtry *et al.,* (1966, 1968, 1969) have also shown the importance of natural enemies, particularly the native lady beetle *Stethorus picipes* Casey in controlling avocado brown mite.

In citrus groves, DeBach (1946) and DeBach and Bartlett (1951) also clearly showed that when DDT was used, both native and introduced natural enemies of mealybugs, scale insects, aphids, and spider mites were being selectively killed, allowing the homopterous pests and spider mites which are less susceptible to DDT to flourish. In addition to the evidence from insecticidal check methods, the importance of native predators such as *Chrysopa* spp. and *Sympherobius californicus* Banks in aiding the biological control of mealybugs was confirmed by correlating the abundance of natural enemies with the densities of the mealybugs (DeBach *et al.,* 1949).

The native walnuts of the southwest, *Juglans rupestris* Engelm., *J. major* Hell, *J. californica* S. Wats. and *J. hindsii* Rehd. harbor native aphids of the genus *Monellia.* Rarely do these aphids become abundant. There are apparently very few parasites that attack them but the common ashy gray lady beetle, *Olla abdominalis* (Say), seems to be the main natural enemy of this aphid in wild areas. Along with the introduction of the Persian walnut, *J. regia* L., from the Old World, the walnut aphid, *Chromaphis juglandicola* (Kltb.) was inadvertently introduced. The frequent outbreaks of this aphid are associated with the scarcity of natural enemies often associated with the use of certain insecticides. The lady beetles, *Hippodamia convergens* Guerin and *O. abdominalis,* when present, are usually the important factors in controlling the aphid (Table 2) (Bartlett and Ortega, 1952; Michelbacher and Ortega, 1958; Sluss, 1967).

[A "race" of the walnut aphid parasite *Trioxys pallidus* (Haliday) recently introduced from Iran, has become established in the hot valleys of central California and is presently having a tremendous impact in reducing walnut

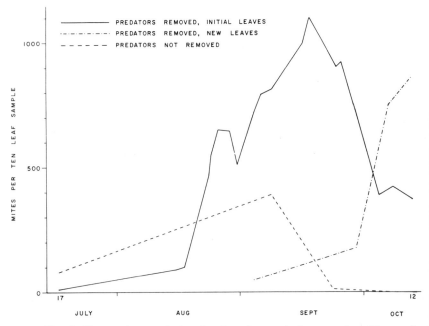

Fig. 2. Changes in population density of avocado brown mite, *Oligonychus punicae* (Hirst), with and without predators. Predators were removed by hand during daylight hours from avocado branches for 84 days. (After Fleschner *et al.*, 1955.)

aphid populations. It is completely overwhelming and replacing other natural enemies, including the French race of *T. pallidus* which is feebly represented in a few coastal northern California areas (van den Bosch *et al.*, in press).]

One of the first distinct upsets of natural biological control occurred in walnut orchards where DDT was being used against the codling moth. A relatively rare native scale insect, the frosted scale, *Lecanium pruinosum*

Coquillett, suddenly appeared in great numbers in insecticide-treated orchards (Table 3) (Michelbacher *et al.,* 1946). Growers not familiar with this coccid, believed the insect to be a new invader. The frosted scale, however, was always present but in inconspicuous numbers, being held under natural

Table 2. Number of coccinellids and their eggs associated with walnut aphid abundance in a northern California orchard. (Data from Sluss, 1967)

	May		June				July				Aug		Sept		
	22	29	5	12	20	26	3	10	24	31	18	28	3	11	18
Aphids/ leaflet	0.6	1.9	4.6	16.5	32.1	14.9	3.8	3.3	6.8	4.9	4.1	11.3	8.9	5.5	5.1
Total coccinellids[†]	6	4	2	28	43	119	129	40	11	5	9	7	20	8	1
Coccinellid eggs	—	15	187	1048	444	165	318	32	26	35	15	240	32	—	37

[†]*Hippodamia convergens* Guerin and *Olla abdominalis* (Say) were the commonest species.

biological control. The main natural enemy of the scale is the encyrtid, *Metaphycus californicus* (Howard) (Bartlett and Ortega, 1952). An accidentally introduced scale insect from the Orient, the calico scale, *Lecanium cerasorum* Cockerell, also appeared in outbreak numbers in response to these DDT treatments. In this case, a native encyrtid parasite *Blastothrix longipennis*

Table 3. Number of frosted scales on twigs of Persian walnuts in California after applications of DDT spray. (Data from Michelbacher, Swanson, and Middlekauff, 1946)

Treatment	No. twigs	x̄ scales per twig
One spray DDT, commercial block	96	7.5
Trees surrounding block	48	0.54
Two sprays DDT, commercial block	96	6.0
Trees surrounding block	48	0.37

Howard which had adapted to the scale was found to be differentially killed by DDT, permitting the scale to reproduce unchecked (Michelbacher and Hitchcock, 1956). These latter entomologists also observed the Audubon's warbler, *Dendroica auduboni* Townsend feeding on the scale insects.

Mealybug populations also "exploded" in other insecticide-treated orchards. Pear orchards treated with DDT had up to 75 more mealybugs, *Pseudococcus maritimus* complex, than unsprayed control trees, and by harvest, 70 per cent of the fruits were infested, contrasted to 3.3 per cent in unsprayed trees. Correlated with the mealybug suppression in the check trees was an abundance of *Chrysopa carnea* Stephens (Table 4) (Doutt, 1948). Mass culture and periodic releases of *Chrysopa* eggs in pear orchards receiving DDT treatments for codling moth control brought the mealybug under control (Doutt and Hagen, 1950). The *Chrysopa* adults were killed by DDT, but eggs and larvae of the *Chrysopa* were not susceptible; therefore, the use of immature stages could be integrated with the DDT treatments.

Table 4. Mealybugs and *Chrysopa carnea* Stephens larvae and pupae trapped in bands on six DDT-sprayed and six unsprayed pear trees. (Data from Doutt, 1948)

Date	Unsprayed		DDT	
	Mealybug	*Chrysopa*	Mealybug	*Chrysopa*
July 9	3	2	63	0
July 15	5	2	61	0
July 21	3	5	50	0
July 30	0	24	40	0
Aug 6	0	36	66	5
Aug 13	0	18	60	4
Aug 21	0	30	61	1
Aug 25	0	8	75	5
Sept 2	0	5	74	4
Sept 8	1	0	59	1
Sept 17	1	2	45	0
Sept 23	6	2	59	0

Since the codling moth became resistant to DDT nearly 20 years ago, a succession of different insecticides has been used against it on apples, pears, and walnuts, as resistance has developed to each insecticide. Most of these insecticides have been of the broad spectrum type; thus, natural enemies are eliminated as well as a number of pest insects. Usually, however, the same pest insects and spider mites resurge quickly, not only because their natural enemies had been eliminated by the direct insecticide kill, but any natural enemies surviving the insecticide treatment would die of starvation or disperse for lack of hosts or prey. The importance of leaving some prey or having some alternative food available for any natural enemies that have survived an insecticide treatment is considered in Chapters 7, 12, and 18.

The influence of direct kill of natural enemies by insecticides has already been mentioned. Observing the impact of insecticides upon natural biological control, Michelbacher (1954) wrote an editorial in the *Journal of Economic Entomology* calling attention to new pest problems which could arise from certain spray programs. Another case of a secondary pest arising from insecticide treatments directed at a key pest, the codling moth, is the pear psylla, *Psylla pyricola* Foerster. Native natural enemies, those that followed the psyllid from eastern United States where it was accidentally imported from the Old World, coupled with endemic predators of western North America, are able to keep the psyllid under control. However, these natural enemies are probably not effective enough to suppress the psyllids to such low levels as to prevent transmission of the disease, pear decline, if this is involved. Trees that are not susceptible to this disease and which are either abandoned or which are not treated with insecticides detrimental to the natural enemies rarely show high levels of pear psylla (Madsen *et al.*, 1963; Madsen and Wong, 1964; McMullen and Jong, 1967*a,b;* Nickel *et al.*, 1965; Westigard and Madsen, 1963).

Shimizu (in press) put five pairs of pear psylla adults per sleeve cage enclosing pear limbs on abandoned trees where psyllids were rare. In closed sleeves having only pear psyllids, the psyllids increased to high numbers, indicating that foliage of abandoned trees can indeed support psyllids. No psyllids were found in sleeve cages on the same trees where psyllids were introduced but which had one end open, allowing predators and parasites to enter (Table 5). In the cages on the same trees where five pairs of psyllids were enclosed with predaceous mirids, the psyllids were considerably reduced compared with psyllids in sleeve cages lacking natural enemies (Table 5).

Grapes

Larvae of the western grape leaf skeletonizer, *Harrisina brillians* (B. and

Table 5. Numbers of pear psyllid eggs and nymphs found in sleeve cages with and without predators. Five males and five females of *Psylla pyricola* Foerster were introduced into five sleeve cages of each treatment on April 30; five ultimate instar predacious mirids were placed in five sleeve cages with developing psyllid populations on May 21, 1969. (From Shimizu, in press)

	Mean numbers of pear psyllids per leaf			
	June 18		July 19	
Treatment	Eggs	Nymphs	Eggs	Nymphs
Check; sleeve cage open	0	0	0	0
Sleeve cage closed + mirids	1.4	0.64	1.2	0.56
Sleeve cage closed; no predators	2.0	1.7	7.7	0.84

McD.), feed on foliage of wild and cultivated grapes. This moth occurs naturally in Arizona, New Mexico, Texas, Utah, Colorado and northern Mexico. It invaded San Diego County, California in 1941, and by 1943 crop losses averaged 40 to 60 per cent, reaching 90 per cent in some cases. A biological control program was initiated in 1950 (Smith, 1953). Explorations for parasites were conducted in both the western states where *H. brillians* naturally occurs and in eastern United States where closely related species or genera exist. Grape growers in Arizona and Mexico had reported the skeletonizer sporadic in its attacks, being abundant for one or two years and almost absent for several years following. The tachinid parasite *Sturmia harrisinae* Coquilett was present in all collections, whereas *Apanteles harrisinae* Muesebeck only emerged from material from Cochise County, Arizona (Clausen, 1961).

Sturmia (Smith *et al.*, 1955), *Apanteles*, and a virus disease were established in San Diego County as a result of importations from Arizona. These agents have been highly effective in reducing existing infestations to low levels, with only occasional short-lived outbreaks occurring in subcommercial and backyard plantings. These agents also aided in preventing spread of the pest (Clausen, 1961).

Grape leaf hopper, *Erythroneura elegantula* Osborn provides both an example of natural biological control and the importance of an alternate host in the ecology of a parasite. A tiny mymarid egg parasite (*Anagrus epos*) of the grape leaf hopper was found to heavily parasitize eggs in many vineyards near streams and rivers where wild vegetation exists. As discussed in Chapter 1, Doutt and Nakata (1965) discovered that this egg parasite also parasitizes eggs of a non-economic leaf hopper, *Dikrella cruentata* Gillette which infests wild blackberries in the riverine situations. It is in the eggs of *D. cruentata* that *Anagrus* overwinters. Since the grape leaf hopper overwinters as an adult, the farther removed are the grape leaf hopper-infested commercial vines from areas where *Anagrus* overwinters the greater the problem. (See especially Chapter 1 and Doutt, 1965.)

Doutt *et al.* (1966) planted small patches of wild blackberries (*Rubus* spp.) in vineyard areas far removed from the riverine haunts to provide overwintering hosts for the egg parasite. These plantings of *Rubus* showed a marked effect in parasitization in vineyards at a distance of 3.5 miles away, and some influence over 4 miles away.

Alfalfa

Two important pest control strategies were crystallized in a central California "crucible" involving native natural enemies and two alfalfa pests. First, the tactic of *supervised control* was generated from the campaign against the alfalfa caterpillar, an endemic species. Secondly, the strategy and methodology of *integrated* control, was developed from primordial beginnings from the confrontation of entomologists and the spotted alfalfa aphid, a recent exotic invader.

The Alfalfa Caterpillar. Michelbacher and Smith (1943) discovered that among the natural factors that limited the abundance of the alfalfa caterpillar, *Colias eurytheme* Boisduval, was a braconid wasp, *Apanteles medicaginis* Muesebeck. Many years earlier, this pristine butterfly-wasp system had moved readily from native clovers to man's new irrigated alfalfa fields. It was not until the early 1940's, however, that the importance of the parasite began to be appreciated. Michelbacher and Smith had observed an alfalfa field in 1939 that was infested with a population of small caterpillars of *C. eurytheme* that appeared sufficiently large that the developing population would inflict serious damage. However, most of the larvae sampled proved to be parasitized by *A. medicaginis.* On the next visit the population was greatly reduced, and no significant injury was caused.

In their four-year study of the problem, Michelbacher and Smith (1943) found that *A. medicaginis* plays the key role in regulating the numbers of

Colias and that without this parasite, alfalfa could not be grown economically in the Central Valley of California. Under field conditions, nearly 100 per cent of the small larvae are often parasitized. Data on the first brood provide an excellent field illustration of the effect of host density on parasitism (Table 6). When the host density is low, correspondingly small parasite populations occur; but where density of small larvae is high, many parasites complete their development.

Detailed studies of the biology and ecology of *Apanteles medicaginis* by Allen (1958); Allen and Smith, (1958) explained in part the reasons why the parasite fails at times to control *Colias* in certain fields. *Apanteles* females emerging from a field must either seek out another field with a *Colias* population in a parasitizable stage or wait for emergence and resulting progeny of a proper age of *Colias* from the same field. This difference in emergence time, coupled with the difference in flight characteristics and host habitat selection, at times results in concentration of hosts and parasites in different fields. Under these circumstances, a small number of *Apanteles* females must cope with a very large number of *Colias* larvae, and control is not effected.

Among the other natural enemies of importance in controlling *Colias*

Table 6.   The number of *Apanteles medicaginis* Meusebeck produced from different densities of the first brood of *Colias eurytheme* Boisduval larvae collected per 100 sweeps in an alfalfa field. (From Michelbacher and Smith, 1943)

Small larvae per 100 sweeps	Parasites produced
0	0.0
1	0.15
2	0.36
3	0.50
4-10	1.33
11-15	2.59
16-25	4.55
26-75	11.80
76+	32.90

eurytheme are an egg parasite in southern California and a polyhedrosis virus disease. Stern and Bowen (1963) found *Trichogramma semifumatum* (Perkins), to be a major biological control agent of *C. eurytheme* only in southern California. They believe that *Colias* populations are not so even-brooded in southern California; therefore, more host eggs may be present in a field through the alfalfa cutting cycle in southern California, in contrast to central and northern California. Such continuity of host eggs would greatly favor *T. semifumatum*. In extreme northern California, *Colias philodice* Godart is also heavily parasitized at times by the gregarious internal parasite *Apanteles flaviconchae* Riley. During late summer, in fields where *Apanteles* fails to control *Colias,* a nuclear polyhedral virus often causes an epizootic, dramatically reducing the caterpillar populations (Michelbacher and Smith, 1943; Steinhaus, 1948; Thompson and Steinhaus, 1950).

The ability of *Apanteles* to control this insect so impressed Michelbacher and Smith that they stressed the importance of not using chemical controls against it where it could be shown that *Apanteles* is abundant enough to check a given infestation.

An economic injury level of 10 nearly mature larvae per sweep was established. They then developed sampling techniques to determine the probability that the parasite would control the caterpillar in a given field during a given cutting of hay. This involved a simple field dissection of small larvae to assess parasitization. If no more than 10 *Colias* larvae per sweep could possibly mature, treatment was not advised. As far back as 1943, Michelbacher and Smith stated that the choice of insecticide should be one that killed the caterpillar and not the parasite.

The establishment of an economic level, coupled with an easy parasite sampling technique of the rather even-brooded, randomly distributed alfalfa caterpillar population set the stage for the development of supervised control.

In 1946, Smith persuaded a group of alfalfa growers in the Dos Palos area of California to hire an entomologist for the summer to follow the development of alfalfa caterpillar and *Apanteles* populations in their fields and to advise them if any insecticide treatment was needed. One of us (K. S. Hagen) was the first supervised control entomologist hired in California, and alone he was able to monitor over 10,000 acres of alfalfa during one summer. Later, the supervised control system was further developed and became a prelude to development of a sophisticated integrated control approach in dealing with insect populations (Smith and Allen, 1954). Today, many crops are under similar pest monitoring services and the integrated control approach is being employed (FAO, 1968; Rabb and Guthrie, eds., in press; Smith, 1969; Smith and van den Bosch, 1967).

The Spotted Alfalfa Aphid. The spotted alfalfa aphid, *Therioaphis trifolii* (Monell), swept through the southwest and through California rapidly causing much damage to alfalfa (Smith, 1959). The presence in alfalfa of native aphid predators capable of reproducing during the summer months and the availability of a selective insecticide (2 oz demeton per acre) made possible the evolution of an integrated control program that has proved highly satisfactory in control of the spotted alfalfa aphid. (Smith and Hagen, 1959; Stern *et al.*, 1959; Stern and van den Bosch, 1959.) Most aphid predators reproduce in the spring and fall months in California valleys, but relatively few are reproductive during the summer months, probably because during their evolutionary history aphids have been scarce on low vegetation during the hot, dry California summers. The main aphid predators in California are lady beetles belonging to the genus *Hippodamia,* and in most valleys of California *H. convergens* and *H. quinquesignata* are the two most important species (Hagen, 1962; Hagen and van den Bosch, 1968; Smith and Hagen, 1965, 1966). In the Imperial Valley, in addition to *H. convergens, Cycloneda sanguinea* L. is active against *T. trifolii* throughout the year (Dickson *et al.,* 1955).

When there are enough aphids available and temperatures are not too cool, *Hippodamia* reproduce, otherwise they enter a facultative, reproductive dormancy (Hagen, 1962; Hagen and Sluss, 1966; Smith and Hagen, 1965, 1966). *H. convergens* responded more than any other coccinellid to the initial invasion of the aphid into the western states and helped reduce the damage. In Utah (Goodarzy and Davis, 1958), in Arizona (Nielsen and Currie, 1960), and in Kansas (Simpson and Burkhardt, 1960), *H. convergens* was also observed to respond most strikingly to the new invader during the late 1950's. In contrast, the common *Coccinella* spp. in California enter an obligatory dormancy during the summer, regardless of aphid density. Photoperiod triggers dormancy in those *Coccinella* species that have been studied. A photoperiod of 18 hours (the light of dawn and dusk added to the daylight hours) induces an estival diapause, 16 hours permits normal reproduction, and 10 or 12 hours of light induces hibernal diapause in *Coccinella novemnotata franciscana* Muls. (McMullen, 1967*a,b*). The obligatory induction of reproductive dormancy in *Coccinella* precludes predation by this group of lady beetles during the summers in California. Fortunately the *Hippodamia* species can be reproductively active in the summer.

An unfavorable aphid-predator ratio usually develops in June or July in central California because of the nearly complete disengagement between predators and aphids in May. During April and May, larvae so decimate the aphid populations in most fields that when the adult predators emerge they usually can not find enough aphids to stimulate ovigenesis. In the first generation of *Hippodamia* spp., about 400 *T. trifolii* or 120 pea aphids per female have to be eaten before eggs are produced (Hagen and Sluss, 1966).

At this time of year, a few adult predators find isolated populations of the aphids sufficient to induce reproduction, but this response is not enough to cope with the soon-exploding aphid populations which begin from random loci of extremely small populations. If at this time when the aphids approach damaging numbers, a broad-spectrum insecticide is applied, the aphids would be suppressed, along with the few predators whose numbers are beginning to increase. Thus, with no reservoir of natural enemies in the field, the aphid population resurges unchecked after this insecticide treatment. Thus, still another insecticide would be required (Stern *et al.*, 1959). As many as 14 treatments of phosphate insecticides were applied sometimes to a single alfalfa field within one year.

If, however, in June or July when the aphid population was approaching damaging levels low and nonsystemic dosages of demeton were applied, the scarce predators would be spared, along with some aphids. The predator populations would again become reproductively engaged with the aphid populations. When demeton was used, often only one insecticide treatment was necessary for the remainder of the season (Smith and Hagen, 1959; Stern and van den Bosch, 1959).

Aphid population crashes resulting from predation are also disruptive, but the resurgence of high aphid populations are long delayed, whereas with use of non-selective materials causing chemically-induced crashes, control is short-lived.

Further developments in the integrated control program against the spotted alfalfa aphid were the introduction and establishment of three species of hymenopterous parasites and the selection of spotted alfalfa aphid-resistant varieties of alfalfa. Because of the wide use of these resistant varieties, the establishment of parasites, the action of native predators and fungi in California alfalfa hay fields, very few insecticide treatments are now applied against this aphid. Even in the event that plant resistance is broken by the aphid, the other components of this integrated control program, i.e. parasites, predators and the use of proper timing and dosage of selective insecticides will still allow alfalfa to be grown economically in California.

The Pea Aphid. The same predator species that were so reactive to the spotted alfalfa aphid when it invaded California have been active against the pea aphid, *Acyrthosiphon pisum* (Harris), in central and northern California ever since the pea aphid invaded the state prior to 1883 (Campbell, 1926; Campbell and Davidson, 1924). Prior to the mid-1960's, the pea aphid was essentially common only during the spring and fall months in alfalfa hay fields (Smith and Hagen, 1966). Aphid predators were naturally attuned to such a phenology, since the native aphids on which they naturally existed in pristine

California had the same phenology. These aphids were virtually absent from low valley vegetation during the hot, dry summer months. These predators thus are adapted by selection to become disengaged from any need for aphids as food in late May and early June and seek aestivating refuge. They remain inactive until the fall or until the spring of the following year.

In recent years, however, the pea aphid in the Central Valley has become so abundant in some fields during the summer as to cause damage to alfalfa. The reasons for this new development are not fully understood, but two possible explanations are suggested. A new biotype of pea aphid may have developed which permits it to tolerate the hot summer temperatures, but it is more likely that its greater success during the summer is tied to new harvesting practices for alfalfa hay. Until the 1960's, the alfalfa plants were cut quite close to the soil surface by a mower with a short-cutting bar. The resulting alfalfa stubble bore few basal leaves, and therefore limited opportunity for pea aphid survival. Today, the mowing operation is combined with raking and a much longer-cutting bar is used. The cutting bar is adjusted higher above the soil and the cutting bar tip often rides on top of the irrigation checks. Therefore, the alfalfa stubble is now several inches taller when cut than in the past. We believe this has altered the microenvironment just enough to provide adequate shelter, or the longer stubble affords quicker plant recovery and adequate nourishment, for greater pea aphid survival during the hot summers whereas in past years most pea aphids in that stubble were killed during this period. However, the stubble is not high enough to retain adult predators or parasites; thus they leave.

Coastal alfalfa fields in California are cut in the same manner, but the introduced pea aphid parasite *Aphidius smithi* Sharma and Rao appears to remain more adequately in the fields during stubble periods, due to the mild coastal conditions (Hagen and Schlinger, 1960). Outbreaks of pea aphids rarely occur during any time of year in coastal fields, because of predators, fungus diseases and *Aphidius smithi.*

An endemic parasite of the pea aphid *Aphidius pulcher* Baker and the imported *A. smithi* have been cultured by the millions in caged field alfalfa plots in Washington. The reared parasites were automatically released from the cages into the open alfalfa fields to help reduce the number of pea aphids migrating to peas (Halfhill and Featherston, 1967).

Alfalfa fields in which strip-cutting has been tried[2] retain far more natural enemies than are found in solid-cut fields. The adult natural enemies of aphids and lepidopterous larvae move from the cut area into the adjacent standing hay strips. Under strip-cutting, not only were there more natural

---

[2]Alternate strips of alfalfa are harvested at two different times in the same field.

enemies (Schlinger and Dietrick, 1960) but the densities of aphids and Lepidoptera were distinctly lower (van den Bosch *et al.,* 1967; van den Bosch and Stern, 1969). The pest insect of cotton, *Lygus hesperus* Knight, is retained in strip-cut alfalfa whereas in solid-cut fields it migrates (to other alfalfa or to adjacent crops including cotton) (Stern, 1969; Stern *et al.,* 1964).

Noctuiid Moths. Of the native western North American parasites of the common noctuid moths that occur in California alfalfa and cotton fields, 11 species attack beet armyworm, *Spodoptera exigua* Hbn., 13 parasitize the western yellow-striped armyworm, *Prodenia praefica* Grote, 13 parasitize the cotton bollworm, *Heliothis zea* (Brodie), and 11 parasitize the cabbage looper, *Trichoplusia ni* (Hbn.). Three common species of parasites attack the above four noctuids, *Apanteles marginiventris* (Cress.), *Hyposoter exiguae* Vier. and the tachinid, *Eucelatoria armigera* (Coq.). Furthermore, *Chelonus texanus* Cress. and *Therion californicum* (Cress.) attack all of these pests except *T. ni,* and *Trichogramma semifumatum* (Perkins) parasitizes the eggs of the above noctuids except *Prodenia* (van den Bosch and Hagen, 1966).

In Arizona, three species of tachinid flies, *Euphorocera tachinomoides* Townsend, *Lespesia archippivora* (Riley) and *Hyphantrophaga hyphantriae* Townsend, were reared from the Great Basin tent caterpillar, *Malacosoma californicum fragile* (Stretch) which commonly attacks cottonwoods and willows in many places almost every year. These tachinid parasites also attack the bollworm, *H. zea,* in agricultural crops; thus the wild native tent caterpillars are an important alternate host or reservoir for some parasites of other caterpillars that occur in agricultural crops (Werner and Butler, 1958).

The importance of conserving the natural enemies in alfalfa cannot be overstressed, for entire areas of many different crops over many square miles may suffer heavier aphid, armyworm, bollworm and looper damage if there is discriminatory mortality occurring against their natural enemies in alfalfa. Only recently has some field research been conducted on the importance of these common, important parasites, but we need a great deal more knowledge concerning their biology and ecology in order adequately to protect them and make better use of them.

Cotton

On the west side of the southern San Joaquin Valley, cotton is often planted in 640-acre sections. These square-mile fields are like lush oases in a desert to prospective phytophagous insects. Every condition favors development of pest insects. The plantings are in existence less than a year, and other crops or their residues surrounding cotton fields are usually plowed under within a year. These vast fallow areas during part of the year are thus weedless, treeless, bushless and barren. Such an environment is not conducive to supporting a complex of insects, yet, a few well-adapted species can enter a field and, unrepressed by natural enemies, can enjoy a rapid population explosion. Fortunately, some parasites and predators of the pest species, particularly from alfalfa, also enter cotton early enough to prevent many moth larvae from completing their development. A diverse complex of natural enemies is sometimes found in cotton (van den Bosch and Hagen, 1966; Whitcomb and Bell, 1964), and because of their different behavior they can sustain themselves even in the absence of host or prey for short periods of time.

The ability of the predaceous *Heteroptera* (*Nabis, Geocoris, Orius,* etc.) to subsist partly on plant sap explains their presence when prey is scarce. The sucking of plant sap, however, can be toxic to them if systemic insecticides are used (Ridgway *et al.,* 1967). In fact, it is the destruction of native natural enemies by broad-spectrum insecticides that often prevents the potential biological control of bollworm (*Heliothis* spp.) (see Chapter 17; Gonzalez *et al.,* 1967; Lingren and Ridgway, 1967; van den Bosch *et al.,* 1969).

The predators that prey upon bollworms (*Heliothis* spp.) have been studied in both laboratory and field (Butler, 1966; Lingren *et al.,* 1968; van den Bosch *et al.,* 1967; Whitcomb, 1967; Whitcomb and Bell, 1964). Caged cotton plants infested with known numbers of bollworm (*Heliothis*) eggs and larvae were subject for 11 or 12 days to different species of native predators to compare the relative effectiveness of each predator species. The predators destroyed from 2/3 to more than 9/10 of the bollworms. *Geocoris pallens* Stal. reduced the bollworm population by 50 per cent and *Nabis americoferus* Carayon and *Chrysopa carnea* caused even greater mortality (van den Bosch *et al.,* 1969; Lingren *et al.,* 1968). Ridgway (1969) obtained a reduction of bollworms by various predators including *Chrysopa carnea* larvae. The results of these tests suggested the possibility of releasing of *C. carnea* eggs and/or larvae in cotton as a control measure. Ridgway and Jones (1968) first tried releases under caged cotton conditions, and were encouraged to test releases in open cotton plots. These tests also gave encouraging results (Ridgway and Jones, 1969; Ridgway, 1969).

Hagen and Tassan (1970) and Hagen *et al.* (in press) experimented for over five years using food sprays simulating artificial "honeydew" on alfalfa and cotton to attract, and stimulate increased egg deposition by, *Chrysopa carnea.* Oviposition by *C. carnea* was often increased over three-fold in areas where the sprays were used in contrast to the control plots, and aphids in alfalfa and bollworms in cotton were reduced in food sprayed plots.

For many years, the native cotton leaf perforator, *Bucculatrix thurberiella* Busck (Braun, 1963), in the Imperial Valley was considered a minor pest of cotton. There were occasional outbreaks, but the moth's native parasites generally kept it under control. There are nine parasites and 10 predators recorded as natural enemies of the perforator in Arizona and California (Clancy, 1946; Rejesus and Reynolds, in press).

Since the invasion of the Palo Verde, Imperial and Coachella valleys by the pink bollworm, *Pectinophora gossypiella* (Saunders), in 1966, wide-scale "eradicative" use of insecticides has been practiced against this new cotton pest. The prediction of Reynolds and Stern (1967) that other potential insect pests in the area would be unleashed as pests by the large-volume use over extensive areas of certain insecticides indeed came true. By the summer of 1968, the cotton leaf perforator became very serious in some areas of the Imperial Valley, causing severe losses (Rejesus and Reynolds, in press). There were severe and destructive outbreaks of this species, and to a lesser extent of spider mites. Natural enemies of many pests were eliminated. Later on, when these "eradicative" treatments were stopped, the most severe outbreak ever recorded in this area of cotton loopers and beet armyworms developed on a variety of fall-planted crops (Smith, in press).

Strawberries

Today, strawberry culture has evolved largely into yearly plantings and this has greatly changed the insect pest, spider mite, and cyclamen mite picture. A few years ago, fields of strawberries were maintained three to four years. Among the pests under the three- to four-year culture was the cyclamen mite, *Steneotarsonemus pallidus* (Banks). This tarsonemid was being routinely treated with mass clumps of parathion—a mass in the crown of each plant. However, from laboratory and field experiments using check-methods and field releases of phytoseiid predators, Huffaker and Spitzer (1951) and Huffaker and Kennett (1953, 1956) showed the striking effectiveness of two native predators in controlling this very serious pest (Fig. 3).

By removing the phytoseiid predators, *Amblyseius aurescens* Athias-Henriot, and *A. cucumeris* (Oudemans) from cyclamen-mite infested plants the mite quickly increased to damaging numbers (Huffaker and Kennett, 1956). Marked differences in densities of this mite were also produced by keeping plots predator-free versus predator-stocked (Fig. 3). The phytoseiid predators were so effective that Huffaker and Kennett (1956) purposely inoculated first-year strawberry plantings with strawberry prunings from old fields which supported both cyclamen mites and the phytoseiid predators. These inoculative releases of both prey and predator species gave sustained fully effective

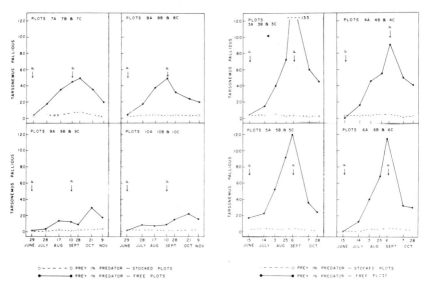

Fig. 3. Changes in population densities of the cyclamen mite, *Tarsonemus pallidus* Banks, in predator-free and predator-stocked (phytoseiid mites) plots in second-year strawberry plantings at different locations in California. The "p"s by arrows indicate dates of parathion treatment in the predator-free plots. (Modified from Huffaker and Kennet, 1956.)

control over the life of the strawberry field.

Present strawberry culture involves field plantings of only one or at most two years duration. The cyclamen mite is not a serious problem today, but the two-spotted spider mite, *Tetranychus urticae* Koch, is a most serious pest. The native phytoseiid *Metaseiulus occidentalis* (Nesbitt) is an important natural enemy of *T. urticae*. It has the capability to control the pest mite on strawberry plants but lags behind the pest in establishment or increase in new fields. A life history and life table study is available on *M. occidentalis* (Laing, 1969*a*), and also for *T. urticae* (Laing, 1969*b*). Laing and Huffaker (1969) compared the efficiency of *M. occidentalis* with the introduced phytoseiid predator, *Phytoseiulus persimilis* A.-H. in regulating *T. urticae* on strawberries in a controlled growth-chamber environment. There were differences between the predators but both species could quickly reduce a population of *T. urticae* when the initial ratio was ten prey to one predator. Laing (*pers. comm.*) has also obtained very promising results in controlling *T. urticae* in field strawberries by releases of *M. occidentalis*. The native *M. occidentalis* is the same predaceous phytoseiid that plays such an important role in controlling spider mites on apples and peaches (Chapter 18) and on grapes (Flaherty *et al.*, 1969; Flaherty and Huffaker, in press).

LITERATURE CITED

Aldrich, J. M. 1912. Note on *Theronia fulvescens*. J. Econ. Entomol. 5: 87-88.

Allen, W. W. 1958. The biology of *Apanteles medicaginis* Muesebeck. Hilgardia 27: 515-541.

Allen, W. W., and R. F. Smith. 1958. Some factors influencing the efficiency of *Apanteles medicaginis* Muesebeck (Hymenoptera: Braconidae) as a parasite of alfalfa caterpillar, *Colias philodice eurytheme* Boisduval. Hilgardia 28: 1-42.

Axelrod, D. I. 1958. Evolution of Madro Tertiary Geoflora. Bot. Rev. 24: 433-509.

Ayre, G. L., and D. E. Hitchon. 1968. The predation of tent caterpillars, *Malacosoma americana* (Lep.; Lasiocampidae) by ants. Can. Entomol. 100: 823-826.

Balch, R. E. 1960. The approach to biological control in forest entomology. Can. Entomol. 92: 297-310.

Bartlett, B., and J. C. Ortega. 1952. Relation between natural enemies and DDT-induced increases in frosted scales and other pests of walnuts. J. Econ. Entomol. 45: 783-784.

Barr, W. F. 1953. *Orthezia annae* Cockerell found in Idaho. Pan. Pac. Entomol. 29: 210.

Bentley, J. R. 1967. Conversion of chaparral areas to grassland: Techniques used in California. U.S. Dept. Agr. Handbook No. 328, 35 pp.

Berryman, A. A. 1966. Studies on the behavior and development of *Enoclerus lecontei* (Wolcott), a predator of the western pine beetle. Can. Entomol. 98: 519-526.

Berryman, A. A. 1967. Preservation and augmentation of insect predators of the western pine beetle. J. Forestry: 260-262.

Berryman, A. A. 1968. Development of sampling techniques and life tables for the fir engraver *Scolytus ventralis*. Can. Entomol. 100: 1138-1147.

Berryman, A. A., I. S. Otvos, D. L. Dahlsten, and R. W. Stark. In press. Interactions and effects of the insectan parasites and predator and avian predator complex. *In* Studies on the population dynamics of the western pine beetle *Dendroctonus brevicomis* LeConte (Coleoptera: Scolytidae), R. W. Stark and D. L. Dahlsten (eds.). Unnumbered publication, Univ. of Calif. Div. of Agr. Sci. (1970)

Braun, A. F. 1963. The genus *Bucculatrix* in America north of Mexico (Microlepidoptera). Mem. Amer. Entomol. Soc. 18. 207 pp.

Bushing, R. W. 1965. A synoptic list of the parasites of Scolytidae in North America north of Mexico. Can. Entomol. 97: 449-492.

Butler, G. D., Jr. 1966. Insect predators of bollworm eggs. Prog. Agr. Arizona 18: 26-27.

Calvert, D. J. 1970. Host specificity of *Monoctonus paulensis* a parasite of certain dactynotine aphids. Ph.D. thesis, Univ. Calif., Berkeley.

Campbell, R. 1926. The pea aphid in California. J. Agr. Res. 32: 816-881.

Campbell, R. E., and W. M. Davidson. 1924. Notes on aphidophagous syrphidae of southern California. Bull. So. Calif. Acad. Sci. 23: 3-9.

Clancy, D. W. 1946. Natural enemies of some Arizona cotton insects. J. Econ. Entomol. 39: 326-328.

Clark, E. C. 1955. Observations on the ecology of a polyhedrosis of the Great Basin tent caterpillar, *Malacosoma fragilis*. Ecology 36: 373-376.

Clark, E. C. 1958. Ecology of the polyhedrosis of tent caterpillars. Ecology 39: 132-139.

Clausen, C. P. 1961. Biological control of western grape leaf skeletonizer (*Harrisma brillians* B. and McD.) in California. Hilgardia 31: 613-638.

Dahlsten, D. L. 1967. Preliminary life tables for pine sawflies in the *Neodiprion fulviceps* complex (Hymenoptera: Diprionidae). Ecology 48: 275-389.

Dahlsten, D. L. In press. Parasites, predators and associated organisms reared from western pine beetle infested bark samples. *In* Studies on the population dynamics of the western pine beetle *Dendroctonus brevicomis* LeConte (Coleoptera: Scolytidae), R. W. Stark and D. L. Dahlsten (eds.). Unnumbered publication, Univ. of Calif. Div. of Agr. Sci. (1970)

Dahlsten, D. L., and R. W. Bushing. In press. Insect parasites of the western pine beetle. *In* Studies on the population dynamics of the western pine beetle *Dendroctonus brevicomis* LeConte (Coleoptera: Scolytidae), R. W. Stark and D. L. Dahlsten (eds.). Unnumbered publication, Univ. of Calif. Div. of Agr. Sci. (1970)

Dahlsten, D. L., E. A. Cameron, and W. A. Copper. 1970. Distribution and parasitization of cocoons of the douglas-fir tussock moth, *Hemerocampa pseudotsugata* (Lepidoptera: Lymantridae), in an isolated infestation. Can. Entomol. 102: 175-181.

Dahlsten, D. L., R. Garcia, J. E. Prine, and R. Hunt. 1969. Insect problems in forest recreation areas. Calif. Agr. 23(7): 4-6.

Dahlsten, D. L., and G. M. Thomas. 1969. A nucleopolyhedrosis virus in populations of the douglas-fir tussock moth (*Hemerocampa pseudotsugata*), in California. J. Invert. Pathol. 13: 264-271.

DeBach, P. 1946. An insecticidal check method for measuring the efficacy of entomophagous insects. J. Econ. Entomol. 39: 695-697.

DeBach, P. (Editor). 1964. Biological Control of Insect Pests and Weeds. Reinhold Publ. Co., N.Y. 844 pp.

DeBach, P., and B. Bartlett. 1951. Effects of insecticides on biological control of insect pests of citrus. J. Econ. Entomol. 44: 372-383.

DeBach, P., C. A. Fleschner, and E. J. Dietrick. 1949. Population studies of the long tailed mealy bug and its natural enemies on citrus trees in southern California. J. Econ. Entomol. 42: 777-782.

DeLeon, D. 1935*a*. A study of *Medetera aldrichii* Wh. (Diptera: Dolichopodidae), a predator of the mountain pine beetle *(Dendroctonus monticolae* Hopk., Coleoptera: Scolytidae). Entomologica Amer. 15: 59-91.

DeLeon, D. 1935*b*. The biology of *Coeloides dendroctoni* Cushman (Hymenoptera-Braconidae) an important parasite of the mountain pine beetle (*Dendroctonus monticolae* Hopk.). Ann. Entomol. Soc. Amer. 28: 411-424.

DeMars, C. J., A. A. Berryman, D. L. Dahlsten, I. S. Otvos, and R. W. Stark. In press *a*. Spatial and temporal variations in the distribution of the western pine beetle, its predators and parasites, and woodpecker activity in infested trees. *In* Studies on the population dynamics of the western pine beetle *Dendroctonus brevicomis* LeConte (Coleoptera: Scolytidae), R. W. Stark and D. L. Dahlsten (eds.). Unnumbered publication, Univ. of Calif. Div. of Agr. Sci. (1970)

DeMars, C. J., D. L. Dahlsten, and R. W. Stark. In press *b*. Survivorship curves for eight generations of the western pine beetle in California, 1962-1965, and a preliminary life table. *In* Studies on the population dynamics of the western pine beetle *Dendroctonus brevicomis* LeConte (Coleoptera: Scolytidae), R. W. Stark and D. L. Dahlsten (eds.). Unnumbered publication, Univ. of Calif. Div. of Agr. Sci. (1970)

Dickson, R. C., and E. F. Laird, Jr. 1955. The spotted alfalfa aphid (yellow clover aphid on alfalfa). Hilgardia 24: 93-118.

Doutt, R. L. 1948. Effect of codling moth sprays on natural control of the Baker mealybug. J. Econ. Entomol. 41: 116-117.

Doutt, R. L. 1961. The dimensions of endemism. Ann. Entomol. Soc. Amer. 54: 46-53.

Doutt, R. L. 1965. New tactics promise grape leafhopper control. Wines and Vines 46: 28-29.

Doutt, R. L., and K. S. Hagen. 1950. Biological control measures applied against *Pseudococcus maritimus* on pears. J. Econ. Entomol. 43: 94-96.

Doutt, R. L., and J. Nakata. 1965. Overwintering refuge of *Anagrus epos* (Hymenoptera: Mymaridae). J. Econ. Entomol. 58: 586.

Doutt, R. L., J. Nakata, and F. E. Skinner. 1966. Dispersal of grape leafhopper parasites from a blackberry refuge. Calif. Agr. 20(10): 14-15.

Evenden, J. C. 1926. The pine butterfly, *Neophasia menapia* Felder. J. Agr. Res. 33: 339-344.

Food and Agriculture Organization (FAO). 1968. Report of the Second Session of the FAO Panel of Experts on Integrated Pest Control. Rome (Sept. 19-24, 1968). 48 pp.

Fillmore, O. O. 1965. The parasitoids of *Aroga websteri* Clarke (Lepidoptera: Gelechiidae) in southern Idaho. M.S. thesis, Univ. Idaho.

Flaherty, D. L., C. D. Lynn, F. L. Jensen, and D. A. Luvisi. 1969. Ecology and integrated control of spider mites in San Joaquin vineyards. Calif. Agr. 23(4): 11.

Flaherty, D. L., and C. B. Huffaker. In press. Biological control of Pacific mites and Willamette mites in San Joaquin Valley vineyards. 1. Role of *Metaseiulus occidentalis*. Hilgardia. (1970)

Fleschner, C. A. 1958. Field approach to population studies of tetranychid mites on citrus and avocado in California. Proc. Xth Intern. Congr. Entomol. Montreal (1956) 2: 669-674.

Fleschner, C. A. 1959. Biological control of insect pests. Science 129: 537-544.

Fleschner, C. A., J. C. Hall, and D. W. Ricker. 1955. Natural balance of mite pests in an avocado grove. Calif. Avocado Soc. Yearbook 39: 155-162.

Gonzalez, D., R. van den Bosch, E. M. Orphanides, L. H. Dawson, and C. White. 1967. Population assessment of cotton bollworm in relation to pest control practices. Calif. Agr. 21(5): 12-14.

Goodarzy, K., and D. W. Davis. 1958. Natural enemies of the spotted alfalfa aphid in Utah. J. Econ. Entomol. 51: 612-616.

Greathead, D. J. 1963. A review of insect enemies of Acridoidea. Trans. Roy. Entomol. Soc. London 114: 451-517.

Hagen, K. S. 1962. Biology and ecology of predaceous Coccinellidae. Ann. Rev. Entomol. 7: 289-326.

Hagen, K. S., E. F. Sawall, Jr., and R. L. Tassan. In press. The use of food sprays to increase effectiveness of entomophagous insects. Proc. Tall Timbers Conf. on Ecol. Anim. Control by Habitat Mgmt. No. 2, Tallahassee, Fla. (1970).

Hagen, K. S., and E. I. Schlinger. 1960. Imported Indian parasite of pea aphid established in California. Calif. Agr. 14(9): 5-6.

Hagen, K. S., and R. R. Sluss. 1966. Quantity of aphids required for reproduction by

Hippodamia spp. in the laboratory. Pp. 47-59. *In* Ecology of Aphidophagous Insects, I. Hodek (ed.). Dr. W. Junk, Publ., The Hague, Netherlands. 360 pp.

Hagen, K. S., and R. L. Tassan. 1970. The influence of food Wheast® and related *Saccharomyces fragilis* yeast products on the fecundity of *Chrysopa carnea* (Neuroptera: Chrysopidae). Can. Entomol. 102: 806-811.

Hagen, K. S., and R. van den Bosch. 1968. Impact of pathogens, parasites, and predators on aphids. Ann. Rev. Entomol. 13: 325-384.

Haldane, J.B.S. 1953. Animal populations and their regulation. New Biology 15: 9-24.

Haldane, J.B.S. 1956. The relation between density regulation and natural selection. Proc. Roy. Soc. London (B) 145: 306-308.

Halfhill, J. E., and P. E. Featherston. 1967. Propagation of braconid parasites of the pea aphid. J. Econ. Entomol. 60: 1756.

Harper, R. W. 1952. Grasshoppers of economic importance in California. Calif. Dept. Agr. Bull. 41: 153-175.

Harville, J. P. 1955. Ecology and population dynamics of the California oak moth, *Phryganidia californica* Packard (Lep.: Dioptidae). Micro-entomology 20: 83-166.

Hewitt, G. B., and W. F. Barr. 1967. The banded-wing grasshoppers of Idaho (Orthoptera: Oedipodinae). Univ. Idaho Agr. Exp. Sta. Res. Bull. 72, 64 pp.

Hopping, G. R. 1947. Notes on the seasonal development of *Medetera aldrichii* Wheeler (Diptera: Dolichopodidae) as a predator of the Douglas-fir bark beetle, *Dendroctonus pseudotsugae* Hopkins. Can. Entomol. 79: 150-153.

Huffaker, C. B., and C. E. Kennett. 1953. Developments toward biological control of cyclamen mite on strawberries. J. Econ. Entomol. 44: 519-522.

Huffaker, C. B., and C. E. Kennett. 1956. Experimental studies on predation: Predation and cyclamen-mite populations on strawberries in California. Hilgardia 26: 191-222.

Huffaker, C. B., and C. H. Spitzer, Jr. 1951. Data on the natural control of cyclamen mite on strawberries. J. Econ. Entomol. 44: 519-522.

Keen, F. P. 1952. Insect enemies of western forests. U.S. Dept. Agr. Misc. Pub. 273, 280 pp.

Kline, L. N., and J. A. Rudinsky. 1964. Predators and parasites of the Douglas-fir beetle: Description and identification of the immature stages. Oregon State Univ. Agr. Exp. Sta. Tech. Bull. 79, 51 pp.

Laing, J. E. 1969a. Life history and life table of *Metaseiulus occidentalis*. Ann. Entomol. Soc. Amer. 62: 978-982.

Laing, J. E. 1969b. Life history and life table of *Tetranychus urticae* Koch. Acarologia 11: 32-42.

Laing, J. E., and C. B. Huffaker, 1969. Comparative studies of predation by *Phytoseiulus persimilis* Athias-Henriot and *Metaseiulus occidentalis* (Nesbitt) on populations of *Tetranychus urticae* Koch. Res. Pop. Ecol. 11: 105-126.

Langston, R. L. 1957. A synopsis of hymenopterous parasites of *Malacosoma* in California. Univ. Calif. Publ. Entomol. 14: 1-50.

La Rivers, I. 1945. The wasp *Chlorion laeviventris* as a natural control of the mormon cricket. Amer. Midland Nat. 33: 743-763.

La Rivers, I. 1948. A synopsis of Nevada Orthoptera. Amer. Midland Nat. 39: 652-720.

Lingren, P. D., and R. L. Ridgway. 1967. Toxicity of five insecticides to several insect predators. J. Econ. Entomol. 60: 1639-1641.

Lingren, P. D., R. L. Ridgway, C. B. Cowan, Jr., J. W. Davis, and W. C. Watkins. 1968. Biological control of bollworm and tobacco budworm by arthropod predators affected by insecticides. J. Econ. Entomol. 61: 1521-1525.

Love, M. 1970. The rangelands of the western U.S. Sci. Amer. 222: 88-96.

Luck, R. F., and D. L. Dahlsten. 1967. Douglas-fir tussock moth (*Hemerocampa pseudotsugata*) egg mass distribution on white fir in northeastern California. Can. Entomol. 99: 1193-1203.

Madsen, H. F., P. H. Westigard, and R. L. Sisson. 1963. Observations on the natural control of the pear psylla, *Psylla pyricola* Forster, in California. Can. Entomol. 95: 837-844.

Madsen, H. F., and T.T.Y. Wong. 1964. Effects of predators on control of pear psylla. Calif. Agr. 18(2): 2-3.

Margalef, R. 1968. Perspectives in Ecological Theory. The Univ. Chicago Press, Chicago, Ill. 111 pp.

McAlpine, J. F., and J.E.H. Martin. 1969. Canadian amber—a paleontological treasure-chest. Can. Entomol. 101: 819-838.

McKenzie, H. L. 1967. Mealybugs of California. Univ. Calif. Press, Berkeley and Los Angeles. 525 pp.

McMullen, R. D. 1967a. A field study of diapause in *Coccinella novemnotata* (Coleoptera: Coccinellidae). Can. Entomol. 99: 42-49.

McMullen, R. D. 1967b. The effects of photoperiod, temperature, and food supply on rate of development and diapause in *Coccinella novemnotata*. Can. Entomol. 99: 578-586.

McMullen, R. D., and C. Jong. 1967a. The influence of three insecticides on predation of the pear psylla, *Psylla pyricola*. Can. Entomol. 99: 1292-1297.

McMullen, R. D., and C. Jong. 1967b. New records and discussion of predators of the pear psylla. *Psylla pyricola* Forster, in British Columbia. J. Entomol. Soc. Brit. Col. 64: 35-40.

McMurtry, J. A., and G. T. Scriven. 1968. Studies on predator-prey interactions between *Amblyseius hibisci* and *Oligonychus punicae:* Effects of host-plant conditioning and limited quantities of an alternate food. Ann. Entomol. Soc. Amer. 61: 393-397.

McMurtry, J. A., and H. G. Johnson. 1966. An ecological study of the spider mite, *Oligonychus punicae* (Hirst) and its natural enemies. Hilgardia 37: 363-402.

McMurtry, J. A., H. G. Johnson, and G. T. Scriven. 1969. Experiments to determine effects of mass releases of *Stethorus picipes* on the level of infestation of the avocado brown mite. J. Econ. Entomol. 26: 1216-1221.

Michelbacher, A. E. 1954. Natural control of insect pests. J. Econ. Entomol. 47: 192-194.

Michelbacher, A. E., and S. Hitchcock. 1956. Calico scale on walnuts. Calif. Agr. 10(9): 6.

Michelbacher, A. E., and J. C. Ortega. 1958. A technical study of insects and related pests attacking walnuts. Univ. Calif. Agr. Exp. Sta. Bull. 764.

Michelbacher, A. E., and R. F. Smith. 1943. Some natural factors limiting the abundance of the alfalfa butterfly. Hilgardia 15: 369-397.

Michelbacher, A. E., C. Swanson, and W. W. Middlekauff. 1946. Increase in the population of *Lecanium pruinosum* on English walnuts following applications of DDT sprays. J. Econ. Entomol. 39: 812-813.

Middlekauff, W. W. 1958. Biology and ecology of several species of California rangeland grasshopper. Pan. Pac. Entomol. 34: 1-11.

Middlekauff, W. W. 1959. Some biological observations on *Sarcophaga faleiformis*, a parasite of grasshoppers. Ann. Entomol. Soc. Amer. 52: 724-728.

Miller, J. M., and F. P. Keen. 1960. Biology and control of the western pine beetle. U.S. Dept. Agr. Misc. Pub. 800, 381 pp.

Nickel, J. L., J. T. Shimizu, and T.T.Y. Wong. 1965. Studies on natural control of pear psylla in California. J. Econ. Entomol. 58: 970-976.

Nicholson, A. J. 1954. Compensatory reactions on populations to stresses, and their evolutionary significance. Austr. J. Zool. 2: 1-8.

Niolson, M. W., and W. E. Curric. 1960. Biology of the convergent lady beetle when fed a spotted alfalfa aphid diet. J. Econ. Entomol. 58: 257-259.

Opler, P. A. 1970. Biology, ecology and host specificity of leaf eating microlepidoptera associated with *Quercus agrifolia*. Ph.D. thesis, Univ. Calif., Berkeley.

Parker, J. R., and C. Wakeland. 1957. Grasshopper egg pods destroyed by larvae of bee flies, blister beetles and ground beetles. U.S. Dept. Agr. Tech. Bull. 1165, 29 pp.

Patterson, J. E. 1921. Life history of *Recurvaria milleri*, the needle miner, in the Yosemite National Park, California. J. Agr. Res. 21: 127-142.

Patterson, J. E. 1929. The pandora moth, a periodic pest of western pine forests. U.S. Dept. Agr. Tech. Bull. 137, 20 pp.

Person, H. L. 1940. The clerid *Thanasimus lecontei* (Wolc.) as a factor in the control of the western pine beetle. J. Forestry 38: 390-396.

Pickford, R., and R. W. Reigert. 1964. The fungous disease caused by *Entomophthora grylli* Fres., and its effects on grasshopper populations in Saskatchewan in 1963. Can. Entomol. 96: 1158-1166.

Pickett, A. D. 1959. Utilization of native parasites and predators. J. Econ. Entomol. 52: 1103-1105.

Rabb, R. L., and F. E. Guthrie (Editors). In press. Concepts of Pest Management, Conference Proceedings. North Carolina State Univ. Press, Raleigh.

Reid, R. W. 1963. Biology of the mountain pine beetle, *Dendroctonus monticolae* Hopkins, in the East Kootenoy Region of British Columbia. III. Interaction between the beetle and its host, with emphasis on brood mortality and survival. Can. Entomol. 95: 225-238.

Rejesus, R. S., and H. T. Reynolds. In press. Impact of the pink bollworm control program upon the cotton leaf perforator-predator-parasite complex in the Imperial Valley, California. (1970)

Reynolds, H. T., and V. M. Stern. 1967. Annual report for the National Cotton Council of America. (Unpublished)

Ridgway, R. L. 1969. Control of the bollworm and tobacco budworm through conservation and augmentation of predaceous insects. Proc. Tall Timbers Conf. Ecol. Anim. Control by Habitat Mgmt. No. 1, Tallahassee, Fla. (1969): 127-144.

Ridgway, R. L., and S. L. Jones. 1968. Field-cage releases of *Chrysopa carnea* for suppression of populations of the bollworm and tobacco budworm on cotton. J. Econ. Entomol. 61: 892-898.

Ridgway, R. L., and S. L. Jones. 1969. Inundative releases of *Chrysopa carnea* for control of *Heliothis*. J. Econ. Entomol. 62: 177-180.

Ridgway, R. L., P. D. Lingren, C. B. Cowan, Jr. and J. W. Davis. 1967. Populations of arthropod predators and *Heliothis* spp. after applications of systemic insecticides to cotton. J. Econ. Entomol. 60: 1012-1016.

Ritcher, P. O. 1966a. White grubs and their allies; A study of North American Scarabaeoid larvae. Oregon State Univ. Press, Corvallis, Ore. 219 pp.

Ritcher, P. O. 1966b. Biological control of insects and weeds in Oregon. Oregon State

Univ. Agr. Exp. Sta. Tech. Bull. 90, 39 pp.

Root, R. B. 1966. The avian response to a population outbreak of the tent caterpillar, *Malacosoma constrictum* (Stretch). Pan-Pac. Entomol. 42: 48-53.

Ryan, R. B. 1962. Durations of the immature stadia of *Coeloides brunneri* at various constant temperatures with descriptions of five larval instars. Ann. Entomol. Soc. Amer. 55: 403.

Ryan, R. B. 1965. Maternal influence on diapause in a parasitic insect, *Coeloides brunneri* Vier. J. Insect. Physiol. 11: 1331-1336.

Ryan, R. B., and J. A. Rudinsky. 1962. Biology and habits of the douglas-fir beetle parasite, *Coeloides brunneri* Viereck in western Oregon. Can. Entomol. 94: 748-763.

Schlinger, E. I., and E. J. Dietrick. 1960. Biological control of insect pests aided by strip-farming alfalfa in experimental program. Calif. Agr. 14(1): 8, 9, 15.

Shimizu, J. In press. The influence of natural enemies on field-caged pear psylla. Calif. Agr. (1970)

Simpson, R. G. and C. C. Burkhart. 1960. Biology and evaluation of certain predators of *Therioaphis maculata* (Buckton). J. Econ. Entomol. 53: 89-94.

Sluss, R. R. 1967. Population dynamics of the walnut aphid, *Chromaphis juglandicola* (Kalt) in northern California. Ecology 48: 41-58.

Smith, O. J. 1953. Grape leaf skeletonizer. Calif. Agr. 7(5): 9.

Smith, O. J., P. H. Dunn, and J. H. Rosenberger. 1955. Morphology and biology of *Sturmia harrisinae* Coquillett (Diptera), a parasite of the western grape leaf skeletonizer. Univ. Calif. Publ. Entomol. 10: 321-358.

Smith, R. F. 1959. The spread of the spotted alfalfa aphid *Therioaphis maculata* (Buckton) in California. Hilgardia 26: 647-685.

Smith, R. F. 1969. Integrated control of insects—a challenge for scientists. Agr. Sci. Rev. 7: 1-5.

Smith, R. F. In press. Pesticides: Their use and limitations in pest management. *In* Concepts of Pest Management, Conference Proceedings, R. L. Rabb and F. E. Guthrie (eds.). North Carolina State Univ. Press, Raleigh.

Smith, R. F., and W. W. Allen. 1954. Insect control and the balance of nature. Sci. Amer. 190(6): 38-42.

Smith, R. F., and K. S. Hagen. 1959. The integration of chemical and biological control of spotted alfalfa aphid. Impact of commercial insecticide treatments. Hilgardia 29: 131-154.

Smith, R. F., and K. S. Hagen. 1965. Modification of the natural regulation of aphids by local climates in California. Proc. XII Intern. Congr. Entomol., London (1964): 372-374.

Smith, R. F., and K. S. Hagen. 1966. Natural regulation of alfalfa aphids in California. Pp. 297-315. *In* Ecology of Aphidophagous Insects, I. Hodek (ed.). Dr. W. Junk, Publ., The Hague, Netherlands. 360 pp.

Smith, R. F., and R. van den Bosch. 1967. Integrated control. Chap. 9. *In* Pest Control: Biological Physical and Selected Chemical Methods, W. W. Kilgore and R. L. Doutt (eds.). Academic Press, N.Y. 477 pp.

Smith, R. W. 1958. Parasites of nymphal and adult grasshoppers in western Canada. Can. J. Zool. 36: 217-262.

Stark, R. W. 1966. The organization and analytical procedures required by a large ecological systems study. Chap. 3. *In* Systems Analysis in Ecology, K. E. F. Watt (ed.). Academic Press, N.Y., 276 pp.

Stark, R. W., and J. H. Borden. 1965. Observations on mortality factors of the fir engraver beetle, *Scolytus ventralis*. J. Econ. Entomol. 58: 1162-1163.

Stark, R. W., and D. L. Dahlsten (Editors). In press. Studies on Population Dynamics of the Western Pine Beetle, *Dendroctonus brevicomis* LeConte. Unnumbered publication, Univ. of Calif. Div. of Agr. Sci. (1970)

Stark, R. W., P. R. Miller, F. W. Cobb, Jr., D. L. Wood, and J. R. Parmeter, Jr. 1968. Photochemical oxidant injury and bark beetle infestation in injured trees. Hilgardia 39· 121-126.

Stehr, F., and E. F. Cook. 1968. A revision of the genus *Malacosoma* Hubner in North America (Lepidoptera: Lasiocampidae): Systematics, biology, immatures and parasites. Smithsonian Inst. U.S. Natl. Mus. Bull. 276, 321 pp.

Steinhaus, E. A. 1948. Polyhedrosis ("wilt disease") of the alfalfa caterpillar. J. Econ. Entomol. 41: 859-865.

Stern, V. M. 1969. Interplanting alfalfa in cotton to control *Lygus* bugs and other insect pests. Proc. Tall Timbers Conf. on Ecol. Anim. Control by Habitat Mgmt. No. 1, Tallahassee, Fla. (1969): 55-69.

Stern, V. M., and W. R. Bowen. 1963. Ecological studies of *Trichogramma semifumatum* with notes on *Apanteles medicaginis* and their suppression of *Colias eurytheme* in southern California. Ann. Entomol. Soc. Amer. 56: 358-372.

Stern, V. M., R. F. Smith, R. van den Bosch, and K. S. Hagen. 1959. The integration of chemical and biological control of the spotted alfalfa aphid. The integrated control concept. Hilgardia 29: 81-101.

Stern, V. M., and R. van den Bosch. 1959. The integration of chemical and biological control of the spotted alfalfa aphid. Field experiments on the effects of insecticides. Hilgardia 29: 103-130.

Stern, V. M., R. van den Bosch, and T. F. Leigh. 1964. Strip cutting of alfalfa for lygus bug control. Calif. Agr. 18(4): 5-6.

Strohecker, H. F., W. W. Middlekauff, and D. C. Rentz. 1968. The grasshoppers of California. Bull. Calif. Insect Survey 10: 1-177.

Struble, G. R. 1943. Laboratory propagation of two predators of the mountain pine beetle. J. Econ. Entomol. 35: 841-844.

Struble, G. R. 1967. Insect enemies in the natural control of the lodge-pole needle miner. J. Econ. Entomol. 60: 225-228.

Struble, G. R., and W. D. Bedard. 1958. Arthropod enemies of the lodgepole needle miner, *Recurvaria milleri* Busck, (Lepidoptera: Gelechiidae). Pan-Pac. Entomol. 34: 181-186.

Struble, G. R., and M. E. Martignoni. 1959. Role of parasites and disease in controlling *Recurvaria milleri* Busck. J. Econ. Entomol. 52: 531-532.

Telford, A. D. 1961a. Lodgepole needle miner parasites: Biological control and insecticides. J. Econ. Entomol. 54: 347-355.

Telford, A. D. 1961b. Features of the lodgepole needle miner parasite complex in California. Can. Entomol. 93: 394-402.

Thompson, C. G., and E. A. Steinhaus. 1950. Further tests using a polyhedrosis virus to control the alfalfa caterpillar. Hilgardia 19: 411-445.

Tilden, J. W. 1951. The insect associates of *Baccaris pilularis* De Candolle. Microentomology 16: 149-188.

Tothill, J. D. 1958. Some reflections on the causes of insect outbreaks. Proc. X. Intern. Congr. Entomol. Montreal (1956) 4: 525-531.

van den Bosch, R., B. D. Frazer, C. S. Davis, P. S. Messenger, and R. C. Hom. In press. An effective walnut aphid parasite from Iran. Calif. Agr. (1970)

van den Bosch, R., and K. S. Hagen. 1966. Predaceous and parasitic arthropods in California cotton fields. Univ. Calif. Agr. Exp. Sta. Bull. 820, 32 pp.

van den Bosch, R., C. F. Lagace, and V. M. Stern. 1967. The interrelationship of the aphid, *Acyrthosiphon*, and its parasite, *Aphidius smithi*, in a stable environment. Ecology 48: 993-1000.

van den Bosch, R., T. F. Leigh, D. Gonzalez, and R. E. Stinner. 1969. Cage studies on predators of the bollworm in cotton. J. Econ. Entomol. 62: 1486-1489.

van den Bosch, R., and V. M. Stern. 1969. The effect of harvesting practices on insect populations in alfalfa. Proc. Tall Timbers Conf. Ecol. Anim. Control by Habitat Mgmt. No. 1, Tallahassee, Fla. (1969): 47-54.

Varley, G. C., and G. R. Gradwell. 1963. The interpretation of insect population changes. Proc. Ceylon Assoc. Adv. Sci. (1962): 142-156.

Wellington, W. G. 1965. Some maternal influences on progeny quality in the western tent caterpillar, *Malacosoma pluviate* (Dyar). Can. Entomol. 97: 1-14.

Werner, F. G., and G. D. Butler, Jr. 1958. Tent caterpillars in breeding useful parasites, they may be valuable friends. Progressive Agr. Ariz., Spring 1958, p. 13.

Westigard, P. H., and H. F. Madsen. 1963. Pear psylla in abandoned orchards. Calif. Agr. 17(1): 6-8.

Whitcomb, W. H. 1967. Field studies on predators of the second-instar bollworm, *Heliothis zea* (Boddie). J. Georgia Entomol. Soc. 2: 113-118.

Whitcomb, W. H., and K. Bell. 1964. Predaceous insects, spiders, and mites of Arkansas cotton fields. Univ. Arkansas Agr. Exp. Sta. Bull. 690, 84 pp.

Wilson, F. 1943. The entomological control of St. John's wort (*Hypericum perforatum* (L.)) with particular reference to the insect enemies of the weed in southern France. Council Sci. Ind. Res. (Aust.) Bull. 169. Pp. 1-87.

Wygant, N. D. 1941. An infestation of the pandora moth, *Coloradia pandora* Blake, in lodgepole pine in Colorado. J. Econ. Entomol. 34: 697-702.

York, G. T., and H. W. Prescott. 1952. Nemestrinid parasites of grasshoppers. J. Econ. Entomol. 45: 5-10.

Chapter 12

NATURALLY-OCCURRING BIOLOGICAL CONTROL IN THE EASTERN UNITED STATES, WITH PARTICULAR REFERENCE TO TOBACCO INSECTS

R. L. Rabb

North Carolina State University, Raleigh

INTRODUCTION

The role of biotic agents (parasites, predators, and pathogens) in the natural control of insect populations tends to be greater and more complex in geographical areas where climatic and edaphic factors favor a diverse and productive flora and fauna. The eastern United States generally is such an area. The southeastern states, in particular, are characterized by a wide diversity of biotic communities, and the agro-ecosystems are comprised of a very complex pattern of cultivated and fallow fields of various shapes and dimensions, many streams, ponds, lakes, and marshes, and an abundance of mixed woodlands. The Insecta permeates this rich biotic matrix, forming an integral part of the intricate food chains and webs involved. A hundred or so of these insects are pests, but the vast majority are essential to man's welfare. Among the latter are thousands of species of parasites and predators.

In recent years, there has been a sharp increase in the study of native natural enemies, and it is apparent that the insect pests in the eastern states generally are attacked by many kinds of parasites, predators and pathogens causing diverse diseases. In spite of this increased tempo of research, however, the role of these biotic agents in the control of specific pests is poorly understood. The observational, rather than the experimental, approach continues to be the rule; thus decisive proof of cause and effect relationships between enemy and host populations, in most cases, is still elusive. However, quantitative research has improved, and various modifications of the life table perspective have enhanced the value of correlative methods. Experimental techniques are also being used more frequently in spite of the intrinsic difficulties of using them in studying populations under representative field conditions. Thus the science and art of evaluating naturally-occurring biological control are still in the developing phase and the examples I shall discuss should be viewed in this context.

Man is failing to reap the full benefit from native natural enemies of insect pests and in most instances those who could find them of most value do not even recognize them. A part of the problem is the need for more research, but in many cases, the importance of natural enemies is overlooked because judgments are made from a too restricted perspective. Many authors also have confounded the problem by classifying natural enemies as important or unimportant without adequately specifying the criteria used (or the pertinent conditions).

In evaluating a native natural enemy in terms of its actual or potential role in the control of a pest species, the following questions seem relevant:

1. To what extent is economic loss reduced by the enemy's action against a local pest population?

2. To what degree does the enemy's action reduce the mean level of abundance of the pest over a large geographic area?

3. Can the actions of the enemy be practically enhanced by environmental manipulations or through mass rearing and release?

4. Can data on the enemy be used in predicting the population trend of the pest?

I shall discuss selected examples in reference to these four questions to illustrate that a natural enemy may be of little significance by one or more criteria but of major importance by another and that all criteria should be used in evaluating its general status. In addition, examples will be given of enemies whose actions are particularly difficult to evaluate but whose potential importance should not be ignored. My examples will be from the agro-ecosystem, and largely as related to tobacco pests, for this is the area of my experience.

MAJOR FEATURES OF TOBACCO CULTURE RELATIVE TO NATURAL ENEMY ACTIONS

The function of a natural enemy cannot be appreciated apart from its ecological setting. In addition, some acquaintance with specific features of an insect's ecosystem must be attained before it is possible to visualize environmental manipulations designed to enhance its beneficial activities. A brief description of tobacco culture, therefore, may enhance the clarity of subsequent discussions of associated pests and natural enemies.

Approximately one million acres of tobacco occurs on the east coast from Connecticut to Florida. The fields are generally small and widely dispersed among other crops. The North Carolina crop, which represents approximately 40 per cent of the total, is produced on over 115,000 farms. This comprises less than five acres per farm. Many varieties are grown, but none was developed on the basis of insect resistance.

Tobacco plants are transplanted from seed beds to fields in early spring and cultivated several times during the early stages of growth. The appearance of flowers signals two important operations: topping and suckering. The inflorescence is removed (topped) so that a greater proportion of photosynthetic products might be available for the production of more leaf tissue. Topping, however, stimulates the growth of adventitious buds in leaf axils which produce suckers of no monetary value and which, when allowed to develop, also diverts nutritional resources from the marketable leaves. Prior to the late 1950's, these suckers were removed by hand, a process requiring great expenditure of labor and money. Within the past 10 years, however, sucker control has been simplified with the introduction of maleic hydrazide (MH-30), one application of which inhibits sucker growth for the remainder of the season, even on the tobacco stalks after harvest.

Harvesting proceeds at about weekly intervals, with the removal of leaves as they ripen from the bottom to the top of the stalks. After harvest many growers immediately cut down their stalks, but others allow them to stand for varying periods of time. Stalks in fields where suckers are controlled before harvest by hand produce an abundance of sucker growth in which late-season populations of certain insects abound. In sharp contrast, post-harvest sucker growth is greatly inhibited and in many cases non-existent in fields receiving applications of MH-30. The advent of MH-30 has thus had a great impact on the populations of certain pests and their natural enemies. In order to avoid confusion, however, it should be emphasized that the motivation for using MH-30 was not insect control; it was to save money and labor involved in removing suckers by hand.

ENEMIES REDUCE THE LOCALIZED ECONOMIC LOSS

The beneficial effects of certain natural enemies are readily recognized because their activities bring about an immediate reduction in the numbers of pests and crop losses. For example: in August of 1969, soybeans in eastern North Carolina attracted large numbers of ovipositing corn earworm moths, *Heliothis zea* (Boddie); consequently, heavy populations of early-instar larvae threatened severe pod damage. In most cases, however, damage was averted, without the necessity of insecticidal applications, by an epizootic of *Spicaria* sp. (identified by W. M. Brooks, North Carolina State University). The large numbers of mummified larvae provided an easily observed and spectacular demonstration of the importance of a biotic agent in reducing the local population of a pest and its harmful effects. Moisture conditions preceding and during the epizootic were particularly favorable for the dissemination and development of the pathogenic fungus. The direct effect of many natural enemies in reducing economic loss, however, is not so readily apparent, although their beneficial effects year after year may be more consistent than is the case with such fungus agents requiring rather precise weather conditions. Two such enemies are (1) *Compoletis perdistinctus* (Viereck), a larval parasite of the tobacco budworm, *Heliothis virescens* (F.), and the corn earworm, *H. zea* (Boddie), and (2) wasps of the genus *Polistes* which prey on free-feeding larvae, including the tobacco hornworm, *Manduca sexta* (Johannson).

The tobacco budworm is one of the most important tobacco pests prior to topping, and the closely related corn earworm at times affects tobacco in a very similar manner. Neunzig (1969) describes the biology of these two pests as associated with tobacco in North Carolina and includes notes and references to their natural enemies. Infestations in terminal buds can cause extensive damage to unfolding leaves, but little damage normally occurs after the full complement of leaves and terminal flowers have been produced. Due to several mortality factors, the survival of budworms varies greatly, and the amount of damage per larva is closely correlated with the length of the feeding period. One important factor killing budworms while they are still small is the hymenopterous parasite *C. perdistinctus*. The adult female lays her eggs in very small budworms which are killed in the 3rd or 4th instar by the developing parasite. Therefore, budworms parasitized by *C. perdistinctus* inflict much less leaf damage than non-parasitized larvae. As shown in Table 1, budworms parasitized by *C. perdistinctus* inflicted little more damage than four-day old non-parasitized invididuals and much less than that caused by older non-parasitized larvae. Another parasite, *Cardiochiles nigriceps* Viereck, does not reduce damage directly because it destroys its budworm host in the prepupal stage after feeding is completed.

C. perdistinctus parasitizes a high percentage of budworms in North Carolina from May to early July prior to flowering when the plants are most susceptible to budworm injury. The data in Fig. 1, which were provided by W. M. Brooks contrasts the seasonal incidence of budworm parasitism by *C. perdistinctus* and *C. nigriceps*. In many fields, *C. perdistinctus* may prevent budworm damage from becoming economically important. Apparently, this parasite has excellent searching ability because parasitism is very high even at

Table 1. Damage to tobacco by parasitized and non-parasitized budworms (*Heliothis virescens* Vier.)[†].

Category of budworms	Number of plants	Number of stalk borings	Number of leaves eaten
Non-parasitized			
4 days old	20	0.3	0.4
12 days old	20	0.5	1.3
Full term	12	1.3	2.8
Parasitized			
Campoletis perdistinctus (Vier.)	11	0.2	0.6
Cardiochiles nigriceps Vier.	11	0.8	3.0

[†]From an unpublished 1960 report by Rabb and Neunzig, North Carolina State University.

low host densities during this pre-flowering period. After flowering and on post-harvest sucker tobacco, however, parasitism by *C. perdistinctus* falls to relatively low levels and parasitism by *C. nigriceps* rises. Thus while *C. perdistinctus* is more important than *C. nigriceps* in directly reducing economic loss from a given infestation, *C. nigriceps* may be of greater importance by reducing overwintering populations, and thus the initial density of the progeny infestation referred to here.

The tobacco hornworm is capable of great destruction of marketable leaves any time before harvest, and, prior to the early 1960's, it was the most important insect pest of tobacco. In order to appreciate the dynamic roles of different natural enemies of the pest, some knowledge of its biology is essential. There are three full generations and a partial fourth in North Carolina, but they are rather ill-defined due to wide overlap caused by the long emergence period of diapausing individuals (Rabb, 1966, 1969). Only the first two generations are of potential importance to a current crop, while the third and partial fourth generations occur after harvest on sucker growth of no economic value. These late season hornworms, however, are of critical importance to the number of overwintering (diapausing) pupae produced.

Most of the damage inflicted by hornworms is due to the feeding of the large 5th instars. Approximately 90 per cent of the total leaf area consumed during the entire larval period is eaten by this stage. In the field, where mortality factors modify the proportionate loss, the 5th instars still inflict over 75 per cent of the damage (Lawson and Rabb, 1964).

In the mid-1950's, Lawson (1959) obtained data on the age-specific

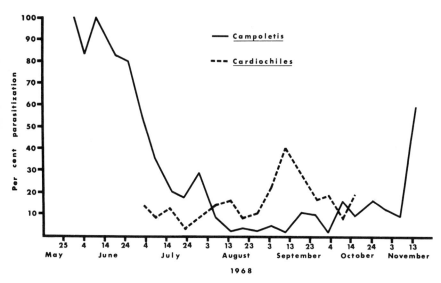

Fig. 1. Seasonal incidence of parasitization of *Heliothis virescens* (F.) by the larval parasites *Campoletis perdistinctus* (Vier.) and *Cardiochiles nigriceps* Vier. in tobacco fields. Clayton, North Carolina.

mortality of hornworm eggs and larvae in relation to specific factors. Data from his paper are shown graphically in Fig. 2 to illustrate the heavy mortality inflicted in the first and second generations by predators of the larvae. Much of this larval mortality was attributed to vespid wasps of the genus *Polistes,* including at least seven species in North Carolina. Rabb (1960) studied the prey complex of the three most common species, *P. annularis* (L.), *P. exclamans* Viereck, and *P. fuscatus* (F.), and found it to consist of free-feeding caterpillars of many kinds, including the tobacco hornworm. In their studies of wasp predation relative to the tobacco hornworm, Rabb and Lawson (1957) placed inverted boxes in the field as nesting sites, which after establishment of the colonies, could be moved to desired locations to augment the wasp's effectiveness. The closer the wasp colonies were located to tobacco the greater was the number of hornworms killed by the hunting workers. Lawson *et al.* (1961) found that the provision of nesting sites near tobacco fields resulted in a reduction in hornworms of about 60 per cent and a significant reduction in leaf loss. Much of the reduction in leaf loss was due to

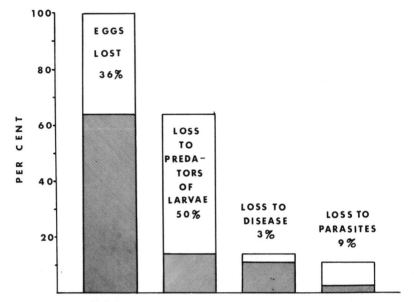

Fig. 2. Mortality and survival of first and second generation hornworms in North Carolina in 1955—mean of five locations (Lawson, 1959).

wasps killing hornworms in the early instars before the most destructive last instar was reached. While *Polistes* reduced leaf loss from the first two hornworm generations, there was little evidence that they greatly affected the production of diapausing pupae, the initial densities of the generations concerned or the year to year population trend.

ENEMIES REDUCE THE GENERAL PEST PROBLEM

Conversely to the preceding example, a natural enemy may be of little importance in reducing crop damage by the mortality it causes in a given pest generation, but it may, nevertheless, have a major influence on the mean level of pest abundance. *Apanteles congregatus* (Say), a small braconid parasite of the tobacco hornworm, is such an enemy. The adult female lays its eggs in early-instar hornworms, and the full-grown larvae emerge from the 4th and 5th instar hosts, spinning white cocoons which remain attached to the backs of their hosts, often until the adults emerge (Fulton, 1940).

Lawson (1959) found *A. congregatus* to be out of synchrony with the tobacco hornworm in early spring and to build up only slowly on this host. He showed that hornworm mortality caused by this parasite during the first two generations was very low as contrasted to that inflicted by the predatory *Polistes*. In addition, the full potential effect of *A. congregatus* was both masked and precluded in part by *Polistes,* since the predator consumed parasitized and non-parasitized hornworms indiscriminantly and consequently also denied the parasite its full numerical response potential. Thus *A. congregatus* has little direct effect on leaf damage by hornworms of the first two broods. It does have an important effect, however, on the mortality and survival of the third and partial fourth generations of hornworms, which give rise to the diapausing pupae. Data obtained in 1962 (Fig. 3) illustrate the seasonal pattern of parasitism by *A. congregatus*. After harvest, *Polistes* cease to rear brood, their colonies disband, and their predation falls sharply. On the other hand, the increased parasitization by *A. congregatus* more than compen-

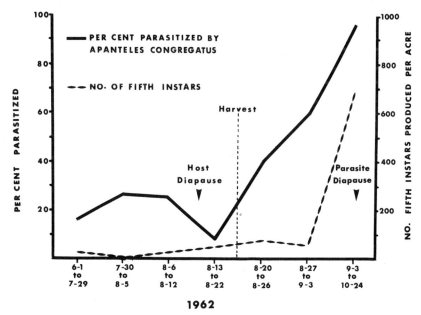

Fig. 3. Seasonal production and parasitization of 5th-instar hornworms in commercial tobacco in 1962—mean of 15 fields near Clayton, North Carolina.

sates for the decline in predation. The great numerical response by the parasite is possible because its diapause is initiated much later in the fall than that of the host (Rabb and Thurston, 1969). In addition, this time-lag effect of *A. congregatus* on the post-harvest populations of hornworms is accentuated by the practice of destroying stalks. Within a period of two to three weeks, the acreage of tobacco stalks and suckers is reduced by about 50 per cent. This reduction in host plant acreage effectively increases the ratio of emerging parasites to the acreage of tobacco in which the parasites must search to find hosts.

The significance of a natural enemy of a crop pest may be modified by a change in production practices. Such has occurred during the past 10 years regarding the enemies of the tobacco hornworm.

Maleic hydrazide (MH-30) was introduced for sucker control in the late 1950's, and since 1964 it has been used on virtually 80 to 95 per cent of the tobacco grown. The residual effect of this systemic chemical has drastically reduced sucker growth after harvest and thus the production of diapausing hornworms (Rabb *et al.,* 1964). Light trap records and larval counts indicate a 10- to 20-fold decrease in populations of *M. sexta* accompanying the change in sucker control practices (Rabb, 1969). Thus, chemical sucker control has accomplished, at least for the time being, the same objective of earlier entomologists (Metcalf, 1909) who recommended early stalk destruction as a means of reducing the hornworm problem.

Some entomologists had reservations about recommending early stalk destruction because they were concerned that reduction of the heavily parasitized post-harvest population of hornworms might in turn drastically reduce the winter carry-over of the parasites, and thus aggravate rather than alleviate the hornworm problem. There was no valid basis for this concern. The use of MH-30 has had the same effect as stalk destruction in reducing host material for parasites, but parasitization remains just as high if not higher than in the early 1950's (Table 2). Before the advent of chemical sucker control, hornworm populations were often above the economic threshold. Now, variations around a much lower equilibrium position are seldom high enough to justify temporary suppression methods. This analysis of the decline in hornworm numbers is of general significance in that it demonstrates the great importance of factors directly reducing the production and survival of diapausing individuals. The importance of *A. congregatus* in controlling the host population now seems to come into sharper focus economically, because without the heavy parasitization, even within the present much lower carrying capacity of tobacco land, hornworm populations again would generally exceed the economic threshold and would bring about expansion of insecticidal applications.

Table 2. Seasonal parasitization of tobacco hornworms by *Apanteles congregatus* (Say) and *Winthemia manducae* Sabrosky & DeLoach at Clayton, North Carolina in 1969.

Months	Per cent hornworms parasitized			
	Based on number hatching[†]		Based on weekly harvest of instars 3-5[‡]	
	Apanteles	*Winthemia*	*Apanteles*	*Winthemia*[§]
May–June	30.5	11.5		
July–Aug	31.0	5.0	80.1	10.8
Sept–Oct	42.5	2.5	97.6	1.2

[†]Each cohort = 100 newly-hatched larvae.
[‡]498 larvae harvested in July–August; 164 in September–October.
[§]Hosts parasitized by *Apanteles* as well as *Winthemia* are not included because they are inevitably killed by *Apanteles*.

In retrospect, the dramatic change in the equilibrium position of the hornworm population presented an unusual opportunity to study host-enemy density relationships. Preliminary investigations utilizing age-specific life tables indicate that larval predators such as *Polistes* are not inflicting as heavy a proportionate mortality in the first two generations as in previous years when populations were much higher. On the other hand, parasitization in the first two generations accounts for a much higher proportionate hornworm loss than in past years. Apparently, the *Polistes* predators have been diverted to other prey having densities higher than those of hornworms in tobacco. The association between the parasites and *M. sexta*, however, has become more pronounced with the decline in host numbers because the parasites do not have the flexibility in host relationships that is available to the *Polistes* wasps.

ENEMIES TRACTABLE TO MANIPULATION

The importance of a natural enemy may be related in part to the ease with which it can be manipulated in managing a pest population. Unfortunately, few native enemies have been the subject of serious manipulation attempts, thus little progress has been made in developing expertise in utilizing this approach. The probability of good judgments being made is proportional to the degree in which the ecology of the enemy and pest is understood.

The provision of nesting sites for *Polistes*, as mentioned earlier, is one example of modifying the habitat to enhance the beneficial effects of a native natural enemy. However, the manipulation of wasp colonies may be of little advantage in the control of many species on which they are known to prey. A brief mention of several factors to consider before utilizing this approach may be useful. Since these wasps attack principally free-feeding larvae, pest species whose larvae bore and tunnel in buds, stems or fruits are not susceptible to control by them. Rabb and Lawson (1957) observed that *Polistes* were not effective in controlling tobacco budworms when they were feeding within the terminal bud. Being general predators, the wasps tend to procure their prey from vegetation bearing the highest prey density, rather than concentrating on any particular prey species. Thus, the presence of alternate prey at relatively high densities can divert the wasps from the pest of concern. In addition, a pest whose depredations occur late in the season would not likely be amenable to control by *Polistes* because its predatory activity declines in late summer as colonies disband, and there is also a great abundance of alternate prey at that time. And finally, although all *Polistes* species are general predators, there are important differences among them in their predatory activities. For example, Rabb (1960) found that *P. annularis* hunted predominantly in wooded areas and thus killed fewer crop pests than the other species studied. Kirkton (1968) utilized artificial nesting sites near cotton fields in Arkansas in an attempt to increase *Polistes* predation on cotton bollworms. His failure to obtain control may be attributed to one or more of the factors mentioned above.

Another obvious approach is to rear enemies in large numbers and release them at appropriate times and places. (See Chapters 8, 10 and 16.) Jaynes and Bynum (1941) reported results of mass releases of a tiny egg parasite *Trichogramma minutum* Riley for control of the sugarcane borer, *Diatraea saccharalis* (F.), in Louisiana. These and many other early attempts to use this native parasite were disappointing, but currently, U.S. Department of Agriculture personnel are studying the potential of this approach from new perspectives, and new techniques may produce practical results (Knipling and McGuire, 1968; Lingren, 1969; and Parker, Chapter 16 herein). Encouraging results have also been reported by Ridgeway and Jones (1969) for inundative releases of a chrysopid predator for the control of cotton bollworms and tobacco budworms in cotton.

The modification of cropping practices to provide greater ecological diversity and critical resources for enemies is being studied in several U.S. Department of Agriculture and State laboratories in the east. The Tall Timbers Conference in Florida this year on "Ecological Control of Animals by Habitat Management" is serving to stimulate this work, and relevant reports may be found in the Conference's "Proceedings."

ENEMIES AS INDICES OF PEST POPULATION TRENDS

Success in predicting population trends of pests requires identifying and monitoring factors responsible for changes in pest populations. Morris (1959) and Varley and Gradwell (1960) refer to factors responsible for *changes* in population size as key factors. Such factors, however, are not to be confused with factors having key roles in the regulation of populations about the equilibrium position. Research, chiefly in Canada, has indicated that natural enemies are key factors for some but not all forest insect pests. Logically, this must also apply to the agro-ecosystems of eastern United States. Studies in this region, however, are too incomplete to be of value for either predicting changes in populations of pests in time or for assigning regulating roles to specific factors. In evaluating the importance of a natural enemy, however, it is relevant to note that a factor may be of relatively little control importance by the more obvious economic criteria but may be of great predictive value where changes in density are the main concern.

CRYPTIC ENEMIES OF IMPORTANCE

Even with criteria of importance clearly in mind, the significance of biotic agents in naturally-occurring biological control often remains unnoticed or unheralded for a variety of reasons.

In some cases, an important enemy may be classed as insignificant because the mortality it inflicts is small as contrasted to that from other factors. A recently described species of tachinid, *Winthemia manducae* Sabrosky and DeLoach (in press), which attacks large last-instar tobacco hornworms, is such an enemy. Lawson (1959) referred to this species as *Winthemia* n. sp. and presented data showing that tachinid parasitization during 1952-1956 never exceeded 3 per cent of the total generation mortality of the tobacco hornworm. DeLoach (1964) also found parasitization of hornworms by *W. manducae* to be low (5 to 12 per cent of large last-instar hornworms in 1962 and 1963). Although the generation mortality attributable to *W. manducae* is relatively small, it always occurs after most other mortality factors have completed their influence on the host generation and thus essentially all of this mortality is indispensable and non-compensated. In addition, observations indicate that parasitization by *W. manducae* has increased in recent years as the general population level of the host has declined. In some fields, *W. manducae* reduced the surviving population of hornworms over 50 per cent in 1969 (Table 3). A full appreciation of the importance of *W. manducae*, however, must include a consideration of its role

Table 3. Effect of *Winthemia manducae* Sabrosky & DeLoach and *Apanteles congregatus* (Say) on survival of hornworms in two fields of tobacco near Clayton, North Carolina in 1969.

| Field no. | Total no. instars 3-5 harvested | Number of hosts with | | | Total | % reduction in surviving hosts by *Winthemia* [†] |
		Apanteles alone	*Winthemia* alone	Both parasites		
D3	278	220	8	35	263	34.8
C9a	387	281	49	21	351	57.6

[†] Based on the number of 3-5 instars which escaped parasitization by *Apanteles*.

as a compensatory mortality factor as mentioned by DeLoach (1964). A part of this potential is represented in Table 3 by the figures on multiple parasitism, but the full potential cannot be evaluated unless mortality from other factors such as *A. congregatus* were removed. To some extent such removal does occur in late October and early November, for parasitization by *A. congregatus* falls sharply during October because its diapause is initiated in late September (Rabb and Thurston, 1969; and Fig. 3). Where fall conditions are mild, an appreciable number of hornworms may be produced after mortality from *A. congregatus* and *Polistes* wasps has fallen essentially to zero. This occurred in 1969 near Clayton in a field which produced an abundance of succulent ground suckers from root stocks remaining after the tobacco stalks were cut. Ten of 21 5th instar larvae collected in this field on October 24 bore tachinid eggs. Only the small 5th instars were not attacked by these flies, which parasitize the largest larvae preferentially. One large host bore over 100 tachinid eggs. Few, if any hornworms in this field reached the prepupal stage without being parasitized.

Among the most unheralded natural enemies are those associated with pest species inhabiting alternate host plants of little if any economic importance but whose action in those other habitats are important in keeping the pest species at low levels in the general environment. Such is the case with regard to egg parasites of the tobacco hornworm (Rabb and Bradley, 1968). Egg parasitism is rare on flue-cured tobacco because the tiny parasites cannot move about on the sticky surface of tobacco foliage. However, a segment of the general tobacco hornworm population occurs on alternate hosts such as Jimson-weed, horse-nettle, and tomato, where foliage is not sticky. Egg parasites, particularly *Telenomus sphingis* (Ashmead) and *Trichogramma minutum* Riley, are quite active on these alternate host plants and at times parasitize over 50 per cent of the hornworm eggs present. The role of parasitism of pests on alternate hosts in determining the pest population levels on our crops is poorly understood, particularly where the mobility of the pest is high, as in the case of the tobacco hornworm, but its potential importance could be great and should not be ignored.

Some important natural enemies go unnoticed because they are so effective. This is the case with natural enemies which keep populations of potential pests below economic levels. Unfortunately, certain of these potential pests have become actual pests because the use of broad spectrum insecticides and fungicides in certain situations has affected the natural enemy populations more adversely than those of the pests. Some previously unimportant plant-feeding mites have become pests in apple orchards (Swift, 1968; Huffaker *et al.* 1969; and unpublished reports of G. C. Rock at North Carolina State University) and citrus groves (Muma, 1958), at least in part because routine spray programs have disrupted the normal function of predators and disease agents. Another example involves the green peach aphid, *Myzus persicae* (Sulzer), on tobacco in North Carolina, where it is seldom a serious pest due to a combination of physical and biotic factors, including a number of predators and diseases. Lawson (1958) considered these predators of little significance in the natural control of this species on tobacco. Instead, he found that populations were drastically reduced by the effects of low temperature on the host plants and by a fungous disease in winter. In late spring, aphid populations increased rapidly and were not controlled by predators but crashed when temperatures rose to 90°F or above for several successive days. He postulated that a virus may have been involved in these crashes.

Aphids are seldom a problem during the heat of the summer, but some problems have been encountered following application of certain insecticides. Thurston (1965) experienced heavy infestations of aphids on tobacco following applications of carbaryl and stated that it was not known whether the increase was the result of some physiological effect on the aphid, or on the plant, or whether it was the result of its killing the predators and parasites. Data (Table 4) obtained by W. J. Mistric (North Carolina State University) substantiate Thurston's report and show a relatively heavy build-up of aphids in plots treated with carbaryl as contrasted to populations in check plots or in plots where methyl parathion was included. The suspected cause and effect relationship between predators and aphid populations is difficult to prove, but this case has much in common with other examples of pest outbreaks following the disruptive effects on natural enemies of insecticidal applications, and carbaryl is known to be highly toxic to many predators. These results

Table 4.   Response of green peach aphid populations
on tobacco to applications of certain insecticides.[†]

Insecticide	Lb./acre active ingred.	Aphids per 40 leaves 5 tests				
		1	2	3	4	5
Carbaryl	2.0	422	284	317	591	1026
Carbaryl + methyl parathion	2.0 0.5	1	1	1	3	31
Check	—	65	32	16	58	106

[†] Unpublished data supplied by Dr. W. J. Mistric. Treatments in each test replicated three times. August 20
through September 18, 1969.

suggest that predators may have an important role in preventing heavy aphid
infestations at certain times. During mid- and late-summer, perhaps their
effect, along with the effect of suboptimal temperature and food conditions,
is sufficient to maintain sub-economic aphid populations unless enemy disruption occurs from such factors as applications of certain insecticides.

Some predators are nocturnal, and many of the diurnal ones are highly
mobile and difficult to observe. In some cases an act of predation can occur
in a few seconds without leaving a clue as to the cause of mortality.
Consequently, current information on the identity and importance of specific
predators and their comparative roles relative to specific crop pests is very
incomplete. One of the most comprehensive studies of predators of a major
agricultural pest in the eastern states was that by W. H. Whitcomb and
colleagues who studied predation in Arkansas cotton fields (Whitcomb and
Bell, 1964; Whitcomb, 1967). They demonstrated that many cotton bollworm
eggs and larvae were destroyed by a large variety of predators (insects, mites,
and spiders), many of which had not been suspected of being involved. One
important aspect of this multiplicity of enemies seems to be the ability of the
total complex to compensate for variations in predation by individual species
of predators. This adds stability in the control experienced. Where broad
spectrum insecticides were used for boll weevil control, the number and
variety of predators was reduced and bollworm populations increased. Reports
on the role of ants and other predators in the natural control of sugarcane
borers, *Diatraea saccharalis* (F.), also serve to illustrate the importance of
easily overlooked enemies in pest management (Hensley *et al.* 1961; Negm and

Hensley, 1967, 1969).

The examples given here of naturally-occurring biological control are not unusual because all insect pests have an array of natural enemies, many of which are important by one or more of the criteria discussed. Many additional examples could have been cited, not only as related to agricultural crops but also as associated with pests of forests, livestock, and of medical importance. Although there is urgent need for additional research on our native biological control agents and for a more astute assessment of their importance, an objective review of the currently available information leads to the inescapable conclusion that native natural enemies comprise one of our most valuable resources in eastern United States. The protection and enhancement of this resource should be a major objective of those utilizing our farm lands, forests, swamps and marshes. Much greater benefits from these biotic agents can be realized if the kinds of agricultural chemicals are chosen more carefully and used more judiciously. In addition, more effort should be made to devise cropping systems and cultural practices more advantageous to the rich and varied natural enemy complex which is characteristic of the eastern United States. (See further Chapters 14, 17, and 19.)

ACKNOWLEDGMENTS

I should like to thank Drs. W. M. Brooks, W. J. Mistric, and H. H. Neunzig for allowing me to use some of their previously unpublished data. In addition, Drs. W. M. Brooks and F. E. Guthrie critically read the manuscript and made a number of helpful suggestions.

LITERATURE CITED

DeLoach, C. J. 1964. Biology, ecology and description of *Winthemia* n. sp. (Diptera: Tachinidae), a parasite of the tobacco hornworm. Ph.D. Thesis. North Carolina State Univ., Raleigh. Unpublished.

Fulton, B. B. 1940. The hornworm parasite, *Apanteles congregatus* Say and the hyperparasite, *Hypopteromalus tabacum* (Fitch). Ann. Entomol. Soc. Amer. 33: 231-244.

Hensley, S. D., W. H. Long, L. R. Roddy, W. J. McCormick, and E. J. Concienne. 1961. Effect of insecticides on the predaceous arthropod fauna of Louisiana sugarcane fields. J. Econ. Entomol. 54: 146-149.

Huffaker, C. B., M. van de Vrie, and J. A. McMurtry. 1969. The ecology of tetranychid mites and their natural control. Ann. Rev. Entomol. 14:'125-174.

Jaynes, H. A., and E. K. Bynum. 1941. Experiments with *Trichogramma minutum* as a control of the sugarcane borer in Louisiana. U.S. Dept. Agr. Tech. Bull. 743, 42 pp.

Kirkton, R. M. 1968. Building up desirable wasp populations. Arkansas Farm Res. 17: 8.

Knipling, E. F., and J. V. McGuire, Jr. 1968. Population models to appraise the limitations and potentialities of *Trichogramma* in managing post insect populations. U.S. Dept. Agr. Tech. Bull. 1387, 44 pp.

Lawson, F. R. 1958. Some features of the relation of insects to their ecosystems. Ecology 39: 515-521.

Lawson, F. R. 1959. The natural enemies of the hornworms on tobacco (Lepidoptera: Sphingidae). Ann. Entomol. Soc. Amer. 52: 741-755.

Lawson, F. R., and R. L. Rabb. 1964. Factors controlling hornworm damage to tobacco and methods of predicting outbreaks. Tobacco Sci. 8: 145-149.

Lawson, F. R., R. L. Rabb, F. E. Guthrie, and T. G. Bowery. 1961. Studies of an integrated control system for hornworms on tobacco. J. Econ. Entomol. 54: 93-97.

Lingren, P. D. 1969. Approaches to the management of *Heliothis* spp. in cotton with *Trichogramma* spp. Proc. Tall Timbers Conf. Ecol. Anim. Control by Habitat Mgmt., No. 1, Tallahassee, Fla. (1969): 207-217.

Metcalf, Z. P. 1909. Insect enemies to tobacco. North Carolina Dept. Agr. Spec. Bull., 72 pp.

Morris, R. F. 1959. Single-factor analysis in population dynamics. Ecology 40: 580-588.

Muma, M. H. 1958. Predators and parasites of citrus mites in Florida. Proc. X. Intern. Congr. Entomol., Montreal (1956) 2: 633-647.

Negm, A. A., and S. D. Hensley. 1967. The relationship of arthropod predators to crop damage inflicted by the sugarcane borer. J. Econ. Entomol. 60: 1503-1506.

Negm, A. A., and S. D. Hensley. 1969. Effect of insecticides on ant and spider populations in Lousiana sugarcane fields. J. Econ. Entomol. 62: 948-949.

Neunzig, H. H. (1969). The biology of the tobacco budworm and the corn earworm in North Carolina with particular reference to tobacco as a host. North Carolina Agr. Exp. Sta. Tech. Bull. No. 196: 1-76.

Rabb, R. L. 1960. Biological studies of *Polistes* in North Carolina (Hymenoptera: Vespidae). Ann. Entomol. Soc. Amer. 53: 111-121.

Rabb, R. L. 1966. Diapause in *Protoparce sexta* (Lepidoptera: Sphingidae). Ann. Entomol. Soc. Amer. 59: 160-165.

Rabb, R. L. 1969. Environmental manipulation influencing populations of tobacco hornworms. Proc. Tall Timbers Conf. Ecol. Anim. Control by Habitat Mgmt., No. 1, Tallahassee, Fla. (1969): 175-191.

Rabb, R. L., and F. R. Lawson. 1957. Some factors influencing the predation of *Polistes* wasps on the tobacco hornworm. J. Econ. Entomol. 50: 778-784.

Rabb, R. L., H. H. Neunzig, and H. V. Marshall, Jr. 1964. Effect of certain cultural practices on the abundance of tobacco hornworms, tobacco budworms, and corn earworms on tobacco after harvest. J. Econ. Entomol. 57: 791-792.

Rabb, R. L., and J. R. Bradley. 1968. The influence of host plants on parasitism of eggs of the tobacco hornworm. J. Econ. Entomol. 61: 1249-1252.

Rabb, R. L., and R. Thurston. 1969. Diapause in *Apanteles congregatus*. Ann. Entomol. Soc. Amer. 62: 125-128.

Ridgeway, R. L., and S. L. Jones. 1969. Inundative releases of *Chrysopa carnea* for control of *Heliothis*. J. Econ. Entomol. 62: 177-180.

Sabrosky, C. W., and C. J. DeLoach. In press. A new *Winthemia* (Diptera, Tachinidae) parasitic on the tobacco hornworm. Proc. Washington Entomol. Soc. (1970)

Swift, F. C. 1968. Population densities of the European red mite and the predacious mite

Typhlodromus (A.) fallacis on apple foliage following treatment with various insecticides. J. Econ. Entomol. 61: 1489-1491.

Thurston, R. 1965. Effect of insecticides on the green peach aphid, *Myzus persicae* (Sulzer), infesting burley tobacco. J. Econ. Entomol. 58: 1127-1130.

Varley, G. C., and G. R. Gradwell. 1960. Key factors in population studies. J. Animal Ecol. 29: 399-401.

Whitcomb, W. H. 1967. Field studies on predators of the second-instar bollworm, *Heliothis zea* (Boddie) (Lepidoptera: Noctuidae). J. Georgia Entomol. Soc. 2: 113-118.

Whitcomb, W. H., and K. Bell. 1964. Predacious insects, spiders, and mites of Arkansas cotton fields. Arkansas Agr. Expt. Sta. Bull. 690, 84 pp.

Chapter 13

CASES OF NATURALLY-OCCURRING BIOLOGICAL CONTROL IN CANADA

A. W. Macphee and C. R. MacLellan

*Research Station, Canada Department of Agriculture
Kentville, Nova Scotia*

INTRODUCTION

There are, in Canada, many documented reports of naturally-occurring biological control of arthropod pests. The examples we have chosen are a few recent cases from forest and orchard entomology. Changes in densities of apple pest populations, initiated through destruction of natural enemies by spray chemicals, have been under investigation in Nova Scotia since 1943, and of peach pests in Ontario since 1946; intensive ecological studies on forest pest populations began about the same time in New Brunswick. We have selected from the many examples available the native black-headed budworm, *Acleris variana* (Fern.), found in forests; and European red mite, *Panonychus ulmi* (Koch), a lecanium scale, *Lecanium coryli* (L.), and codling moth, *Laspeyresia* [=*Carpocapsa*] *pomonella* (L.), commonly found as pests in orchards. These orchard pests apparently have reached North America from Europe since the apple was introduced early in the 17th century. They are well established, widely distributed species.

Intensive studies have been carried out in Canada on the ecology of these pests, with particular emphasis on factors controlling or regulating their populations. In many of these cases, the factors bringing about and maintaining control are native natural enemies which have received close evaluation.

Many observed outbreaks of forest pests have resulted from the failure of native natural enemies and corrections have been sought through introduction of parasites or predators. A different kind of problem was posed in apple orchards, for here many species owe their status as serious pests to the interjection of crop production methods into a relatively stable natural plant/animal community. In the early years of apple growing most of the important pests of today were relatively scarce, presumably because they were more in balance with their natural enemy populations. Intensive crop production and introduction of spraying as a new factor in the environment altered the faunal relationships and fewer natural enemies survived. Research aimed at restoring these biotic control agents to their former or potential effectiveness has, therefore, had priority over the introduction of exotic parasites and predators. Employment of the latter approach has been limited to new serious introduced pests, such as the winter moth, *Operophtera brumata* (L.) (see Chapter 9).

Apple trees potentially carry a very rich and complex fauna. Cycles and fluctuations are the rule but the use of pesticides has accentuated the variability and has often adversely influenced the natural enemy complex. Any number of the many available predators or parasites may attack and control some pests, such as lecanium scale or phytophagous mites, but one rarely finds, for example, the same predator complex actively controlling an outbreak of European red mite in two different orchards. On the other hand, efficient natural enemies of oyster shell scale, *Lepidosaphes ulmi* (L.), are limited practically to two species, a parasitic aphelinid, *Aphytis mytilaspidis* LeB., and a predaceous hemisarcoptid mite, *Hemisarcoptes malus* (Shimer) (Lord, 1947).

CASE HISTORIES

The case histories of the four pest species, European red mite, a lecanium scale, codling moth, and the black-headed budworm, are dealt with under the respective sub-headings below.

European Red Mite

Phytophagous mites, and specifically the European red mite, *Panonychus ulmi* (Koch), have been held to low densities in a large number of instances in apple orchards in Nova Scotia by predaceous mites and insects. This has occurred on susceptible and resistant varieties alike and under a wide range of growing conditions. It has also occurred at both high and low fertility levels (Lord and Stewart, 1961). The composition of the predator complex may vary widely but it almost invariably includes several mirid species and one or more phytoseiid mite species. When populations of predators were limited to a single species in experimental orchards the P. *ulmi* population was contained, but it underwent large fluctuations (Sanford and Lord, 1962; Lord, Herbert and MacPhee, 1958). Under more normal conditions where the destruction of predators is less, as in most commercial orchards, the numerous predator species maintain mite populations at low steady densities (with fluctuations of small amplitude) and actively destroy incipient outbreaks when they occur.

Most of the predaceous species are polyphagous but many respond numerically, as well as functionally, to increases and decreases in numbers of prey mites. The interrelations between mites and polyphagous predators that attack them are, therefore, important in the dynamics of other pests on which these predators also feed.

Studies made on mites to date do not lend themselves to analyses of a life table nature, but judgments have been substantiated in a number of ways in the course of faunal studies by entomologists at the Agricultural Research Station in Kentville. The subsequent size of any mite population is predictable from information derived from measuring the extent of destruction by noxious sprays of the predator populations and the future course of the population can be forecast if proper materials are consistently used in the spray program. The data presented in Tables 1 and 2 give more formal evidence of the mechanism of maintenance of low densities of the red mite for a succession of years. The data in Table 1 embrace a ten-year period in an orchard where pest control was achieved through biological control and

Table 1.   Numbers of phytophagous mites and their predators in an experimental, integrated control orchard during a 10-year period.

Year	Average number of adults, nymphs and eggs per leaf			Insect predators Ave. no. per tray
	European red mite	Brown mite	Typhlodromid mites	
1950	75	0.5	1	64
1951	4	1	8	8
1952	0.5	2	1	38
1953	0.5	2	1	26
1954	0.5	1	1	14
1955	Trace	1	1	34
1956	Trace	1	2	12
1957	Trace	1	1	51
1958	Trace	1	1	65
1959	Trace	0.5	1	22

Table 2. Number of European red mites per cluster in a non-insecticide orchard - Sheffield orchard.[†]

Date	European red mites per cluster				
	1965	1966	1967	1968	1969
Delicious					
June 15	1.0	0.2	0.0	0.0	0.01
July 1	—	0.3	0.0	0.05	0.01
July 15	7.0	0.3	0.0	0.04	0.0
August 1	25.0	0.2	0.0	0.03	—
August 15	12.0	0.8	0.0	0.05	0.0
August 30	—	0.04	0.02	0.05	0.05
McIntosh					
June 15	0.1	0.01	0.01	0.0	0.0
July 1	—	0.02	0.0	0.02	0.02
July 15	0.8	0.07	0.0	0.0	0.01
August 1	1.3	0.02	0.0	0.01	—
August 15	0.5	0.05	0.03	0.04	0.0
August 30	—	0.0	0.02	0.02	0.01

[†]Data from unpublished records of a faunal study headed by F. T. Lord, Research Station, Canada Department of Agriculture, Kentville, Nova Scotia.

selective use of chemicals (Patterson, 1966). The high level of mites in the first year was the consequence of previous use of DDT. Predators soon overtook the mite population and thereafter were the main factor in preventing any subsequent increase. This result, following a change from broad-spectrum pesticides to more selective treatments, shows that indigenous predators can maintain a mite population at a low steady density. These relationships were further observed over a five-year period, beginning in 1965, in a young orchard planted in 1961 (Table 2). Here, a long-term study is being made of the total foliage fauna, including *P. ulmi*. The orchard, grown to commercial standards, has been treated with fungicides only and has not received insecticide or miticide treatments. The *P. ulmi* records, made on 24 trees of each of the varieties Delicious and McIntosh, clearly demonstrate that, if uninhibited, resident natural enemies in Nova Scotia can maintain mites on apple at a low steady density indefinitely and that the mechanism begins when the orchard is young. The only significant mite population was in the first year of the study and this occurred on a few trees in only one section of the orchard. As the predator population increased, the incipient outbreak collapsed.

A picture of the results of interactions of mites and predators in 46 commercial orchards under three treatment programs was obtained in 1954 from a survey of phytophagous mites, predaceous mites and predaceous insects (Fig. 1) (Patterson and MacLellan, 1954). No insecticides were used in 16 of the orchards, 20 received ryania as a selective control for codling moth and 10 were treated with DDT. The drastic reduction of predaceous species by DDT initiated increases in mite populations to damaging levels in the orchards so treated. Due to adequate predator activity, mites caused no problem in most orchards not treated with insecticides or ones treated with ryania. Ryania, used in one or two applications, kills some predators but enough survive to give effective mite control.

Putnam and Herne (1966) described the role of native natural enemies of *P. ulmi* on peaches. They concluded that populations of *P. ulmi* are maintained at low endemic levels by predators, chiefly *Typhlodromus caudiglans* Schuster, that subsist to a considerable extent on other sources of food. On the other hand, epidemics of *P. ulmi* are reduced largely by other predators, chiefly *Haplothrips faurei* Hood and *Stethorus punctillum* Weise,

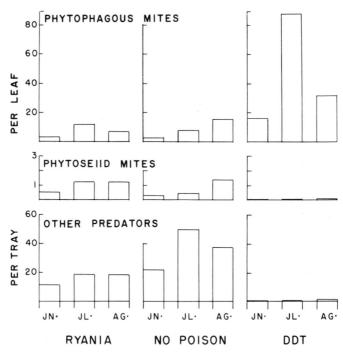

Fig. 1. Numbers of phytophagous and predaceous phytoseiid mites and other predators in commercial orchards under treatments of (1) ryania; (2) no insecticide; and (3) DDT in 1954.

that increase by feeding on the mite during its period of rapid population growth. After increasing in numbers they exert their greatest effect through destruction of winter eggs. *H. faurei* and *S. punctillum* have been observed to move into orchards in sufficient numbers to exercise some control on endemic mite populations. Also, *T. caudiglans* preys on higher densities of the mite though it appears to have an unexplained upper limit to its own density, and it does not destroy winter eggs.

Certain important differences occur in the mechanism regulating the density of *P. ulmi* in peach orchards from that in apple orchards. The predator complex differs, chiefly in the virtual absence of mirids on peach, due possibly to peach being an unsuitable host plant for the nymphs, or it may be unattractive to ovipositing females. The lack of pubescence on peach leaves, in contrast to apple, undoubtedly causes differences in the predator-prey relationship by affecting their behavior and motility (Putnam and Herne, 1966).

Lecanium Scale

Lecanium coryli (L.) is, potentially, a serious pest common to the apple growing areas of Nova Scotia but is not present in large numbers. Sporadic outbreaks occur only in localized orchards. When these occur they are invariably associated with the destruction of natural enemies by a toxicant that is innocuous to the scales but poisonous to the biotic agents. When balance is thus destroyed the scales increase with great rapidity, for although there is but one generation per year, each female may lay up to 500 or more eggs. The eggs hatch in July and the nymphs crawl to the leaves where they feed until October and then return to the bark of small twigs to spend the winter in an immature stage. The scales mature in the spring. Only the males are winged. There is one generation per year.

The most obvious and immediate damage is smutting of the fruit by a black fungus which grows on the honeydew produced by the scales, but heavy infestations will cause stunting and, eventually, death of tree branches.

The role of parasites and predators in the population dynamics of lecanium scale was investigated in a four-year study in a block of mature Spy trees during the period from July, 1963 to July, 1967 (McPhee and Phillips, unpublished data).

The first phase of the study was to reproduce the outbreak conditions postulated from observations of commercial orchards by applying a chemical spray destructive to natural enemies. For this purpose the orchard was treated twice with DDT, once in 1963 and again in 1964. The second phase was to allow the orchard to return to conditions suitable for natural control. Periodic observations and counts *in situ* and supplementary laboratory counts were made on 12 to 18 small limbs of the orchard.

The data for the four-year period are shown in Fig. 2. The scale density, which is not shown numerically in the figure, was light in the summer of 1963, moderate in 1964, sufficiently heavy to cause damage in 1965, and light in 1966, while in 1967 scales were difficult to find. In other words, the scale went through a cycle from low to high density and to low again in the four generations. The percentage survival of scales (black areas) based on egg populations each year under the natural biotic and prevailing weather conditions and the artificially modified chemical (spray) environment in the orchard, are given in Fig. 2. The predator and parasite populations (black

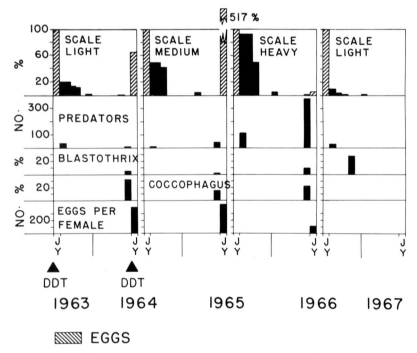

Fig. 2. Percentage survival of eggs (second cross-hatch) and nymphs (solid black) of lecanium scale from a basic egg population [adjusted annually (first cross-hatch) to 100 per cent]; numbers of predators per tray; percentage parasitism by *Blastothrix sericea* (Dalm.) and *Coccophagus* sp.; and number of eggs per female scale, for the years 1963-1967.

areas) were low the first two years when DDT was applied but relatively large in the next two when DDT was not used.

Blastothrix sericea (Dalm.), a common multiparasite of *L. coryli,* has two generations a year in Nova Scotia and has been observed to cause 30 to 40 per cent mortality of overwintering scale nymphs in orchards generally. When DDT was used in the experimental orchard in June, 1963 parasitism by *B. sericea* was not over 5 per cent by fall. Following the DDT treatment in 1964, this parasite was further reduced to about the 1 per cent attack level. Because no additional DDT was applied, mortalities by this parasite in the 1965-1966 and 1966-1967 generations were 10 per cent and 28 per cent, respectively, eventually contributing significantly to control. Unfortunately, the first spring generation of *B. sericea* develops too late to contribute to spring mortality of the scale; thus, parasitized scales are still feeding and causing damage. It is believed that the pesticide application in late June kills the parasite adults as they emerge from adult scales and thus reduces the summer generation of parasites and the consequent scale mortality. The data in Fig. 2 on *B. sericea* are from counts of nymphs attacked in the fall, supplemented by data on percentage of nymphs killed in the spring.

The parasite *Coccophagus* sp., also causes significant mortality of scale nymphs and this species was not appreciably suppressed by DDT applications. It may or may not be susceptible for the treatments were made after its emergence. The behavior and life history of this species are not well known but it is not well synchronized; there is a long gap between the time of its emergence as an adult and the period in which the nymphs of the next scale generation are available for attack.

Predator populations, chiefly mirids, were low in 1963 and 1964 and relatively high in the summer of 1965 and 1966. Some predation of scale nymphs was observed in 1965, but due to the high density of scales the percentage was not great. Circumstantial evidence suggests that predation may have been high in the spring of 1966. Firstly, the percentage of "under-developed" scales (i.e., there was no egg production) rose to nearly 50 per cent from the usual 20 per cent, and secondly, egg production fell from the usual 500 eggs per overwintered female to just over 100. The end result was a scale egg population trend index of 0.05 (95 per cent reduction) in the 1965-1966 generation contrasted to that of the previous generation, 1964-1965, which was 5.17 (five-fold increase).

Parasitism and predation combined nearly annihilated the 1966-1967 scale generation. There was no survival on the experimental limbs and very little on the trees generally. Since mortality from both predators and parasites was high in these last two years and since such agents possess compensatory abilities, it is not meaningful to estimate which enemy species, or combination of species, regulates scale numbers. Observations in commercial orchards do not resolve this difficulty since both parasites and predators have been causing significant mortality in all cases noted. It seems that the presence of both parasites and predators speeds up the control of outbreaks.

Codling Moth

The codling moth, *Laspeyresia pomonella* (L.), is one of the worst pests of apple in Nova Scotia despite the fact that it has but one generation a year. The larva lives only in the fruit, with relatively low numbers per tree causing economic loss. Consequently, the peaks of natural fluctuations of populations controlled by natural enemies may be serious. The maintenance of a stable, low endemic situation is the desired objective. However, in commercial situations selective chemical sprays are often necessary but only rarely must we resort to the last-ditch methods of using broad spectrum insecticides.

The mean population levels and the factors determining them were measured periodically during a 10-year study in an orchard relatively undisturbed by pesticides. The summarized results are shown in "life-table" formats in Table 3. There are two hazardous periods in the life of the codling moth: (1) there is a loss of almost 70 per cent in the period from egg deposition until the larva enters the fruit, and (2) during winter when the mature larvae are in cocoons on the tree trunks they may be significantly attacked by bird

Table 3. The mean population level of codling moth in life-table format from 10 years' data obtained in an experimental integrated control orchard.

Developmental stage	Number alive	Number killed	Cause of mortality
Eggs	100[†]	20	Predators, parasites
1st instar larvae	80	48	Predators, weather, other
Larvae in fruits	32	3	Predators, drowning
Larvae leaving fruits	29	4	Dispersal
		12	Woodpeckers
Wintering larvae	25	2	Parasites
		1	Other
Pupae	10		
Adults	10		
Females	5		
Eggs next generation	100		

[†]Approximate number per 700 fruit clusters.

predators and other natural enemies. A decrease in effectiveness of natural enemies in either period will cause a significant increase in codling moth within one generation. However, rarely do large changes in any one factor have any pronounced influence on the pest population for the year, for such are the potentials for compensation.

A detailed two-year study in an orchard under integrated control practices gave results that are typical for orchards with moderate predator populations (Table 4) (MacLellan, 1963). The 17 and 21 per cent losses of eggs in the two years were attributed to predation by mirids and to parasitism by *Trichogramma*. The predators were mostly of the species: *Hyaliodes harti* Knight; *Diaphnocoris pellucida* (Uhler); *Blepharidopterus angulatus* (Fall.); and *Phytocoris* spp. Data of Table 5 reveal that 80 and 84 per cent of the codling moths, respectively, were lost in the larval stages either because of predation,

Table 4. Fate of codling moth eggs in an experimental, integrated control orchard with moderate predator populations.

Year	Number of eggs found	Mortality caused by		Number of eggs hatched	Per cent mortality
		Predation[†]	Parasitism[‡]		
1960	1310	190	30	1090	17
1961	1460	290	15	1155	21

[†]Eggs preyed by: *Hyaliodes harti* Knight, *Diaphnocoris pellucida* (Uhler), *Blepharidopterus angulatus* (Fall.), *Phytocoris* spp.
[‡]Eggs parasitized by: *Trichogramma* sp.

Table 5. Fate of codling moth larvae which hatched from eggs in an experimental, integrated control orchard with moderate predator populations.

Year	Number of hatched eggs	Number of larvae				
		Killed or lost[†]	Establishing in fruits	Preyed on in fruits	Surviving in fruits	Per cent mortality
1960	1090	850	240	25	215	80
1961	1155	955	200	15	185	84

[†]Larvae preyed on by: *Anystis agilis* Banks, *Diaphnocoris pellucida* (Uhler), *Hyaliodes harti* Knight, and *Blepharidopterus angulatus* (Fall.).

failure to find fruits, unsuccessful fruit penetration, dispersal or adverse weather. The predominant larval predators were the mite *Anystis agilis* Banks and the mirids *D. pellucida*, *H. harti* and *B. angulatus.* Insect and mite predators are thus capable of effecting a high mortality of codling moth populations even though this action is rarely sufficient in itself to cause a significant decrease from high densities.

We call attention here to a challenging, and at the same time frustrating, aspect of this phase of control. Firstly, these important predaceous species exhibit no numerical response to codling moth density for other prey must form the major part of their food in the apple orchard or elsewhere. (See also Chapters 11 and 12.) Secondly, when pesticides are used against pests other than codling moth, these natural enemies may be reduced in numbers. Therefore a selective pesticide should be chosen and harmful effects minimized by use of optimum dosage and care in timing the applications.

The effectiveness of natural enemies on mature codling moth larvae in cocoons during the fall and winter was measured for 17 years by sampling a number of commercial orchards (MacLellan, 1958, 1966). Fig. 3 shows the density of codling moth larvae entering winter quarters per 10 ft^2 of tree trunk for 10 years of this study and Fig. 4 shows the percentage mortality attributed to bird predation (chiefly by woodpeckers), larval parasites, and

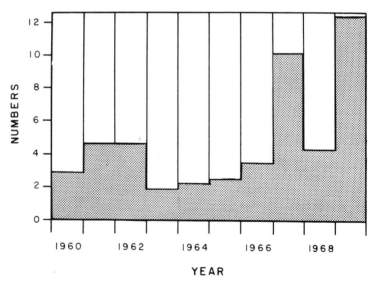

Fig. 3. Density of codling moth larvae entering winter quarters per 10 ft^2 of tree trunk in 16 commercial orchards.

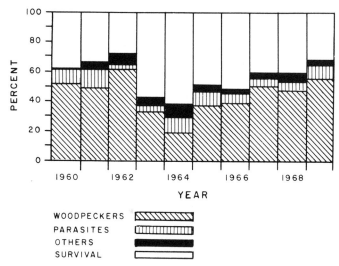

Fig. 4. Percentage mortality of codling moth wintering larvae attributed to various natural enemies in 16 commercial orchards.

other causes, including low temperatures, diseases, and predation by *Tenebroides corticalis* Melsh. Woodpecker predation appears to exhibit a somewhat density-related response but, more significantly, it averages about 50 per cent for most years.

The fecundity of codling moth appears to be lower in Nova Scotia than in some areas. Females lay about 20 eggs each. This low egg production, combined with the fact that there is only one generation a year, results in a slow population response. Neither of these factors, however, limits in any way the levels attainable over a few generations, as has been commonly observed. The slow response does, however, ameliorate problems in pest management and enhances biological control. Thus evidence that a combination of naturally-occurring enemy species suppress, and are capable of containing, codling moth is almost overwhelming. Aside from the use of insecticides, there is little to suggest that anything other than these natural enemies has any important bearing on codling moth density in Nova Scotia. In areas where winter temperatures go much lower, such as in New Brunswick, observations indicate that winter mortality may be an important factor in codling moth density and this modifies the problems of control.

Black-Headed Budworm

The black-headed budworm, *Acleris variana* (Fern.), a forest pest native to North America has been recorded in periodic outbreaks in eastern Canada since the late 1920's on its preferred host, balsam fir. As a rule, an outbreak causes severe defoliation for two years but there is no tree mortality and only a slight loss in radial increment.

The population dynamics of this species have not been worked out in detail, so no series of life tables are available (Miller, 1966). Miller, therefore, appraised the various factors by a process of exclusion in which he used miscellanous evidence, including survivorship curves and key-factor analysis (Morris, 1959) of the indicated important factors. A key-factor analysis of data for 17 years by Miller (1966) suggests that population release is associated with a number of years of low parasitism and favorable weather, while population decline is largely determined by late larval parasitism.

Four species of primary parasites, *Ascogaster* sp. nr. *provancheri* Prov., *Microgaster peroneae* Walley, *Phytodietus* sp. nr. *vulgaris* Cress., and *Actia diffidens* Curr., were shown to cause, collectively, the highest and most variable degree of mortality of late instar larvae. In the Green River area of New Brunswick, the four species showed a high percentage attack correlation with changes in population levels of the budworm. As shown in Figure 5 (Miller, 1966), the annual rate of population change was less than 1.0 (i.e., a decrease occurred) when parasitism exceeded 30 per cent. Parasitism estimates were as high as 97 per cent in one year and 88 per cent in another. At the highest rates of parasitism, the annual rate of change in density tended to flatten out at about 0.30. This suggests that as parasitism rises above 70 per cent there is a tendency for some compensating factor or factors to act at reduced intensity. Not all changes were accounted for by parasitism. There was a correlation between population change and weather with high degree-day years (above 1169 degree-days from June 1 to August 20) related to increases and low degree-day years related to decreases in population density. There were also a number of instances when high parasitism was associated with decreases in population density despite favorable weather. Conversely, there were instances when increases in population density were associated with low parasitism and unfavorable weather.

Populations showed an elliptical sequential pattern (Fig. 6, from Miller, 1966), especially from 1947 to 1959, when the population density in one generation was plotted against the density in the preceding generation. When the population density was plotted against the density surviving parasitism in the preceding generation (not illustrated here), the ellipse was flattened but not eliminated, indicating that some unknown factor was having an effect on budworm population regulation.

Miller therefore concluded that the initial release of black-headed budworm populations in the Green River area of New Brunswick was associated with low parasitism and favorable weather and that population declines during the two gradations studied were due to high parasitism during the large-larval age interval.

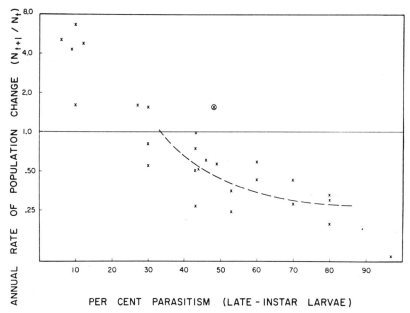

Fig. 5. The relationship between the annual rate of change of the black-headed budworm in the Green River area and percentage parasitism among late instar larvae; one generation N_{t+1}; previous generation N_t (after Miller, 1966).

SUMMARY AND CONCLUSIONS

The data presented show some examples of the effect of native biological control agents in the control of serious pests in Canada. These

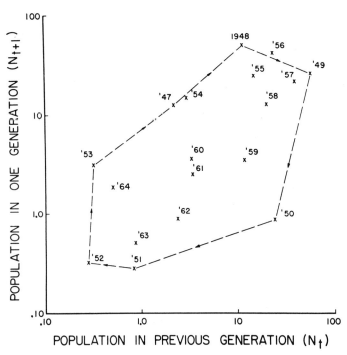

Fig. 6. A logarithmic relationship showing the density of black-headed bud-worm in one generation (N_{t+1}) as a function of density in the previous generation (N_t). Green River area, New Brunswick (after Miller, 1966).

examples represent only a small proportion of the pest species in orchards and forests. However, they illustrate the kind of problems involved and indicate the necessity of obtaining sound ecological data to design a faunal management program. In the Nova Scotia research program, a stepwise approach is followed. That is, each pest species is studied as a species ecosystem and the pieces fitted together later into a projected integrated program leading to later modification as a commercial, integrated control program. The tools of management of such a complex fauna are many, but in Nova Scotia orchards we have come to rely chiefly upon selective pesticides or broad spectrum pesticides rendered selective by timing and reduced dosage.

Pests which flare up only occasionally give us some trouble since their suppression is necessary and few selective pesticides are available. The resulting treatment usually disturbs the native natural enemies of other pests, so that these pests begin to increase and may require treatment by still additional pesticides. Until such time as selective pesticides are more readily obtainable, we must devise ways and means of applying broad spectrum pesticides in orchard ecosystems with as little interference as possible to the beneficial fauna (MacPhee and Sanford, 1956).

The report on scales and mites illustrates a type of potentially serious pests which are amenable to management since their biological controls are very effective and not easily disturbed.

The codling moth is in many ways a different kind of problem since the natural controls are not individually highly effective but collectively give good control. In this situation we have found that we must occasionally resort to a supportive control, such as a selective chemical which will not lessen biological control, to obtain good commercial management. Many orchards operate for years without any such supportive chemical control. There are other related pests which are managed in a similar way, but most of these require artificial control only rarely in time or place, and they are of less economic importance.

The effectiveness of native natural controls on forest pests has been established on a few species but, as in orchard pests, scant attention has been paid to pest species continuously at endemic levels. Extensive surveys of forest fauna, underway for some years, may lay the groundwork for clarifying this situation.

LITERATURE CITED

Lord, F. T. 1947. The influence of spray programs on the fauna of apple orchards in Nova Scotia: II. Oystershell scale, *Lepidosaphes ulmi* (L.). Can. Entomol. 81: 202-230.

Lord, F. T., H. J. Herbert, and A. W. MacPhee. 1958. The natural control of phytophagous mites on apple trees in Nova Scotia. Proc. X Intern. Congr. Entomol., Montreal (1956) 4: 617-622.

Lord, F. T., and D. K. R. Stewart. 1961. Effects of increasing the nitrogen level of apple leaves on mite and predator populations. Can. Entomol. 93: 924-927.

MacLellan, C. R. 1958. Role of woodpeckers in control of the codling moth in Nova Scotia. Can. Entomol. 90: 18-22.

MacLellan, C. R. 1963. Predator populations and predation on the codling moth in an integrated control orchard—1961. Mem. Entomol. Soc. Can. 32: 41-54.

MacLellan, C. R. 1966. Present status of codling moth populations. 103rd Rep. Nova Scotia Fruit Growers' Assoc. Pp. 77-83.

MacPhee, A. W., and K. H. Sanford. 1956. The influence of spray programs on the fauna of apple orchards in Nova Scotia. X. Supplement to VII. Effects on some beneficial arthropods. Can. Entomol. 88: 631-634.

Miller, C. A. 1966. The black-headed budworm in Eastern Canada. Can. Entomol. 98: 592-613.

Morris, R. F. 1959. Single-factor analysis in population dynamics. Ecology 40: 580-588.

Patterson, N. A. 1966. The influence of spray programs on the fauna of apple orchards in Nova Scotia. XVI. The long-term effect of mild pesticides on pests and their predators. J. Econ. Entomol. 59: 1430-1435.

Patterson, N. A., and C. R. MacLellan. 1954. Control of the codling moth and other orchard pests with ryania. 85th Rep. Entomol. Soc. Ont. Pp. 25-32.

Putman, W. L., and D. H. C. Herne. 1966. The role of predators and other biotic agents in regulating the population density of phytophagous mites in Ontario peach orchards. Can. Entomol. 98: 808-820.

Sanford, K. H., and F. T. Lord. 1962. The influence of spray programs on the fauna of apple orchards in Nova Scotia. XIII. Effects of perthane on predators. Can. Entomol. 84: 928-934.

SECTION IV

BIOLOGICAL CONTROL AS A KEY ELEMENT IN THE SYSTEMS APPROACH TO PEST CONTROL

Chapter 14

SYSTEMS ANALYSIS AND PEST MANAGEMENT

R. W. Stark* and Ray F. Smith

Department of Entomology and Parasitology
University of California, Berkeley

INTRODUCTION

Previous chapters have dealt with various aspects of biological control theory and practice and with natural control factors, particularly the role of predators and parasites. The following chapters continue the same theme but with greater emphasis on integrated control and the role biological control plays in the application of integrated control in pest management.

The intent in this chapter is to discuss the part systems analysis should play in this increasingly complex function of applied ecology in pest management. This may be the first treatment of systems analysis which omits formulae of any kind. Rather than reproduce complicated mathematical and statistical formulae (or even elementary ones) found elsewhere, we have chosen to discuss the basic assumptions of systems analysis, the techniques involved, the real and potential use, and the pitfalls. Included is advice (sound, we hope) on the input of the entomologist and ecologist, where they should withdraw and bow to the skills of the statistician and mathematician, and where the statistician and mathematician should be jettisoned.

THE SYSTEMS ANALYSIS APPROACH

The following general discussion of systems analysis is borrowed freely from many sources (Giese, 1966; Holling, 1966*a*, 1966*b*; Margalef, 1968; Oettinger, 1966; Southwood, 1966, 1969; Strackey, 1966; Sutherland, 1966)

*Now Dean of the Graduate School, University of Idaho, Moscow.

but predominantly the works of Watt (1962, 1966, 1968). Little attempt is made to separate their contributions since they now resemble a Christmas pudding—most of the ingredients are too well-blended to distinguish.

Systems analysis is a set of techniques and procedures, mathematical, statistical and mechanical, for analyzing complex systems such as agro-ecosystems. It has now been demonstrated that such techniques and procedures can and should be applied to the solution of large-scale problems such as ecological pest management.

The rationale for such optimism is that the objective of pest management, minimizing losses caused by a pest, maximizing productivity and minimizing environmental degradation has an impressive body of pure and applied mathematical theory and techniques by which optimal management can be recognized. This has long been suggested but it was not until the development of the larger electronic computers that an investigation could reasonably encompass the complex detail and problems of broad scope such as found in almost any ecosystem.

Systems analysis for pest management is based on ecological concepts which have long been recognized but which have had limited application because of the limited technological means at our disposal. These concepts can be summarized as follows:

a) Complex ecological processes can be separated into a large number of relatively small and simple components, although these subsystems or parts may still involve complexity and interaction with other components of the full system.

b) Historical processes involving time (past, present and future) can be dealt with or simulated in terms of existing recurrence formulae.

c) Complex multi-dimensional, interacting processes can be explained or simulated by existing formulae, or are amenable to mathematical description when (a) and (b) are known.

The major contributions of systems analysis to the application of these concepts to actual pest management practice depend almost wholly on the modern electronic computer and are in the area of data processing, the testing of complex concepts and the roles of factors. The insights gained are then used to design field research aimed at still further insight until the whole is understood. Lest this be construed as minimizing or belittling the concept of "systems analysis," we must add that the formalization of these procedures and techniques is having, and will have, a profound effect on ecological research and pest management practice.

Data Processing

Every ecologist knows of the vast accumulation of data necessary in ecological research. We hope all know the advantages of the modern computers in compiling, abstracting, storing and executing routine statistical procedures on such data. A simple comparison of one extensive study using a desk calculator, an IBM 1620 computer (now old fashioned) and an IBM 7094 will suffice (Giese, 1966). Using the desk calculator the analysis would involve 50-100 man-hours; using the 1620, only 30 minutes; and using the 7094, less than three minutes. (This does not, of course, include the many man-hours necessary to amass the informational input.) New computers are now being developed which further increase efficiency and multi-programming capability. These are machines which can perform more than one operation at a time.

Programs, written in appropriate machine languages, less difficult to learn than a foreign tongue, are available for most routine statistical procedures and expert programers are available either in large public institutions or as a private service. There are several "languages," according to the computer used, but there is now a "universal" computer language which can "translate" all previous machine languages.

The data processing function of systems analysis, incidentally, is of great potential use in the international aspects of biological control. Consider, for example, the world collection records and biologies of entomophagous and phytophagous organisms. A teletype-linked data storage center could provide on very short notice, the locations and limitations of any known parasite, predator, or disease pathogen as these become known. Also, programs could be written whereby the investigator describes the biology of his pest, the attributes of the biological control agent he desires, the use area (climate, altitude, floristic characters, et cetera), and the computer could name those biological control agents which most closely meet those requirements and where they could be found. This is an ideal goal for the future since at present we do not have the biological information, e.g., a high proportion of natural enemies are not even described.

Execution of Ecological Research

The capability of large, modern digital computers to incorporate large masses of data in their "memory" and to perform complicated functions at a high rate of speed permits, in principle, the ecologist to incorporate all the components, and their interactions, of an ecosystem and to subject them to analysis and experimentation as to their roles. This will be discussed in detail below.

The computer can also be a real contributor in the development of ecological theory. It is now possible to program a theory, e.g., as regards host-parasite relationships (Griffiths and Holling, 1969), explore the structure of the theory, confront it with field or experimental data and interpret the implications, then change the theory as indicated and reinterpret without leaving the console of the computer! However, lest we are accused of having adopted the "computer mystique" and attributing the human power of thought to this hardware, let it be understood that we subscribe to the following: "Computers are capable of profoundly affecting science by stretching human reason and intuition much as telescopes or microscopes extend human vision. The ultimate effects of this stretching will be as far-reaching as the invention of writing. Whether the product is truth or nonsense, however, will depend more on the user than the tool." (Oettinger, 1966).

Although the relief presented by the computer in assuming the onerous chore of data processing is great, the greatest contribution thus far is in this ability to construct and examine "sub-models," or "models" of parts, or of the whole ecosystem. There have been many works of this nature (Chapman, 1969; George, 1965; Morris, 1969; Mott, 1969; Southwood, 1966; Varley and Gradwell, 1968; Watt, 1961, 1962, 1963, 1969, to name a few) and we will only discuss this function briefly in the context of pest management.

As stated earlier, optimization of productivity or minimization of losses with minimal environmental degradation is the prime goal of pest management. Programs to attain this goal are often too complex to be worked out with a desk calculator. The natural regulatory factors, the complexities of their interactions and the many potential methods of artificial, biological, and cultural control present so many possible combinations that use of armchair logic alone, or together with the trial and error method, is totally impractical for spatial, temporal and economic reasons. [An example of how these complexities relate to pest management in an agro-ecosystem is given by Smith and Reynolds (1970).] Also, many of the possible combinations of control attempts might be severely disruptive ecologically or even totally destructive. Here is where model-building or simulation theory offers the greatest reward. The procedure goes something like this:

1) From field observations, laboratory experimentation and control pilot studies, a comprehensive list of the organisms and their interactions that seem to be of importance in regulating or affecting the pest population is compiled.

2) When a reasonable list of the regulatory mechanisms and other population influencing factors has been compiled, then each alone and in interaction with others must be measured as to their influences on the

population density of the pest. This implies and introduces experimental design, sampling theory and application so that limits may be expressed for the effects of each control or influencing factor. Of course, it must not be forgotten that imaginative, carefully-designed and executed experiments are an integral part of this process. Biological interactions of factors must be determined by biological experimentation. The nature of the interactions cannot be assumed from a computer treatment of extensive and complex multiple regressions and analyses of variance unless the factors have been tested biologically in various combinations, with single factors varied, pairs of factors varied together, et cetera. For example, the magnetic force in homing of pigeons was not suggested in experiments conducted on clear days, but on overcast days use of magnets on the wings disturbed the birds' orientation process. The relative importance of each factor must be determined in relation to the whole, which brings in extensive multiple regression, analysis of variance, et cetera—the calculations for which are too complex for all but high-speed digital computers.

3) When the workings of the existing pest ecosystem is reasonably well understood, it can then be described by a model which includes all the known, measured regulatory and influencing factors operating in a known, measured physical environment.

4) Lastly, having a descriptive model described in a computer program, one may, at will, change the various components of that system to determine the optimum way in which the pest density may be regulated or maintained at any level set by the simulator. The desired level might be a pest population below the economic level. The relative cost of alternative programs for achieving this level could be compared. In another situation, it might be useful to determine just how close to the desired population level we might reach by altering a single factor as others are held constant. The advantages of such simulation techniques in the use of integrated control in pest management are obvious. Again, this is the ideal goal, for, to our knowledge, no program of systems analysis has developed to this final stage for any pest management program.

It is not necessary to follow the four ideal steps outlined above which describe the procedure for systems analysis as well as that for model-building. A competent, intuitive ecologist can construct an abstract conceptual model based on processes such as reproduction, competition, predation, and parasitism and then compare his largely theoretical model with events in the real world. Such a model will at first be crude, but it will reveal gross features of the pest system under study. It will aid in determining what are the most valuable data which should be collected, which phenomena should be investigated more thoroughly and even dictate or suggest what experiments should be conducted to bring realism to the model. Most models which have been developed to date are inadequate in that they have neglected to incorporate adequately the movements of individuals or populations of individuals. This is a very important but little understood question. Not only are immigration and emigration involved, but also the inherent vagility of different species and the dispersion patterns of their populations. These are areas which the ecologist must explore thoroughly to estimate values for constants for incorporation in more realistic future models.

Regardless of which method is used to construct the model, interpretable results can be achieved by "manipulating" the factors of the simulated system, or even by posing appropriately formed questions to the machine. The answers given will either suffice or indicate what additional data are required. The questions asked are no different than those posed by the ecologist in the field (Solomon, 1964), e.g., in what way is the action of a given predator related to the density of a particular species?

The tremendous potential of computers in this work can also lead to serious pitfalls. Confronted by the awesome capacity of the computer, it is easy to forget that what is fed into the machine must be realistic, supported by real data; otherwise, the computer gives results without meaning. The output of intuitive models can be grossly misleading when applied to existing complex biological processes. For example, at the First International Symposium of Statistical Ecology, such a model was presented. It was developed to investigate a particular predator-prey relationship. The "machine prey" was assigned its niche and the "machine predator" was let loose against it, subject to various mathematical and statistical stipulations. The model worked beautifully, but not a single "prey" was caught! Without including knowledge of the real searching power of the predator and the dispersive or other protective behavior of the prey, based on field and laboratory studies, such "games" are meaningless. There is a basic computer warning which should be posted on every computer, i.e., "G.I.G.O.," garbage in, garbage out. There must be the most intimate mating of theory, observation and experiment in a real world (Smith, 1969).

Lastly, it must always be kept in mind that the computer is dependent on the outside world for communication. Real data and programs have to be put into the machine (input) before it can work and its answers (output) must then be interpreted and put to use. The ultimate in computer communication was recently recounted by Herb Caen, a San Francisco columnist. It seems the Pentagon had, over a lengthy period, programed a massive war game. The final question to be answered, given innumerable probabilities of international and domestic events, was whether the final outcome of the hypothetical war would be beneficial to the future of the United States. The answer, naturally, was classified so only the Commander-in-Chief of the Armed Services was present at the final print-out of the calculator. He, of course, had had little, if anything, to do with the program. The answer came

. . . . YES . . . .

"Yes, what?" cried the baffled President. After many electronic burps and groans came the read-out

. . . . YES *SIR* . . . .

SYSTEMS ANALYSIS AND PEST MANAGEMENT

Pest management may be defined as management of a negative resource, or perhaps most clearly, pest management is part of resource management. The resource we are managing may be fields of alfalfa or cotton, or forests of ponderosa pine, or the entire forest cover of a state or nation. We are concerned with reducing the losses inflicted by a pest or pests on the particular resource. To that end, we have developed and are developing control methods for our pest species. Various ways and means of biological and integrated control are the theme of this book and we need not review them here.

Integrated control, the integration and exploitation of all feasible methods which are practical and which will cause the least disruption of ecosystems, is now reasonably well documented (Geier, 1966; Smith and van den Bosch, 1967; Stern *et al.,* 1959; van den Bosch and Stern, 1962; Voute, 1964) and examples are discussed in other chapters of this book. In our view, there is no doubt that successful pest management of the future will depend on the use of integrated control. With perhaps a few exceptions it is becoming more and more difficult to think of a single effective, continuous control method for pests of a crop, especially if one views the crop as a whole (i.e., the agro-ecosystem) and the complex of pest species that attack the crop. Although many single pest species may be reliably controlled by a highly efficient parasite or predator, other pests will commonly occur on the same crop for which such reliable biological control agents are not at hand.

Methods and Procedures for Use in Pest Management

We discuss here the relationships of systems analysis to integrated control, which should further elucidate the "input" required of the ecologist, the role of the statistician, mathematician, and computer and the possible application of the "output" or results.

The methodology of resource management described by Watt (1968) can be transposed almost directly to pest management. It is exemplified here by an actual study (Stark, 1966; Stark and Dahlsten, eds., 1970) of the methods and procedures being developed for management of western pine beetle populations.

The western pine beetle, *Dendroctonus brevicomis* LeConte, is the most destructive insect enemy of ponderosa pine in the western United States and has been the subject of investigation for over 60 years. The sheer magnitude and complexity of the pine beetle ecosystem and the analytical techniques available restricted research to relatively small segments of the problem (Miller and Keen, 1960).

Application of the systems analysis approach with all it implies has greatly accelerated the pace and influenced the nature of these studies. It is now possible to develop a systems analysis that offers the likelihood of managing this major pest in an optimum way according to a long-term plan. The key evaluation studies, impossible without high speed computers, were those which permitted accurate evaluation of the extent of the problem and which provided techniques for measuring changes in insect abundance as a result of treatments. The six steps in the formulation of a management policy for this bark beetle are treated under the following topics. The last four of these steps represent our estimate of the future role of systems analysis in the management of western pine beetle populations.

1. *Determination of Pest Status.* The forest area under management is compartmentalized. Pertinent information, such as timber types present, age classes, volume, road systems and other features which affect management, is catalogued and an "information" program is established (Amidon, 1964). From this bank of data the value of any portion of any compartment can be calculated "on the stump" and the cost to harvest, process, and deliver, computed without cutting a single tree. The annual cut by species and volume can be determined quickly and incremented yearly, if the growth rates of each species on each site are known. Priorities can be assigned on the basis of stand maturity and market demands. "Machine management" of commercial forest properties is now a reality and eminently feasible using systems analysis with the aid of the computer.

The losses inflicted by unforeseeable events such as windthrow, fire and some insect damage can be calculated and their over-all impact on the management plan calculated. The feasibility of salvage and the changes in plans necessary to achieve salvage can be computed in the office and the decision made to proceed or write off the loss. Although much of the information still must be obtained on the ground, a very large portion of it can now be obtained from high altitude photography (Caylor and Thorley, 1970; Dolph and Wear, 1963; Langley, 1959; Pope and Wear, 1961; Stark, 1966).

The techniques for determining generation by generation tree kill by the western pine beetle and, hence, the capability of including this information in forest management planning have now been developed. Remote sensing techniques (Caylor and Thorley, 1970; Stark, 1966) now permit precise estimates of the number, location, size and volume of infested trees. This information, entered into the pool of "compartment" data, permits precise estimates of potential loss and the setting of "economic thresholds." The values recoverable through salvage can be weighed against the increased costs of logging, changes in cutting schedules, et cetera (on the computer), and a decision made whether to accept the loss or proceed with salvage.

This step in systems analysis has engaged the expertise of forester, statistician and entomologist. Decisions made at this level are only in terms of impact of the current beetle population on current forest crop. For pest management, we must determine what the future holds.

2. *Analysis of the System.* Determination of population changes requires precise sampling techniques for various stages of the insect, evaluation of the mortality factors and some measure of trend. This can be done in bark beetle work to some extent simply by comparisons of numbers of trees killed by each generation as described above. For some management problems this may be sufficient; if flexible enough, the forest manager can accept the cutting dictated by the movements of the insects. However, if the trees killed begin to exceed annual allowable (or planned) cut, then control measures may be indicated and these will be inefficient unless the ecology of the system is understood.

Precise sampling techniques have been developed for all stages of western pine beetle (except *in-flight* populations) (Berryman *et al.,* 1970; Berryman, 1970*a*; Dudley, 1969), and for parasitism (Dahlsten and Bushing, 1970), insectan predation (Berryman, 1970*b*) and avian predation (Otvos, 1970). These techniques and an increasing understanding of the interactions of the mortality complex (Berryman *et al.,* 1970) and population fluctuations (DeMars *et al.,* 1970) now permit accurate calculation of within-tree populations and, coupled with the infested-tree counts from remote sensing, the formulation of population trend curves. Thus, we are a short time away from a capability of short-term predictions essential to management of this insect's populations. Completion of these studies to their present level was possible only with the combined efforts of biologists and statisticians.

3. *Determination of The Most Important Factors.* In determining the most important factors, particularly in regulating pest density, but also in greatly affecting survival at a given time, the statistician and the computer again play significant roles. Experimental methods and techniques using regressions and analysis of variance will expose the most significant regulatory features (Varley and Gradwell, Chapter 4, this book; DeBach and Huffaker, Chapter 5). It is the ecologist who must interpret the output and apply it in the field. For example, knowledge that drought is the most important limiting factor, or the one having the greatest predictive value of western pine beetle fluctuations in most situations (Hall, 1958) is, however, of little value in management of this pest where periodic lack of rain is the norm. However, in most ponderosa pine areas, variations in this factor are certainly of value for predictive purposes and management must be altered accordingly. We must understand the hierarchy of density-influencing factors and their roles, i.e., whether truly regulatory or simply density-influencing. An ecologically-oriented pest management specialist is required to determine which of these may be most useful in pest management and are capable of manipulation (Smith, 1969).

For example, we now know that the most important mortality factors in western pine beetle populations at endemic levels are insect and avian predation (Berryman, 1970b; Otvos, 1970) and that it is possible to augment insect predator populations through cutting practices and mass rearing (Berryman, 1967). It may also be possible to augment avian predation through strategic placement of nesting boxes. Knowledge of the aggregating behavior of western pine beetle has led to manufacture of a synthetic lure which may be used as an applied control (Silverstein *et al.*, 1968; Bedard *et al.*, 1969).

More significantly, detailed ecological analysis of the western pine beetle ecosystem has shown that many of the "outbreaks" arise from poor silvicultural practices (Miller and Keen, 1960; Stark, 1966) and the presence of root diseases (Stark and Cobb, 1969). In the majority of stands, there are areas of recurrent infestations, the cause of which can now be determined. Such conditions as overstocking, low quality of site, and excessive organic debris from previous logging, are amenable to correction. Only intensive investigation by foresters, physiologists, pathologists and entomologists can determine the relative importance of these factors.

4. *Modelling The Pest Ecosystem.* The next step is to build the model by taking all variables shown to be important (or even thought to exist) and incorporating them into a system. We have not progressed to this step completely for the western pine beetle, but statisticians, ecologists, and programers are working to this end.

5. *Simulation.* When the systems model is developed it will then be possible to evaluate the consequences of changes in the components of the system. For example, knowing a predator's efficiency, we can simulate an "altered" predator density and determine the effect on western pine beetle emergence. Or, we can experimentally "thin" the forest to a stocking density more suited to the site and determine the "effect" on the vigor of the resultant stand. Or, knowing the drawing power of pheromones, we can "install" traps, "measure" the depletion of the *in-flight* population and determine the potential attacking populations. The possibilities are many and depend on the marriage of the statistician (including mathematician) and biologist. It is the biologist who must ask the relevant questions, the statistician who must incorporate them into the model.

6. *Field Testing, Strategy and Tactics Determination.* Lastly, the sole province of the ecologist is the verification of machine simulation by field testing and the determination of strategy and tactics. When dealing with a mature forest crop, the latitude for experimentation is much less than when dealing with an annual crop. On the other hand, the relatively slow growth of western pine beetle populations permits more time for implementation of reasonable treatments as deduced from ecological consideration of the system. The techniques described above (not yet fully modelled) now provide a basis for predicting population trend, and the biological studies show that the way to management of western pine beetle is in the area of proper silviculture, augmented by encouragement of natural controls. The danger of this approach lies in too great a faith in the computer. With regard to simulation modelling and determining control strategy and tactics, many assumptions are usually made. These assumptions must be tested critically by observation or experiment.

CONCLUDING REMARKS AND SUMMARY

What we have described is not in the category of Walter Mitty's dream world. It is here now. Although the roots of the concepts described here lie deep in scientific history, the flowers have but recently bloomed and few have been plucked. A glance at our scanty bibliography shows their recent origin. Many ecologists have shied away from use of the techniques for various reasons, the "mystique" of the computer set being but one.

However, it is becoming increasingly obvious that the unilateral approach of pesticide applicators must be curtailed and even classical biological control has its limitations. Unilateral solutions will not be widely acceptable or applicable in a world increasingly aware of ecological action and reaction.

The techniques of systems analysis in pest management are the tools of the forward-looking ecologist concerned with integrated control.

In summary, systems analysis involves a set of techniques and procedures, mathematical, statistical and mechanical, for analyzing the influence of factors and factor interactions of complex systems such as agro-ecosystems. Its tremendous potential in pest management rests on the development of high speed electronic computers which permit analysis and storage of vast amounts of data, interpretation of complex interacting processes and actual experimentation with simulated factors of the environment. A generalized methodology is envisaged which demonstrates the respective roles of the farmer or forester, ecologist, mathematician and statistician. It is emphasized that the decisions reached by analysis or simulation via computer technology should be the product of actual and representative biological data supplied by the biologist. The soundness of the decisions arrived at, or output, will depend upon the completeness and accuracy of their informational input. These decisions, even then, mean nothing until they are implemented in the field by an ecologically oriented pest management program.

LITERATURE CITED

Amidon, E. L. 1966. MIADS2...An alphanumeric map information assembly and display system for a large computer. U.S. Dept. Agr. For. Serv. Res. Paper PSW-38, 12 pp.

Bedard, W. D., P. E. Tilden, D. L. Wood, R. M. Silverstein, R. G. Brownlee, and J. O. Rodin. 1969. Western pine beetle: Field response to its sex pheromone and a synergistic host terpene, myrcene. Science 164: 1284-1285.

Berryman, A. A. 1967. Preservation and augmentation of insect predators of the western pine beetle. J. Forestry 65: 260-263.

Berryman, A. A. 1970a. Procedures employed in sampling the populations of insect predators attacking developing broods of the western pine beetle. Pp. 66-74 in Studies on the Population Dynamics of the Western Pine Beetle, Dendroctonus brevicomis LeConte (Coleoptera: Scolytidae), R. W Stark and D. L. Dahlsten (eds.). Univ. of Calif. Div. of Agr. Sci.

Berryman, A. A. 1970b. Evaluation of insect predators of the western pine beetle. Pp. 102-112 in Studies on the Population Dynamics of the Western Pine Beetle, Dendroctonus brevicomis LeConte (Coleoptera: Scoiytidae), R. W. Stark and D. L. Dahlsten (eds.). Univ. of Calif. Div. of Agr. Sci.

Berryman, A. A., C. J. DeMars, Jr., and R. W. Stark. 1970. The development of sampling methods for "within-tree" populations of the western pine beetle. Pp. 33-36 in Studies on the Population Dynamics of the Western Pine Beetle, Dendroctonus brevicomis LeConte (Coleoptera: Scolytidae), R. W. Stark and D. L. Dahlsten (eds.). Univ. of Calif. Div. of Agr. Sci.

Caylor, J. A., and G. A. Thorley. 1970. Sequential aerial photography as an aid in the evaluation of bark beetle population trends in west-side Sierra forests. Pp. 8-32 in Studies on the Population Dynamics of the Western Pine Beetle, Dendroctonus brevicomis LeConte (Coleoptera: Scolytidae), R. W. Stark and D. L. Dahlsten (eds.). Univ. of Calif. Div. of Agr. Sci.

Chapman, R. C. 1969. Modeling forest insect populations—the stochastic approach. Proc. Forest Insect Population Dynamics Workshop, New Haven, Connecticut (Jan. 23-27, 1967). U.S. Dept. Agr. For. Serv. Res. Paper NE-125. Pp. 73-87.

Dahlsten, D. L., and R. W. Bushing. 1970. Insect parasites of the western pine beetle. Pp. 113-118 in Studies on the Population Dynamics of the Western Pine Beetle, *Dendroctonus brevicomis* LeConte (Coleoptera: Scolytidae), R. W. Stark and D. L. Dahlsten (eds.). Univ. of Calif. Div. of Agr. Sci.

DeMars, C. J., Jr., D. L. Dahlsten and R. W. Stark. 1970. Survivorship curves for eight generations of the western pine beetle in California. Pp. 134-146 in Studies on the Population Dynamics of the Western Pine Beetle, *Dendroctonus brevicomis* LeConte (Coleoptera: Scolytidae), R. W. Stark and D. L. Dahlsten (eds.). Univ. of Calif. Div. of Agr. Sci.

Dolph, R. E., and J. F. Wear. 1963. A survey of western pine beetle damage on the Fremont National Forest using color photographs. U.S. Dept. Agr. For. Serv., Pac. NW Region Report. 18 pp.

Dudley, C. O. 1969. A sampling design for the egg and first instar larval populations of the western pine beetle, *Dendroctonus brevicomis* LeConte (Coleoptera: Scolytidae) in ponderosa pine. Univ. of Calif. M.Sc. Thesis. 115 pp. (Unpubl.)

Geier, P. W. 1966. Management in insect pests. Ann. Rev. Entomol. 11: 471-490.

George, F. H. 1965. Cybernetics and Biology. University Reviews in Biology. W. H. Freeman and Co., San Francisco. 130 pp.

Giese, R. L. 1966. From cradle to computer—the Columbian timber beetle. Proc. North Central Branch, Entomol. Soc. Amer. 21: 85-92.

Griffiths, K. J., and C. S. Holling. 1969. A competition submodel for parasites and predators. Can. Entomol. 101: 785-818.

Hall, R. C. 1958. Environmental factors associated with outbreaks by the western pine beetle and the California five-spined engraver beetle in California. Proc. X Intern. Congr. Entomol., Montreal (1956) 4: 341-347.

Holling, C. S. 1966a. The strategy of building models of complex ecological systems. Chap. 8. In Systems Analysis in Ecology, K. E. F. Watt (ed.). Academic Press, N. Y. 276 pp.

Holling, C. S. 1966b. The functional response of invertebrate predators to prey density. Mem. Entomol. Soc. Can. 48: 1-86.

Langley, P. G. 1959. Aerial photography as an aid in insect control in western pine and mixed conifer forests. J. Forestry 57: 169-172.

Margalef, R. 1968. Perspectives in Ecological Theory. Univ. of Chicago Press. 111 pp.

Miller, J. M., and F. P. Keen. 1960. Biology and control of the western pine beetle. U.S. Dept. Agr. For. Serv. Misc. Publ. 800, 381 pp.

Morris, R. F. 1969. Approaches to the study of population dynamics. Proc. Forest Insect Population Dynamics Workshop, New Haven, Connecticut (Jan. 23-27, 1967). U.S. Dept. Agr. For. Serv. Res. Paper NE-125. Pp. 9-28.

Mott, D. G. 1969. Dynamic models for population systems. Proc. Forest Insect Population Dynamics Workshop, New Haven, Connecticut (Jan. 23-27, 1967). U.S. Dept. Agr. For. Serv. Res. Paper NE-125. Pp. 53-72.

Oettinger, A. S. 1966. The uses of computers in Sciences. Sci. Amer. 215:161-172.

Otvos, I. S. 1970. Avian predators of the western pine beetle. Pp. 119-127 in Studies on the Population Dynamics of the Western Pine Beetle, *Dendroctonus brevicomis* LeConte (Coleoptera: Scolytidae), R. W. Stark and D. L. Dahlsten (eds.). Univ. of Calif. Div. of Agr. Sci.

Pope, R. B., and J. F. Wear. 1961. Results of a study to test the accuracy of large scale vertical and oblique photographs for the appraisal of ponderosa pine mortality. U.S. Dept. Agr. For. Serv., Pac. NW For. and Range Exp. Sta. Unpubl. Rept. 19 pp.

Silverstein, R. M., R. G. Brownlee, T. E. Bellas, D. L. Wood, and L. E. Browne. 1968. Brevicomin: Principal sex attractant in the frass of the female western pine beetle. Science 159: 889-891.

Smith, R. F. 1969. The new and the old in pest control. Proc. Acad. Nazion. dei Lincei, Rome (1968), 366(128): 21-30.

Smith, R. F., and H. T. Reynolds. 1970. Effects of manipulation of cotton agro-ecosystems on insect pest populations. In Ecology and International Development, M. T. Farvar and J. P. Milton (eds.). Natural History Press.

Smith, R. F., and R. van den Bosch. 1967. Integrated control. Chap. 9. In Pest Control: Biological, Physical and Selected Chemical Methods, W. W. Kilgore and R. L. Doutt (eds.). Academic Press, N. Y. 477 pp.

Solomon, M. E. 1964. Analysis of processes involved in the natural control of insects. Adv. Ecol. Res. 2: 1-58.

Southwood, T. R. E. 1966. Ecological Methods. Methuen & Co. Ltd., London. 391 pp.

Southwood, T. R. E. 1969. The abundance of animals. Inaugural Lecture, Imp. Coll., Univ. London, Sci. and Technol. 8: 1-16.

Stark, R. W. 1966. The organization and analytical procedures required by a large ecological systems study. Chap. 3. In Systems Analysis in Ecology, K. E. F. Watt (ed.). Academic Press, N. Y. 276 pp.

Stark, R. W., and F. W. Cobb, Jr. 1969. Smog injury, root diseases and bark beetle damage in ponderosa pine. Calif. Agr. 23(9): 13-15.

Stark, R. W. and D. L. Dahlsten (Editors). 1970. Studies on Population Dynamics of the Western Pine Beetle, Dendroctonus brevicomis LeConte. Unnumbered publication, Univ. of Calif. Div. of Agr. Sci.

Stern, V. M., R. F. Smith, and R. van den Bosch. 1959. The integration of chemical and biological control of the spotted alfalfa aphid. Pt. I. The integrated control concept. Hilgardia 29: 81-101.

Strackey, C. 1966. Systems Analysis and Programming. Sci. Amer. 215: 112-124.

Sutherland, I. E. 1966. Computer inputs and outputs. Sci. Amer. 215: 86-96.

van den Bosch, R., and V. M. Stern. 1962. The integration of chemical and biological control of arthropod pests. Ann. Rev. Entomol. 7: 367-386.

Varley, G. C., and G. R. Gradwell. 1968. Population models for the winter moth. Pp. 132-142. In Insect Abundance, T. R. E. Southwood (ed.). Symposia Roy. Entomol. Soc. London, No. 4.

Voute, A. D. 1964. Harmonious control of forest insects. Intern. Rev. Forestry Res. 1: 326-383.

Watt, K. E. F. 1961. Mathematical models for use in insect pest control. Can. Entomol. 93, Suppl. 19. 62 pp.

Watt, K. E. F. 1962. Use of mathematics in population ecology. Ann. Rev. Entomol. 7: 243-260.

Watt, K. E. F. 1963. Mathematical population models for five agricultural crop pests. Mem. Entomol. Soc. Can. 32: 83-91.

Watt, K. E. F. 1966. The nature of systems analysis. Chap. 1. In Systems Analysis in Ecology, K. E. F. Watt (ed.). Academic Press, N. Y. 276 pp.

Watt, K. E. F. 1968. Ecology and Resource Management—A Quantitative Approach. McGraw-Hill, N. Y. 450 pp.

Watt, K. E. F. 1969. Methods of developing large-scale systems models. Proc. Forest Insect Population Dynamics Workshop, New Haven, Connecticut (Jan. 23-27, 1967). U.S. Dept. Agr. For. Serv. Res. Paper, NE-125. Pp. 35-51.

Chapter 15

MICROBIAL CONTROL AS A TOOL IN INTEGRATED CONTROL PROGRAMS

L. A. Falcon

Department of Entomology and Parasitology
University of California, Berkeley

INTRODUCTION

Microbial control includes all aspects of the utilization of microorganisms or their by-products in the control of pest species. Applied to insect populations, many of the principles governing the role of insect parasites and predators to control insects also apply to entomogenous microorganisms. In nature, viruses, bacteria, protozoa, fungi, rickettsiae and nematodes, perform important roles in the dynamics and natural regulation of insect and mite populations. Their effect is most evident in epizootics of disease which occur at intervals and under some circumstances may decimate the host populations. Less obvious is the effect of insect pathogens in the enzootic stage or those which cause chronic or low grade infections. Other than causing outright death, pathogens may interfere with insect development, alter reproduction, lower insect resistance to attack by parasites, predators, and other pathogens and influence the susceptibility of insects to control by chemical insecticides or other artificial methods.

The major emphasis in the application of microbial control has been to field-collect or artificially mass culture a specific insect pathogen and disseminate it when the host is most susceptible to its effect. One approach is to introduce and colonize pathogens as a permanent mortality factor in the host population. The success of this method has been demonstrated for: (1) the milky spore disease bacteria (*Bacillus popilliae* Dutky and *B. lentimorbus* Dutky) to control the Japanese beetle, *Popillia japonica* Newman, in the

United States (White and Dutky, 1940); (2) the accidental introduction and establishment of a polyhedrosis virus which led to control of the devastating European spruce sawfly, *Diprion hercyniae* (Hartig), in Canada (Balch and Bird, 1944); and (3) the importation and successful dissemination of a polyhedrosis virus for control of the European pine sawfly, *Neodiprion sertifer* (Geoffroy), also in Canada (Bird, 1953).

Another microbial control technique is to make repeated applications of a pathogen for temporary suppression of an insect pest. The best illustration of this is the development and use of the bacterium, *Bacillus thuringiensis* Berliner. This bacterium is mass produced by fermentation and formulated as a dust, wettable powder, or emulsion. The material is handled and applied in a manner similar to that for chemical insecticides. It is short-lived and several applications are often required.

Commercial preparations containing *B. thuringiensis* first became available in the United States in 1958. Since then, the bacterium has been produced by six companies in this country and at least six in other areas of the world (Table 1). *B. thuringiensis* is the only microbial control agent registered for use on food crops in the United States. Current registrations

Table 1. Insect pathogens developed by industry and various agencies in recent years.

Group	Pathogen	Product name	Source
Bacteria	*Bacillus lentimorbus* Dutky	Japidemic	Ditman Corp., USA
	B. popilliae Dutky	Doom	Fairfax Biological Labs.,USA
	B. sphaericus Neide	—	International Minerals Chemical Corp. (IMC), USA
	B. thuringiensis Berliner	Agritrol®	Merck and Co., USA
		Bakthane® L69	Rohm and Haas Co., USA
		Bactospeine	Pechiney Progil Lab. Roger Bellon, France
		Bathurin	Chemapol, Biokrma, Czechoslovakia
		Biospor 2802	Farbwerke Hoechst, Germany
		Biotrol® BTB	Nutrilite Products, Inc.,USA
		Dendrobacillin	Moskovs. zavod. bakt. prepavatov. atov., USSR
		Entobakterin 3	All-Union Institute Plant Protection, USSR
		HD-1 (Experimental)	Abbott Laboratories, USA
		Parasporin®	Grain Processing Corp., USA
		Sporeine	Laboratoire L.I.B.E.C., France
		Thuricide®	IMC, USA
Fungi	*Beauveria bassiana* (Balsamo) Vuillemin	—	IMC, USA
		Biotrol FBB	Nutrilite Products, Inc., USA
	Metarrhizium anisopliae (Metch)	—	IMC, USA
Nematodes	DD - 136	Biotrol NCS	Nutrilite Products, Inc., USA
Polyhedrosis viruses	*Heliothis*	Biotrol VHZ	Nutrilite Products, Inc., USA
		Viron/H ®	IMC, USA
	Neodiprion	Polyvirocide	Indiana Farm Bureau Co-op Assoc., USA
	Prodenia	Biotrol VPO	Nutrilite Products, Inc., USA
	Spodoptera	—	IMC, USA
	Trichoplusia	Biotrol VSE	Nutrilite Products, Inc., USA
		—	Biological Control Supplies, USA
		—	IMC, USA
		Biotrol VTN	Nutrilite Products, Inc., USA

permit its application on more than 20 agricultural crops, in addition to shade trees, ornamentals and in forests, for the control of at least 22 insect pests (Table 2). Its use has increased markedly in recent years primarily because it is exempt from residue tolerances and can be applied for pest control right up to the harvest of a food crop.

Several insect viruses, including the respective nuclear polyhedrosis viruses of the corn earworm, *Heliothis zea* (Boddie), the cabbage looper, *Trichoplusia ni* (Hubner), and the beet armyworm, *Spodoptera exigua* (Hubner), are under experimental development by the same firms which produce *B. thuringiensis* in the United States.

Aside from the commercially developed products, field collected insect viruses have also been used successfully for temporary suppression of several insect pests, including the alfalfa caterpillar, *Colias eurytheme* Boisduval (Steinhaus, 1951), the cabbage looper and the beet armyworm (Falcon, unpublished).

Most of the data on safety and the utility for insect pathogens has been obtained for *B. thuringiensis* and the nuclear polyhedrosis virus of *H. zea,* with a scattering of information available for other insect pathogens. Their

Table 2. Some registered uses for *B. thuringiensis* products in the United States.

Pest		Crop
Vegetable and Field Crops		
Alfalfa caterpillar	*Colias eurytheme* Boisduval	Alfalfa
Artichoke plume moth	*Platyptilia carduidactyla* (Riley)	Artichokes
Bollworm	*Heliothis zea* (Boddie)	Cotton
Cabbage looper	*Trichoplusia ni* (Hübner)	Beans, broccoli, cabbage, cauliflower, celery, collards, cotton, cucumbers, kale, lettuce, melons, potatoes, spinach, tobacco
Diamondback moth	*Plutella maculipennis* (Curtis)	Cabbage
European corn borer	*Ostrinia nubilalis* (Hübner)	Sweet corn
Imported cabbageworm	*Pieris rapae* (L.)	Broccoli, cabbage, cauliflower, collards, kale
Tobacco budworm	*Heliothis virescens* F.	Tobacco
Tobacco hornworm	*Manduca sexta* (Johannson)	Tobacco
Tomato hornworm	*Manduca quinquemaculata* (Haworth)	Tomatoes
Fruit Crops		
Fruit-tree leaf roller	*Archips argyrospilus* (Walker)	Oranges
Orange dog	*Papilio cresphontes* Cramer	Oranges
Grape leaf folder	*Desmia funeralis* (Hübner)	Grapes
Forests, Shade Trees, Ornamentals		
California oakworm	*Phryganidia californica* Packard	
Fall webworm	*Hyphantria cunea* (Drury)	
Fall cankerworm	*Alsophila pometaria* (Harris)	
Great Basin tent caterpillar	*Malacosoma fragile* (Stretch)	
Gypsy moth	*Porthetria dispar* (L.)	
Linden looper	*Erannis tiliaria* (Harris)	
Salt Marsh caterpillar	*Estigmene acrea* (Drury)	
Spring cankerworm	*Paleacrita vernata* (Peck)	
Winter moth	*Operophtera brumata* (L.)	

From information supplied, in part, through the kindness of International Minerals and Chemical Corp., Libertyville, Ill. and Nutrilite Products Inc., Buena Park, Calif., USA specificity is restricted to the class Insecta; the nuclear polyhedrosis viruses are generic- or species-specific (Ignoffo, 1968), while *B. thuringiensis* is a "broad spectrum" microbial agent, reportedly infecting more than 137 insect species from the orders Lepidoptera, Hymenoptera, Diptera and Coleoptera (Heimpel, 1967). These pathogens are highly compatible with other forms of control and can be used concurrently with most chemical insecticides as well as autocidal, attractant and behavioral manipulation control methods.

For additional information on the development and use of microbial agents for insect control, the reader is referred to Steinhaus (1963, 1964), Hall (1964), Martignoni (1964), Tanada (1964a, 1967), Anonymous (1969), and Burges and Hussey (in press).

USE OF MICROBIAL CONTROL
IN INTEGRATED CONTROL PROGRAMS

Major Considerations

Many factors can influence effectiveness of microbial agents and ultimate success is dependent on satisfying four basic requirements:

(1) Detailed knowledge of the biology, ecology, phenology and behavior of the target insect is necessary to determine when and where to apply a pathogen for maximum effectiveness.

(2) The pathogen chosen must be safe to use, easy to handle, reasonably selective and sufficiently virulent to effectively control its host.

(3) The method of application must result in a persistent, evenly distributed lethal deposit of the pathogen which will provide adequate suppression of the pest.

(4) The benefits derived must justify the use of a microbial agent.

In applying a pathogen, both physical and biotic factors of the environment may affect its virulence, stability, persistence, dispersal and transmission, as well as the resistance and susceptibility of the target pest, and the interaction between pathogen and pest. Among the physical factors of the environment, sunlight may deactivate the pathogen; temperature and humidity can influence the virulence, infectivity and pathogenicity of the pathogen; and the effects of rain and wind may aid in spreading a pathogen or, conversely, may dilute its concentration in the target area. For biotic factors, the morphological characteristics of the plant and the density of the foliage may influence wetting, adhesion and coverage by an artificially disseminated pathogen; rapid plant growth may dilute deposits of a pathogen; plant produced substances or the pH of deposits of the plant surface may deactivate a pathogen; and the nutritional value, pH of the plant tissue and plant-produced substances may alter the susceptibility of the insect host to a pathogen. In addition, artificially disseminated pathogens may interfere with or supplement the control by other organisms such as parasites, predators and naturally-occurring pathogens. Conversely, parasites and predators may contribute to dispersal and transmission of a pathogen.

Selection of the pathogen, method of propagation, formulation, and use of adjuvants (stickers, spreaders, wetting agents, et cetera) can influence the effectiveness of a microbial agent. Standardization of preparations containing insect pathogens is difficult and potency can vary from batch to batch. The procedures and equipment employed for dissemination of a pathogen will affect coverage, adhesion and drift. Chemical pesticides may affect a pathogen directly if applied together, or indirectly if the pathogen is applied over existing chemical deposits or if chemicals are applied over existing deposits of a pathogen.

The benefits derived from using a pathogen must exceed the initial investment. This is often determined solely on the basis of cost of application and degree of crop protection or yields obtained. Additional aspects which should always be considered are the impact of the pathogen on the abundance and vigor of subsequent life stages and generations of the pest, the value of preserving beneficial organisms which may prevent resurgence of the target pest or outbreaks of other pests, the lessened contribution to environmental pollution, and the being able to use the pathogen up to harvest of a crop.

For additional information on factors influencing the natural and artificial dispersion of pathogens and their use in integrated control the reader is referred to Steinhaus (1954), Cameron (1963, 1969), Hurpin (1966), van der Laan (1967) and Burges and Hussey (in press).

Examples of Practical Application

Although the literature contains many reports of how microbial pathogens may be employed for integrated control, they have not been used extensively in this manner. One reason may be that *B. thuringiensis* is the only microbial agent generally available for use in insect control. Another is the shortage of trained and interested researchers willing to investigate the use of insect pathogens under field conditions, and lastly, the actual development of practical integrated control programs has been slow in coming.

Most demonstrations and actual use of microbial agents in integrated control programs have involved the bacterium, *B. thuringiensis*. Steinhaus

(1951) was one of the first to successfully demonstrate its use by employing the bacterium alone or in conjunction with a nuclear polyhedrosis virus for control of the alfalfa caterpillar, *Colias eurytheme*. This bacterium has been used extensively both alone and in combination with chemical insecticides for the control of cabbage looper in crucifer and lettuce crops (Hall, 1968). More recently, Jensen (1966, 1969) demonstrated its usefulness as a tool in the integrated control of grape pests. Good control of the grape leaf folder, *Desmia funeralis* (Hubner) was obtained. Gentry *et al.* (1969) used *B. thuringiensis* to aid in reducing populations of the tobacco budworm, *Heliothis virescens* (F.), and cabbage looper in an integrated control study for tobacco pests. In preliminary investigations on the integrated control of the pest complex on cowpea in Nigeria, Taylor (1969) achieved superior control of the lepidopteran *Maruca testulis* Geyer with three applications of *B. thuringiensis* followed by three applications of the chemical insecticide, benzene hexachloride (BHC), compared to six applications of BHC used alone. In cotton, Bullock and Dulmage (1969) reported that early season applications of *B. thuringiensis* applied for control of the pink bollworm, *Pectinophora gossypiella* (Saunders), may reduce populations low enough so that other techniques, such as sterile-male releases may be used more effectively. Examples of the use of insect viruses in integrated control programs include the work by Allen *et al.* (1966) on the integration of the nuclear polyhedrosis virus of *H. zea* into a biological control program on cotton. The nuclear polyhedrosis virus of cabbage looper has been used for the control of this pest on several crops in California (Falcon, unpublished).

MICROBIAL CONTROL RESEARCH IN CALIFORNIA

Considerable emphasis has been given in California to the development of microbial control as a tool in the integrated control arsenal. Certain aspects of this program will be reviewed in an effort to demonstrate approaches taken and results obtained, as well as the complexities and problems involved in developing such a program. Research has been conducted mainly in two areas, (1) on cotton, wherein several insect pathogens are being investigated as part of a developing integrated control program, and (2) with a granulosis virus as a selective tool to aid in the control of the codling moth, *Laspeyresia pomonella* (L.).

Cotton Insects

In the San Joaquin Valley, cotton is attacked by a variety of insect pests including the mirid, *Lygus hesperus* (Knight), the bollworm, *H. zea,* the cabbage looper, and the beet armyworm. The major effect of the microbial control research program has been to evaluate and develop the use of *B. thuringiensis,* and the respective nuclear polyhedrosis viruses of bollworm, cabbage looper and beet armyworm, as selective control agents for use in integrated control. A team approach involving several disciplines has been employed and extensive information developed. The over-all aspects of this program are discussed in Chapter 17, so I will restrict my comments to the microbial control portion of the program. However, some duplication and repetition is unavoidable.

Nuclear polyhedrosis viruses of cabbage looper and beet armyworm occur naturally in the cotton growing areas and are important in the dynamics and population regulation of their respective hosts. Populations of both species are periodically greatly reduced by virus-caused epizootics. During four years of field studies conducted in the southwest area of the San Joaquin Valley, each species attained peak population densities in alternate years, i.e., the cabbage looper in 1966 and 1968, and the beet armyworm in 1967 and 1969. Upon reaching these high densities virus epizootics occurred in each species and greatly reduced their numbers. Unfortunately, this happened too late in the year to protect cotton from extensive feeding by the larvae.

It has long been known that cabbage looper virus can be collected in the field and applied during the early stages of population increase to prevent major outbreaks. On cotton in California, both cabbage looper and beet armyworm viruses have been used successfully in this manner. In recent years, industrial firms have produced both cabbage looper and beet armyworm viruses on living larvae grown under laboratory conditions. Limited field trials have shown these preparations to be as effective as field collected viruses.

Of the two viruses, cabbage looper virus appears to be the most effective under field conditions. One application with the amount of virus produced by 10 mature virus-killed larvae (10 larval equivalents) applied per acre during the population growth phase will destroy an infestation in a matter of days. In certain situations the virus has appeared to be self-perpetuating and one application provided adequate suppression of the population for the remainder of the growing season. With the beet armyworm virus, repeated applications using 25 larval equivalents per acre, have provided effective suppression of larval populations in cotton.

Larvae of both species feed in aggregations on open foliage, a behavior pattern which favors ingestion and rapid spread of virus. Consequently, the viruses can be applied with standard application equipment without elaborate preparations. This factor, along with the natural availability of the viruses, makes them inexpensive and easy to use. As the cabbage looper and beet armyworm have become more difficult to kill with chemical insecticides, use of field collected viruses has increased on cotton and other crops in California.

In contrast to the cabbage looper and beet armyworm viruses, the nuclear polyhedrosis virus of *Heliothis zea* has not been found to be of major importance as a naturally-occurring mortality agent for the bollworm in California. To obtain sufficient quantities for field use, the virus must be propagated in the living host under artificial conditions. In the mid-1960's unrefined preparations of bollworm virus showed considerable promise for control of this species in laboratory and small-scale field tests (Ignoffo *et al.*, 1965). However, subsequent experimental formulations developed by industry were more refined, and field results with these preparations have been erratic. In California, consistent, effective control of the bollworm has not been obtained (Falcon *et al.*, 1965, 1966, in press).

According to Ignoffo and Garcia (1969) there are no significant differences in the susceptibility of beet armyworm, bollworm and cabbage looper to their respective viruses; yet, under field conditions control efficiency differs markedly. One major factor limiting effectiveness of bollworm virus is larval behavior. Unlike cabbage looper larvae, bollworm larvae are not gregarious and feed in or on squares, flowers or bolls under cover of the plant parts. Thus, optimum dosages, good coverage, and maximum penetration of the plant canopy with virus applications are needed. Most application equipment and procedures used in cotton do not meet these requirements. Another factor limiting effectiveness is ultraviolet radiation from sunlight which inactivates bollworm virus deposited on the surface of the leaves (Bullock, 1967; Gudauskas and Canerday, 1968). In the San Joaquin Valley, the pH of the cotton leaf surface ranges from 8.0 to 11.0. Since a pH of 8.5 or higher may deactivate the virus (Ignoffo and Garcia, 1966) the pH of cottonleaf surfaces may contribute markedly to the degradation of virus under field conditions.

In an effort to improve persistence of bollworm virus deposits in the field, one firm formulated several virus mixtures, with various agents included to inhibit the adverse effects of ultraviolet radiation and to improve wetting, dispersion and adhesion of deposits. In a series of laboratory and field experiments conducted in 1969, the newer formulations demonstrated a high degree of pathogenicity to 1st-instar bollworm larvae up to six days after application, contrasted to less than one day for virus with no protectants added (Fig. 1). The same preparations were tested against bollworm in a replicated field experiment. In addition, one preparation (VH 691) was also buffered. Only this combination resulted in significant reductions in bollworm numbers (Fig. 2). The addition of the buffer lowered the pH of cotton leaf surfaces which may explain the effectiveness of the buffered mixture over the

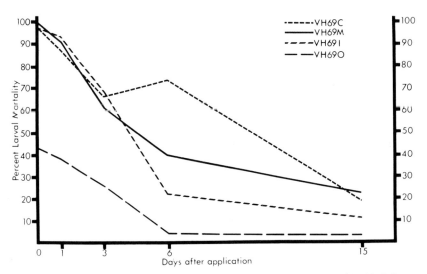

Fig. 1. Mean per cent virus dead 1st instar *Heliothis zea* (Boddie) larvae fed on cotton leaves 0 to 15 days after treatment with Viron/H®formulations of nuclear-polyhedrosis virus.

unbuffered preparations.

Field tests with commercial preparations of *B. thuringiensis* have shown, (1) frequent applications of moderate dosages (2 qts. per acre) significantly reduced defoliation by cabbage looper (Falcon *et al.*, 1968*b*); (2) it has not provided effective suppression of beet armyworm; and (3) recent improved preparations show promise for control of the bollworm (Dulmage *et al.*, in press; Falcon *et al.*, unpublished data; McGarr *et al.*, in press).

In summary, three nuclear polyhedrosis viruses and the bacterium *B. thuringiensis* are being developed for use in integrated control on cotton in California. Thus far, the cabbage looper virus is the most useful, followed by the beet armyworm virus. Both viruses provide effective and selective control of their respective hosts and are readily available if field collected. Obtained in this manner, individual growers can use the viruses on their respective fields. Even though effective experimental commercial preparations of these viruses are available, they cannot be sold until registration has been approved. The bollworm virus is a promising selective control agent, but until efficient methods of dissemination are developed, field persistence improved, registration for use obtained and cost established, its full potential will not be

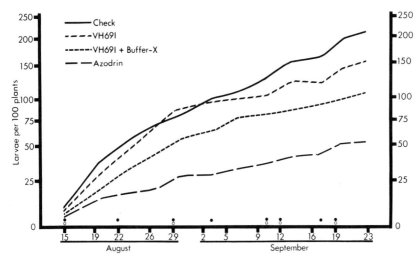

Fig. 2. Cumulative totals of live *Heliothis zea* (Boddie) larvae over 1/2" in length on cotton treated with Viron/H®formulations and Azodrin®. Dots along ordinate designate dates of application, ●- virus, ○- Azodrin.

realized. *B. thuringiensis* is the only microbial material readily available for use in cotton. It provides adequate suppression of cabbage looper and looks increasingly promising for control of bollworm. However, it is very costly and this factor, more than any other, has thus far restricted its use in cotton.

Employment of cotton as a field laboratory to test and develop microbial control agents as tools in integrated control is also providing valuable information for developing their use in other field crops.

The Granulosis Virus of Codling Moth

The codling moth, *Laspeyresia pomonella,* is an important pest of apples, other deciduous fruits, and walnuts in many parts of the world. Currently, broad spectrum organic chemical insecticides such as azinophosmethyl and carbaryl are the primary means of control of this pest. While these chemical insecticides are highly effective, routine and frequent applications result in the destruction of beneficial species and outbreaks of other pests (Clancy and McAlister, 1956; Pickett, 1959; LeRoux, 1960; Hoyt, 1961).

Because of their use, phytophagous mites have become among the most important and difficult pests to control in deciduous fruits. This factor, probably more than any other, has encouraged research to find selective methods for control of the codling moth. The isolation of a granulosis virus in larvae of the codling moth and subsequent laboratory studies which showed the virus to be highly virulent and amenable to quantity production, indicated that it had considerable potential as a selective control for this species (Falcon et al., 1968a).

In laboratory studies conducted to assess the potential of the virus as a control agent for the codling moth, results of bioassays showed that about 4 virus capsules fed per os is a lethal dose for 1st-instar larvae, with death occurring in two to three days. Placed on virus-treated fruits, about 30 per cent of 1st-instar larvae died before damaging the fruits and 60 per cent died after feeding on the epidermis, leaving little more than a "sting" on the surface of the apple. Of the remaining 10 per cent, most died in the apple, while a few succumbed to virus as pre-pupae or pupae. In an effort to determine its usefulness as a control agent for codling moth, a laboratory and field research program was initiated in 1965. The initial approach involved assessment of several factors including: (1) the incidence and importance of naturally-occurring granulosis virus in field populations of the codling moth; (2) susceptibility of other species to the virus; (3) methods for mass propagation of the virus; (4) impact of the virus on codling moth survival and fruit damage under field conditions; (5) dosages and timing of application needed for maximum suppression of codling moth populations; and (6) establishment of artificially disseminated virus as a natural mortality factor in field populations.

Thus far, the virus has not been found occurring naturally in field populations of codling moth in California. The original virus was isolated by our laboratory in 1963 from codling moth larvae collected from apple and pear trees in Mexico by L. E. Caltagirone (Division of Biological Control, University of California, Berkeley) (Tanada, 1964b). Since 1963, the presence of granulosis virus has been confirmed in laboratory cultures of codling moth maintained in the United States, Canada, Europe and the U.S.S.R. Currently, it represents one of the major problems to be overcome in the mass rearing of codling moth for sterile insect release programs underway in Washington (U.S.A.) and British Columbia (Canada). From available information it is not possible to determine its range of distribution and importance as a natural mortality factor for codling moth populations in nature.

In testing for host specificity, the virus was not infective to larvae of the bollworm, and the navel orangeworm, Paramyelois transitella (Walker) when high concentrations were administered per os. However, larvae of the oriental fruit moth, *Grapholitha molesta* (Busck), which is considered a close relative of the codling moth, succumbed to the virus, with the same symptoms as appear in virus-infected codling moth larvae.

To obtain sufficient quantities of virus for field studies it was propagated using 5th-instar codling moth larvae reared on a semi-synthetic diet modified after Ignoffo (1965). To initiate infection a concentration of about 3×10^7 capsules per larva were administered *per os*. Upon death, the virus-killed larvae yielded about 1×10^9 virus capsules per larva (1 larval equivalent = 1 LE) or approximately a 100-fold increase over the amount of virus used to initiate the infection. In 1967, about 1×10^{14} virus capsules were obtained from nearly 100,000 larvae reared at a cost of 0.3 cents per larval equivalent of virus (cost estimates for diet ingredients and rearing containers).

For application purposes, the virus was suspended in water with a wetting agent added (Triton® X-100 or Colloidal Biofilm®) and applied with a handgun from a portable orchard sprayer (150 lb. pressure at a rate of 4 to 5 gals. of total spray per tree). As nearly as possible, applications of virus were made to coincide with periods of maximum codling moth egg-laying activity, as indicated by collections of moths in light-traps and sex-lure traps. All field work was conducted in a one-acre apple orchard (Golden Delicious variety) situated in the Sierra foothills near Placerville, California (elevation - 2000 ft.).

Each year, the area treated, the number of applications and the dosages used were varied: (1) in 1966, 5 per cent of the orchard was treated (4 trees), the dosage used was about 3.4×10^{11} capsules per gallon, five applications were made, and the estimated amount of virus sprayed for the season was 1×10^{13} capsules per tree; (2) for 1967, when 35 per cent of the orchard was treated (28 trees), the dosage used was 1.2×10^{11} capsules per gallon, seven applications were made and about 4×10^{12} capsules were applied per tree; and (3) in 1968, the entire orchard was sprayed with a dosage of 9.5×10^9 capsules per gallon, and 12 applications were made for an average of 7×10^{11} virus capsules per tree. The effectiveness and persistence of the virus spray programs were monitored by laboratory bioassay. The bioassay results showed a similar degree of effectiveness one hour after application for all dosages, with the higher concentrations providing slightly better residual effectiveness 7, 14 and 21 days after application.

To assess the impact of the virus on codling moth in the orchard ecosystem, several sampling methods were employed including: (1) ultraviolet light traps and pheromone traps to measure moth density; (2) systematic serial sampling to assess oviposition and egg survival on leaves and fruit; (3) examination of fruits to determine larval survival and the nature and incidence of fruit damage; and (4) trunk bands to obtain an estimate of larval survival.

The trunk bands were one of the best measures of impact by the virus on survival of codling moth larvae. The bands, 3-inch wide corrugated cardboard strips, were placed around individual tree trunks from 12 to 18 inches above the soil. During the growing season the bands were checked at weekly intervals from early May to late August. In the winter period, one-half of the trunk bands were examined in November, and the other half in March, to determine population densities of the overwintered larvae. All larvae found were collected and held in the laboratory until death or adult emergence.

Trunk bands were maintained in the orchard for three consecutive growing seasons in both the check and virus-treated areas. As shown in Fig. 3,

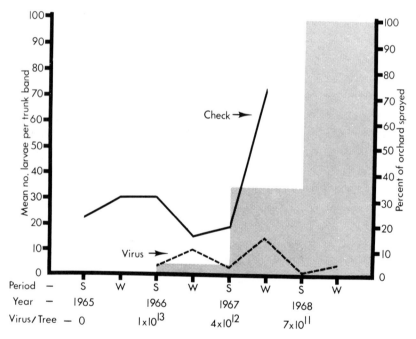

Fig. 3. Impact of granulosis virus on survival of codling moth larvae as determined by the trunk band sample method for the summer periods (S) (May to August) and winter periods (W) (November and March) 1965 to 1968.

bands were used prior to the introduction of virus to obtain an estimate of population density in the orchard and to survey for natural occurrence of granulosis virus. During this period, an average of 22.8 larvae per band were found in the summer period (S) and 30.8 during the winter period (W). Subsequently, larval recoveries in the check areas ranged from 15.3 to 30.8 larvae per band until the 1967-1968 winter period when they increased to 73.4. These data are useful to demonstrate the fluctuations of codling moth larval populations where parasites, predators and natural disease (such as from the fungus, *Beauveria bassiana* (Balsamo) Vuillemin) were primary naturally-occurring population regulatory agents. When the virus was first introduced into the orchard in 1966, larval recoveries from the virus-treated trees averaged 4.0 per band during the summer and 10.3 the following winter. These values represented 87 and 33 per cent less larvae, respectively, than found in the check area for the same periods. In 1967, 81 per cent less larvae (3.8 per band) were recovered from the virus-treated area during the summer and 79 per cent less (15.1) during the 1967-1868 winter contrasted to the check area.

In 1968, the entire orchard was treated. Larval abundance in the trunk bands averaged 1.9 for the summer and 5.0 the following winter. This was the lowest average number of larvae recovered for either period during the three years of virus applications. Although a check area was not available in 1968, larval recoveries from neighboring unsprayed areas indicated that codling moth population densities were as great during the 1968 season as for any of the previous three growing seasons. These results suggest that the virus was most effective when employed on an area basis. In previous years the survival of codling moth in the check plots apparently provided a continual reservoir of moths which would migrate into virus-treated areas and contribute to the over-all infestation.

In addition to the lower survival of codling moth in the virus-treated areas as reflected by the trunk bands, further larval mortality due to virus occurred after the larvae were collected and held in the laboratory. During the summer periods about 40 per cent of the collected larvae succumbed to virus and for the winter periods about 10 per cent.

In 1968, fruit damage in the virus-treated area was the lowest for the three-year virus study period. Of 2,400 fruits examined at harvest, 2.3 per cent were wormy or vacated. By contrast, fruit damage at harvest in the check areas during the 1965-1968 growing seasons, ranged from 45 to 70 per cent wormy or vacated fruits.

Assessment of the results of this study demonstrated that frequent applications of virus applied during the period of codling moth larval activity produced marked reductions in larval abundance. Best results were obtained in

1968 when the entire orchard was treated 12 times during the growing season, with virus applied at a dosage of 9.5 x 10^9 capsules per gallon of finished spray. This amount of virus represented about 8 larval equivalents per gallon, and at a cost of 0.3 cents per larval equivalent, this is about 2.4 cents per gallon of finished spray.

Throughout the entire study, there was no evidence that the virus became established as a natural mortality agent in the codling moth population. Also, its use did not appear to interfere with other organisms in the orchard ecosystem. However, additional research is needed to fully substantiate these observations.

The major shortcoming of the virus is its lack of effective residual persistence. Consequently, to be effectively employed for suppression of codling moth, it must be applied at frequent intervals throughout the growing season. With current methods of application such an approach is impractical. The next phase of the research program, which is already underway, is directed towards developing practical, inexpensive methods for virus dissemination and improving residual effectiveness of virus deposits.

The potential uses of the granulosis virus of codling moth as a tool in the integrated control arsenal are many. One approach may be to combine the virus with sub-lethal dosages of chemical insecticides to extend residual effectiveness. Another is to use the virus in conjunction with sterile insects and thereby possibly reduce the quantity of irradiated moths needed in area-wide control programs. With the development of efficient methods of dissemination, the virus could possibly be used in place of sterilized moths, with a similar degree of effectiveness and at lower cost. This procedure would consist of disseminating larval equivalents of granulosis virus over large areas at frequent intervals (2 to 3 days) during the period of codling moth activity. Mass propagation of virus would probably be cheaper than the autocidal method since only the larval stages are required and both sexes could be used. This approach should also be considered for other insect pathogens.

CONCLUSIONS

In this discussion I have attempted to show how microbial control is being developed for use in integrated control programs in California. The emphasis of these programs is on the development of the bacterium, *B. thuringiensis,* the respective nuclear polyhedrosis viruses of the bollworm, the cabbage looper, the beet armyworm and the granulosis virus of the codling moth. These pathogens were chosen primarily because they seem to offer the greatest potential for microbial control at this time. Also, all can be obtained in sufficient quantities for large-scale field studies.

In developing the use of microbial control, an intricate knowledge of the population dynamics and a thorough understanding of the interactions within the ecosystem are required. Ingenuity is needed to employ microbial control effectively, efficiently and economically. Currently, the major problems in developing the use of *B. thuringiensis* and the viruses for insect control in the field are related to (1) the short residual effectiveness of the pathogens under field conditions, and (2) inadequate methods for dissemination. These problems are of great economic importance, for even though a pathogen may be effective and inexpensive to produce, it still may be impractical to use if the cost for the repeated applications are too high. In attempting to remedy this situation, research should move in two directions, (1) to find ways to improve the persistence of pathogens in the field and (2) to develop inexpensive, simple, efficient methods for their dissemination. These problems will be solved in time; however, their solution would be hastened with more intensive research in these areas.

There are several other problems which may be more difficult to solve and are also in need of greater research efforts—for example, the question of the safety of insect pathogens, which has not been sufficiently demonstrated, in some points of view. There is also the matter that all microbial agents intended for use in some form of pest control currently come under the regulations of the 1947 "Federal Insecticide, Fungicide, and Rodenticide Act." As a result, both the Food and Drug Administration and the Pesticides Regulation Division (USDA) must accept and clear insect pathogens for use in insect control. It is also possible that clearances with the Public Health Service may be needed.

To date, only two entomogenous pathogens have been registered for use in the United States (milky spore disease bacteria for control of Japanese beetle grubs in the soil and products containing the spores of *B. thuringiensis*). No insect viruses have been registered (Heimpel, 1968). Efforts to obtain virus registration were initiated by the U.S. Department of Agriculture in 1962. Since 1965, two commercial firms have petitioned for registration of the nuclear polyhedrosis virus of *Heliothis zea*. However, its status is still uncertain. From these efforts it is hoped that a basic pattern for developing information required for decisions regarding safety of insect pathogens will emerge. Only in this way, it seems to me, will it be possible to reduce the time and financial investment required to develop the information needed by the regulatory agencies. This is of even greater importance for microbial control agents than for chemical pesticides since few if any exclusive patent rights can be applied to insect pathogens.

The need for creating a climate favorable for private industry to research and produce insect pathogens is essential to the future of microbial control. It is primarily private industry which can develop and employ the efficiency needed for mass production and handling of microbial control agents. This must be accomplished without the benefits of exclusive patent rights and with the probability of low dollar return because of the limited markets available for any one pathogen. These are tremendous obstacles and certainly indicate that other means for developing such products must be found. One possible approach may be for state or federal institutions to provide subsidies and to otherwise aid private industry to develop the microbial materials needed for use in insect control programs. Furthermore, integrated control programs themselves must be developed and the entire approach to pest control changed, not only at the merchandising and use levels but, in many cases, at the basic and applied research levels.

LITERATURE CITED

Allen, G. E., B. G. Gregory, and J. R. Brazzel. 1966. Integration of the *Heliothis* nuclear polyhedrosis virus into a biological control program on cotton. J. Econ. Entomol. 59: 1333-1336.

Anonymous. 1969. Microbial control of insects. Pp. 165-195. *In* Principles of Plant and Animal Pest Control. Vol. 3. Insect Pest Management and Control. Nat. Acad. Sci. Publ. No. 1695. Washington, D. C.

Balch, R. E., and F. T. Bird. 1944. A disease of the European spruce sawfly, *Gilpinia hercyniae* (Htg.), and its place in natural control. Sci. Agr. Ottawa 25: 65-80.

Bird, F. T. 1953. The use of a virus disease in the biological control of the European pine sawfly, *Neodiprion sertifer* (Geoffr.). Can. Entomol. 85: 437-446.

Bullock, H. R. 1967. Persistence of *Heliothis* nuclear-polyhedrosis virus on cotton foliage. J. Invert. Pathol. 9: 434-436.

Bullock, H. R., and H. T. Dulmage. 1969. *Bacillus thuringiensis* against pink bollworms on cotton in field cages. J. Econ. Entomol. 62: 994-995.

Burges, H. D., and N. W. Hussey (Editors). In press. Microbial Control of Insects and Mites. Academic Press, New York. (1970)

Cameron, J. W. M. 1963. Factors affecting the use of microbial pathogens in insect control. Ann. Rev. Entomol. 8: 265-286.

Cameron, J. W. M. 1969. Problems and prospects in the use of pathogens in insect control. Proc. Entomol. Soc. Ontario (1968): 73-79.

Clancy, D. W., and H. J. McAlister. 1956. Selective pesticides as aids to biological control of apple pests. J. Econ. Entomol. 49: 196-202.

Dulmage, H. T., M. J. Lukefahr, H. M. Graham, and N. S. Hernandez. In press. Studies on the use of HD-1, a new formulation of the *Bacillus thuringiensis*-δ-endotoxin, against *Heliothis* species on cotton. J. Econ. Entomol. (1970)

Falcon, L. A., W. R. Kane, and R. S. Bethell. 1968*b*. Preliminary evaluation of a granulosis virus for control of the codling moth. J. Econ. Entomol. 61: 1208-1213.

Falcon, L. A., T. F. Leigh, R. van den Bosch, J. H. Black, and V. E. Burton. 1965. Insect diseases tested for control of cotton bollworm. Calif. Agr. 19(7): 12-14.

Falcon, L. A., R. van den Bosch, C. A. Ferris, L. K. Stromberg, L. K. Etzel, R. E. Stinner, and T. F. Leigh. 1968a. A comparison of season-long cotton-pest-control programs in California during 1966. J. Econ. Entomol. 61: 633-642.

Falcon, L. A., R. van den Bosch, T. F. Leigh, L. K. Etzel, and R. E. Stinner. 1966. Microbial control of the bollworm. Pp. 17-22. *In* Cotton Insect Control, A Progress Report on Research. Div. Agr. Sci. Univ. Calif.

Gentry, C. W., W. W. Thomas, and J. M. Stanley. 1969. Integrated control as an improved means of reducing populations of tobacco pests. J. Econ. Entomol. 62: 1274:1277.

Gudauskas, R. T., and D. Canerday. 1968. The effect of heat, buffer, salt and H-ion concentration, and ultraviolet light on the infectivity of *Heliothis* and *Trichoplusia* nuclear-polyhedrosis viruses. J. Invert. Pathol. 12:405-411.

Hall, I. M. 1964. Use of microorganisms in biological control. Chap. 21. *In* Biological Control of Insect Pests and Weeds. P. DeBach (ed.). Reinhold Publ. Co., N. Y. 844 pp.

Hall, I. M. 1968. Integrated microbial control and chemical control of insect pests. Pp. 165-167. *In* Proc. Joint U.S.-Japan Seminar on Microbial Control of Insect Pests, Fukuoka (1967).

Heimpel, A. M. 1967. A critical review of *Bacillus thuringiensis* and other crystalliferous bacteria. Ann. Rev. Entomol. 12: 287-322.

Heimpel, A. M. 1968. Progress in developing insect viruses as microbial control agents. Pp. 51-61. *In* Proc. Joint U.S.-Japan Seminar on Microbial Control of Insect Pests, Fukuoka (1967).

Hoyt, S. C. 1961. Evaluation of new materials for codling moth control in Washington. J. Econ. Entomol. 54: 1127-1130.

Hurpin, B. 1966. Possibilite d'utilisation des microorganisms entomopathogenes dans la lutte integree. Part 2. Pp. 149-165. *In* Proc. FAO Symp. on Integrated Control (1965).

Ignoffo, C. M. 1965. The nuclear-polyhedrosis virus of *Heliothis zea* (Boddie) and *Heliothis virescens* (Fabricius). I. Virus propagation and its virulence. J. Invert. Pathol. 7: 209-216.

Ignoffo, C. M. 1968. Specificity of insect viruses. Bull. Entomol. Soc. Amer. 14: 265-276.

Ignoffo, C. M., A. J. Chapman, and D. F. Martin. 1965. The nuclear-polyhedrosis virus of *Heliothis zea* (Boddie) and *Heliothis virescens* (Fabricius). III. Effectiveness of the virus against field populations of *Heliothis* on cotton, corn, and grain sorghum. J. Invert. Pathol. 7: 227-235.

Ignoffo, C. M., and C. Garcia. 1966. The relation of pH to the activity of inclusion bodies of a *Heliothis* nuclear-polyhedrosis. J. Invert. Pathol. 8: 426-427.

Ignoffo, C. M., and C. Garcia. 1969. Relative susceptibility of four noctuids to their respective nucleopolyhedrosis viruses. J. Invert. Pathol. 14: 282-283.

Jensen, F. L. 1966. Grape leaf folder control with *Bacillus thuringiensis*. Calif. Agr. 20(7): 2-3.

Jensen, F. L. 1969. Microbial insecticides for control of grape leaf folder. Calif. Agr. 23(4): 5-6.

Le Roux, J. 1960. Effects of "modified" and "commercial" spray programs on the fauna of apple orchards in Quebec. Ann. Soc. Entomol. Quebec 6: 87-103.

Martignoni, M. E. 1964. Mass production of insect pathogens. Chap. 20. *In* Biological Control of Insect Pests and Weeds, P. DeBach (ed.). Reinhold Publ. Co., N. Y. 844 pp.

McGarr, R. L., H. T. Dulmage, and D. A. Wolfenbarger. In press. Tests with HD-1, a formulation of the δ-endotoxin of *Bacillus thuringiensis* var. *alesti,* and other insecticides for control of the tobacco budworm and the bollworm. J. Econ. Entomol. (1970)

Pickett, A D. 1959. Utilization of native parasites and predators. J. Econ. Entomol. 52. 1103-1105.

Steinhaus, E. A. 1951. Possible use of *Bacillus thuringiensis* Berliner as an aid in the biological control of the alfalfa caterpillar. Hilgardia 20: 359-381.

Steinhaus, E. A. 1954. The effects of disease on insect populations. Hilgardia 23: 197-261.

Steinhaus, E. A. (Editor). 1963. Insect Pathology, An Advanced Treatise. Vols. 1, 2. Academic Press, New York. 661 pp. + 689 pp.

Steinhaus, E. A. 1964. Microbial diseases of insects. Chap. 18. *In* Biological Control of Insect Pests and Weeds, P. DeBach (ed.). Reinhold Publ. Co., N. Y. 844 pp.

Tanada, Y. 1964*a*. Epizootiology of insect diseases. Chap. 19. *In* Biological Control of Insect Pests and Weeds, P. DeBach (ed.). Reinhold Publ. Co., N. Y. 844 pp.

Tanada, Y. 1964*b*. A granulosis virus of the codling moth, *Carpocapsa pomonella* (Linnaeus) (Olethreutidae, Lepidoptera). J. Insect Pathol. 6: 378-380.

Tanada, Y. 1967. Microbial pesticides. Pp. 31-88. *In* Pest Control: Biological, Physical and Selected Chemical Methods, W. W. Kilgore and R. L. Doutt (eds.). Academic Press, N. Y. 477 pp.

Taylor, T. A. 1969. Preliminary studies on the integrated control of the pest complex on cowpea, *Vigna unguiculata* Walp. in Nigeria. J. Econ. Entomol. 62: 900-902.

van der Laan, P. A., (Editor). 1967. Insect Pathology and Microbial Control. Pp. 252-286. Proc. Intern. Colloq. on Insect Pathol. and Microbial Control (1966). North-Holland Publ. Co., Amsterdam.

White, R. T., and S. R. Dutky. 1940. Effect of introduction of milky diseases on populations of Japanese beetle larvae. J. Econ. Entomol. 33: 306-309.

Chapter 16

MANAGEMENT OF PEST POPULATIONS BY MANIPULATING DENSITIES OF BOTH HOSTS AND PARASITES THROUGH PERIODIC RELEASES

F. D. Parker

Entomology Research Division
Agricultural Research Service
U. S. Department of Agriculture
Columbia, Missouri

INTRODUCTION

Before a biological control program utilizing the concepts of population manipulation can be undertaken, a great deal of ecological information about the pest species must be known. Of primary concern are such basic, but often unknown, factors such as pest abundance and seasonal distribution and the role natural enemies play in the population regulation of the pest. In-depth studies were undertaken at the USDA Biological Control of Insects Research Laboratory at Columbia, Mo., on the population dynamics of the imported cabbageworm, *Pieris rapae* (L.) (Parker, 1970). Results of these studies indicated that the indigenous parasites were ineffective in reducing or holding the host population below the economic injury level. Only one parasite, *Apanteles glomeratus* (L.), was found numerous enough to contribute even partial control of larval stages of this pest, and this species was introduced into the United States from Europe around 1890. Parker (*op. cit.*) attributed the cause of parasite ineffectiveness to several facts:

(1) Populations of *A. glomeratus* were not synchronized with early spring populations of *P. rapae,* as parasites emerged approximately two weeks before the host. Apparently, this braconid merely hibernates, with portions of the overwintering population emerging during warm periods in the winter;

cocoon masses brought into the laboratory during the winter yielded adults within a few days. The host, on the other hand, enters diapause in late Sept.-Oct., and does not emerge until the following spring.

(2) Host density in the first two generations was too low for parasites to increase soon enough and at a rate which could suppress subsequent host populations. In 1967 first generation egg densities of *P. rapae* were only 1 per 27 plants at its peak and only 1 per 4 plants during the peak of the 2nd egg generation. These peaks were separated by periods that extended to 25 days, with an even lower egg density. Host egg densities in the spring of 1969 were similar (Fig. 1).

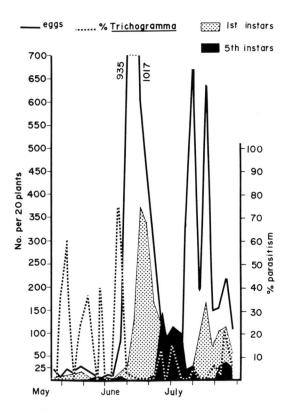

Fig. 1. Egg and larval populations of *Pieris rapae* (L.) and percentage of eggs parasitized by *Trichogramma* on spring planted cabbage near Columbia, Mo. (Check Site—South Farm).

(3) Parasite densities were too low because of overwintering mortalities in general and hyperparasite attacks. Leaves from cabbage plants that were left in the field for the winter were brought into the laboratory during early spring. Only three egg parasites emerged from approximately 100 leaves collected. In addition, normal cultural practices destroy the overwintering egg parasite populations when the land is plowed for spring planting. In 1967 *A. glomeratus* emerged from only 8 per cent of marked cocoons, although 56 per cent yielded hyperparasites and 36 per cent failed to emerge.

(4) Eggs of the primary larval parasite, *A. glomeratus*, were frequently encapsulated. In laboratory studies (unpublished data) 80 per cent of the *Apanteles* eggs were encapsulated when laid in late 2nd- to 5th-instar *P. rapae* larvae. Eggs deposited in very early instars were not encapsulated. Therefore, the opportunity for successful parasitism in the spring was curtailed by gaps between 1st-instar larval populations. The *Apanteles* increased and became an important parasite only when populations of early instar larvae overlapped.

IMPORTED CABBAGEWORM EXPERIMENTS

With the knowledge as to why the existing parasites failed to control the host, it was evident that host and parasite populations might be manipulated to favor the parasites if three conditions were met: (1) more effective parasites were introduced, (2) mass releases were made to synchronize and increase parasite density, and (3) host density were increased by mass release of fertile hosts, thereby insuring an adequate host supply for maintaining continuity of the parasites at a level sufficient to assure a prompt and adequate response.

The first step was accomplished with the introduction of two wasp parasites from Europe: an egg parasite, *Trichogramma evanescens* Westwood (Trichogrammatidae) (Parker, 1970), and an early larval parasite, *Apanteles rubecula* Marshall (Braconidae) (Puttler *et al.,* 1970). The remaining conditions were met by mass releases of hosts and both introduced species of parasites.

The initial experiment was conducted on a 1-acre plot of cabbage. Here, nearly 75,000 butterflies were released during the first three host generations; approximately one million egg parasites, and 6,000 larval parasites were released during the 1st host generation (Table 1, Ent. Farm).

The results of this experiment indicated that release of the host did increase host density. There were three times as many eggs on this plot as on the check plot in the 1st host generation. A continuous egg population of at least two per plant was available to the parasites the entire season except for one short period when the activity of the butterflies was reduced by inclement weather. Parasite releases increased parasitism of the eggs during the 1st host generation to more than 50 per cent contrasted to 1 per cent on 2 check plots, and 70 per cent of the larvae, as contrasted to 12.5 and 0.0 per cent on the check plots.

Periodic releases of hosts and parasites early in the season did synchronize their generations. During the 2nd and 3rd host generations at this location, the parasites were able to control the host population with only minor additional releases (Table 1). Egg parasitism by *T. evanescens* averaged 95.4 and 99.0 per cent and larval parasitism by *A. rubecula* averaged 80.7 and 100.0 per cent during these generations.

The combined effect of releasing both the host and the two parasites caused the host population to collapse. Survival of the host population to the 5th instar stage during the 1st host generation was 0.72 per cent at the release site, whereas 16.8 per cent survived this long at the check site (Table 2, Zubers). Even though the initial 1st generation host population at the release site was 3 times that of the check, 7.2 times as many larvae survived at the check site. During the 2nd host generation, 16.1 times as many larvae survived at the check site. No larvae survived at the release site during the 3rd host generation, whereas the check site was flooded during this generation; however, before the flooding the plants hosted 62 mature larvae per 100 plants. Total egg-larval mortality during the 90 day growing season was 99.67 per cent at the release site, whereas 89.08 per cent of these stages were killed at the check site (Table 2).

Table 1. Numbers of hosts and parasites released on cabbage plots at Columbia, Missouri.

Release sites		Plot size (acres)	Butterflies released	*Trichogramma evanescens* released	*Apanteles rubecula* released
1968					
Ent. Farm	s	1	74,712	1,142,000	7,570
Woods	s	1/3	30,002	416,000	1,900
Judahs	s	1/10	19,008	524,000	1,900
1969					
Ent. Farm	s	1	39,019	2,800,000	10,000
	f	1	9,091	372,000	7,850
Longs	s	1	28,977	1,560,000	6,750
	f	1	7,433	192,000	7,650
Lab	s	1/3	39,416	2,310,000	0
	f	1/3	8,694	525,000	6,350
Regans	s	1-1/2	0	4,090,000	16,950
	s	1-1/2	0	3,056,000	6,400

s - plot planted in spring; f - plot planted in fall.

Table 2. The effect of mass release of fertile hosts and two parasites on the natural population of three generations of *Pieris rapae* (L.) (80-90 days) on cabbage at Columbia, Missouri.

Release sites		Plot size (acres)	Total eggs counted/100 plants (3 months)	% eggs parasitized by *Trichogramma*	1st-instar larvae/100 plants	% larvae parasitized by *A. rubecula*	5th-instar larvae/100 plants	Total % egg-larval mortality
1968								
Release Sites								
Ent. Farm	s	1	9,340	93.53	274	84.87	31	99.67
Woods	s	1/3	6,594	97.88	56	57.14	24	99.64
Judahs	s	1/10	10,244	80.34	570	28.70	472†	95.40
Check Sites								
Zubers	s	1/3	4,649	5.27	1,109	2.30	508‡	89.08
So. Farm	s	1/3	14,268	7.59	3,164	3.14	3,823	73.21
1969								
Release Sites								
Ent. Farm	s	1	22,285	92.32	854	78.94	53	99.77
	f	1	19,995	75.73	1,655	67.25	356	98.22
Longs	s	1	14,825	95.67	315	61.53	39	99.74
	f	1	9,260	88.02	695	60.75	170	98.17
Lab	s	1/3	34,400	98.01	355	20.37	110	99.69
	f	1/3	26,530	89.47	1,135	62.50	377	98.58
Regans	s	1-1/2	1,920	41.34	305	78.94	41	97.87
	s	1-1/2	1,285	57.81	135	− §	15	98.84
Check Sites								
Zubers	s	1/2	13,930	28.17	2,435	10.52	995	92.86
	f	1/2	16,800	64.66	2,115	5.42	810	95.18
So. Farm	s	1/2	28,440	3.79	8,870	0.00	2,955	89.61
	f	1/2	16,480	18.07	4,715	0.00	1,165	92.94

s - plot planted in spring; f - plot planted in fall.
†Only 16 larvae on plot during last host generation when heads matured.
‡Plot was flooded during first part of 3rd host generation.
§Plot was sprayed with *Bacillus* - per cent larval parasitization not calculated.

The important result of these experiments was that a marketable crop of cabbage was produced; 96 per cent of the plants produced Grade A, No. 1 heads, whereas at the check site none of the plants produced marketable heads. The majority of the cabbage matured during the last part of the 2nd host generation and the beginning of the 3rd. After the initial success of releasing fertile hosts and parasites, additional experiments were conducted at two more release sites. Similar results were obtained; the host population was reduced below levels of economic injury after fertile hosts and parasites were released (Table 2).

Fertile butterflies were released at three sites during the spring of 1969 in order to build up a host population that could support a high parasite density early in the season (Fig. 2). Results were essentially the same as in the initial experiment, although more egg parasites were released because of abnormally heavy and prolonged periods of cold and rainy weather. Total

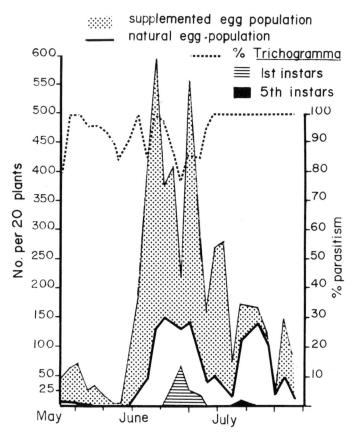

Fig. 2. Natural and supplemented egg and larval populations of *Pieris rapae* (L.) and percentage of eggs parasitized by *Trichogramma* on fall planted cabbage near Columbia, Mo. (Release Site—Entomology Farm).

egg-larval mortality of the host population exceeded 99 per cent at all three sites (Table 2). Crop damage by the imported cabbageworm was negligible.

At another site (Regan's), parasites but not hosts were released. This site was at the farm of a commercial vegetable grower. The host population was controlled, but not without difficulty. First, over four million *T. evanescens* were released but they only contributed 41 per cent control of the egg population. Second, an initial release of 12,000 *A. rubecula* produced no parasitized larvae. The cause of parasite failure was due to two factors:

insecticide drift and low host density. Leaves collected from plants were analyzed and found to contain a residue of 0.28 ppm carbaryl. This grower made frequent applications of this insecticide on nearby turnips and tomatoes. It is likely that the liberally applied insecticides killed many of these parasites before they had an opportunity to be effective. The host egg population was quite low during the experiment and frequently there were no eggs on which to build up a parasite population. At Regan's, nearly twice as many parasites were necessary to kill the same percentage of a much smaller host population (Table 2). At other sites the host population was 5 to 9 times greater.

It was found that releases of both species of parasites were necessary because cold or rainy weather inhibited the activity of *Trichogramma* (Parker *et al.*, 1970). When daily average temperatures fell below 15.5°C egg parasitization declined. Such periods of cool weather are frequent during the

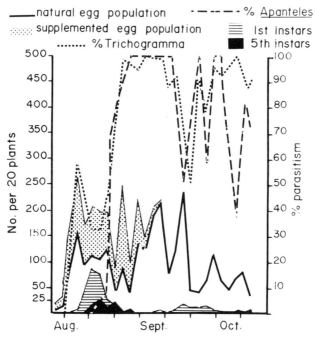

Fig. 3. Natural and supplemented egg and larval populations of *Pieris rapae* (L.) and percentage of eggs parasitized by *Trichogramma* and percentage of larvae parasitized by *Apanteles rubecula* Marshall on fall planted cabbage near Columbia, Mo. (Release Site—Entomology Farm).

spring and fall, the time when cabbage is generally grown. Therefore, releases
of the larval parasite *A. rubecula* were necessary to prevent the escaping host
population from causing damage. In laboratory food consumption tests, larvae
parasitized by this braconid ate only 1/10 the leaf surface as did unparasitized
larvae. Results of field tests confirmed these results; high larval populations
caused little plant damage if they were parasitized by this braconid. On the
other hand, larvae parasitized by *A. glomeratus* ate significantly more than did
unparasitized larvae (unpublished data). Therefore, releases of this parasite
could cause more plant damage.

In Missouri, *A. rubecula* normally enters winter diapause in early
September, nearly a generation before its host. The longevity of this parasite
is approximately six weeks. In order for a parasite population to be available
during the last host generation when cabbage matures, the natural parasite
population must be manipulated. It was postulated that control of the last
larval population could be obtained if a large adult population of *Apanteles*
were produced just prior to the onset of parasite diapause. Since adult
parasites have a longevity of about six weeks, those produced at this time
would be available during the last host generation. To test this theory, fertile
hosts were released at three sites during the 4th host generation.
Trichogramma were purposely not released so that a high larval population
could build up on the young plants. After a larval population was established,
(Fig. 3), releases of *Apanteles* and *Trichogramma* were made. Egg parasitism

Fig. 4. Young cabbage plants at Entomology Farm, 2nd planting, 1969.
Damage by natural and artificial populations of *Pieris rapae* (L.) larvae.

Fig. 5. Same plants as in Fig. 4, but at maturity well after releases of *Trichogramma evanescens* Westwood and *Apanteles rubecula* Marshall were made. Note the extensive damage on older leaves and the amount of injury young plants can sustain and still produce No. 1 heads.

rose and remained at a high level the rest of the season, except during periods of cooler weather, and larval parasitism reacted similarly, but averaged slightly lower toward the end of the season (Fig. 3). Although these plants were extensively damaged initially (Fig. 4), they produced Grade A, No. 1 heads at maturity (Fig. 5).

OTHER CABBAGE PESTS

Populations of seven other cabbage pests, excluding aphids, were found on the fall planting in 1968. These pests included the following Lepidoptera: the cabbage looper, *Trichoplusia ni* (Hubner); the diamondback moth, *Plutella xylostella* (Curtis); the cross-striped cabbageworm, *Evergestis rimosalis* (Guenee); the yellow-striped armyworm, *Spodoptera ornithogalli* (Guenee); the southern cabbageworm, *Pieris protodice* Boisduval and LeConte; the corn earworm, *Heliothis zea* (Boddie); and the variegated cutworm, *Periodroma saucia* (Hubner). Populations of *P. protodice* were controlled by the method outlined above.

Populations of the cabbage looper were controlled by releasing *Trichogramma evanescens* and spraying with a virus. Egg populations of this pest reached as high as 500-800 per 20 plants and egg parasitism averaged between 71-79 per cent. However, no parasite releases were made during the peak host oviposition period; had parasites been released a greater rate of egg parasitism may have resulted. Larval survival to 1st-instar ranged betwen 281-408 per 20 plants as opposed to 1014-2565 at the check sites. After three applications of a polyhedrosis virus at the release sites, larval survival decreased to 18-33 and 205-360 per 20 plants at the check sites.

Populations of the diamondback moth never reached economic levels; larval counts ranged between 1-5 per 20 plants. High rates of larval parasitism were observed and the following parasites were common: *Diadegma plutellae* (Viereck), *Tetrastichus sokolowskii* Kurdjumov, and *Micropletis plutellae* Muesebeck.

Although larval populations of the cross-striped caterpillar were sporadic, parasitism was significant, averaging 70 per cent. Two species of *Apanteles* were reared from this host, *A. orobenae* Forbes, and *A. marginiventris* (Cresson).

Smaller infestations of the four other species of Lepidoptera were found on these cabbage plots. No population estimates of these pests were made and rates of parasitism were not calculated. However, small larval samples were periodically taken and high rates of parasitism were observed, mostly by *A. marginiventris*. None of these Lepidopterans reached an economic level.

The use of laboratory produced biological agents and those that naturally occurred in the cabbage plots resulted in control of all eight species of pests and the production of a high quality crop. None of the three release sites had economic larval populations and most plants produced Grade A, No. 1 heads. One field was harvested, and of the 1,500 heads sampled, only 1 was unmarketable due to worm damage. None of the three check plots produced marketable cabbage (Fig. 6, 7).

CONCLUSIONS

The results of these experiments indicate that two critical requirements for suppression of a host population by parasites are: (1) the maintenance of an adequate supply of hosts to maintain the parasite population, and (2) the

Fig. 6. Field of cabbage damaged by natural populations of *Pieris rapae* (L.) and *Trichoplusia ni* (Hubner) at Zuber's, 2nd planting, 1969.

Fig. 7. Field of cabbage devastated by natural populations of *Pieris rapae* (L.) and *Trichoplusia ni* (Hubner), South Farm, 2nd planting, 1969.

release of effective parasites when existing ones are inadequate. In these experiments the host population was maintained at high levels by mass release of fertile cabbage butterflies. The feasibility of employing this method as a practical and acceptable means of control of *P. rapae* or other insects needs further investigation. However, the results of these investigations could eventually lead to more effective and more dependable means of controlling insects by artificially manipulating both the host and prey densities. This method was highly successful in controlling a pest of cole crops where host and parasite populations were continuously disturbed by farming practices.

LITERATURE CITED

Parker, F. D. 1970. Seasonal mortality and survival study of *Pieris rapae* (L.) in Missouri and the introduction of an egg parasite, *Trichogramma evanescens* Westwood. Ann. Entomol. Soc. Amer. 63: 985-999.

Parker, F. D., F. R. Lawson, and R. E. Pinnell. Suppression of *Pieris rapae* using a new control system: mass releases of both the pest and its parasites. J. Econ. Entomol. 64: 721-735.

Puttler, B., F. D. Parker, R. E. Pinnell, and S. E. Thewke. 1970. Introduction of *Apanteles rubecula* into the United States as a parasite of *Pieris rapae.* J. Econ. Entomol. 63: 304-305.

Chapter 17

THE DEVELOPING PROGRAM OF INTEGRATED CONTROL OF COTTON PESTS IN CALIFORNIA[1]

R. van den Bosch,[2] T. F. Leigh,[3] L. A. Falcon,[2]
V. M. Stern,[4] D. Gonzales,[4] and K. S. Hagen[2]

INTRODUCTION

During the years immediately following World War II, cotton pest control in California's San Joaquin Valley was dominated by the organochlorine insecticides (e.g., DDT, *toxaphene, endrin*). The persistence and high toxicity of these materials made them effective killers of the two key pests, *Lygus hesperus* Knight and *Heliothis zea* (Boddie). As a result, normally only one or two treatments per season were needed to effect satisfactory pest control at relatively low cost.

But the persistence of the organochlorines, their movement away from the sites of application, and their tendency to accumulate in the fatty tissues of warm-blooded animals created a serious pollution problem and health hazard. This led to increasing restrictions on the use of the materials and to their substantial replacement by organophosphate insecticides.

The organophosphates are also highly toxic to insects. Some, in fact, have broader toxicity spectra than the organochlorines. But the organophosphates are generally ephemeral; they do not persist in the environment, nor do they move great distances from the site of application or have an affinity for lipids. This is why they have been favored over the organochlorines.

[1]This paper relates entirely to cotton pest control in the San Joaquin Valley where approximately 90 per cent of California's cotton is grown.
[2]Department of Entomology and Parasitology, University of California, Berkeley.
[3]Department of Entomology, University of California, Davis.
[4]Department of Entomology, University of California, Riverside.

It is unfortunate, therefore, that the ephemerality of the organophosphates and their broad toxicity to arthropods have led to serious problems
which rob the materials of much of their usefulness and have made them
increasingly pollutive. Three major problems have developed in the wake of
organophosphate use: (1) pest resurgence, wherein the target species rapidly
rebound to equal or greater abundance than before treatment, often because
the insecticides eliminate natural enemies and thus release the pests from
biotic repression (Fig. 1), (2) secondary pest outbreaks, wherein the insecticides applied for control of target pests, eliminate the natural enemies of
non-target species, which permits damaging eruptions of the latter (Fig. 2),
and (3) pest resistance to pesticides, wherein repeated use of the materials
causes rapid genetic selection of populations which can tolerate the pesticides
(Fig. 3). As a result of these problems, in California cotton (and in other
crops too) there has been a proliferation of pest problems, an upsurge in pest
control costs, and an increase in environmental pollution.

This situation clearly indicated a need for a fresh approach to cotton
pest control, involving reduced reliance on chemical pesticides. The tack
chosen was integrated control, which involves the coordinated use of a variety
of techniques, agents and materials.

Integrated Control Defined

Integrated control is a pest population management system that utilizes
all suitable techniques either to reduce pest populations and maintain them at
levels below those causing economic injury, or to so manipulate the populations that they are prevented from causing such injury. Integrated control
achieves this ideal by harmonizing techniques in an organized way, by making
the techniques compatible, and by blending them into a multifaceted, flexible
system (Smith and Reynolds, 1966). In other words, it is an holistic approach
aimed at minimizing pest impact while simultaneously maintaining the integ-
rity of the ecosystem.

DEVELOPMENT OF THE INTEGRATED CONTROL PROGRAM

A variety of factors must be considered in the development of an
integrated pest control system. In California cotton the primary factors
include: (1) pest sampling and population prediction methods, (2) pest
economic thresholds, (3) naturally-occurring biotic mortality agents (natural
enemies) and their role in restraining or suppressing pest and potential pest

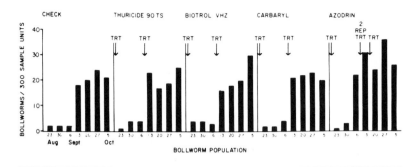

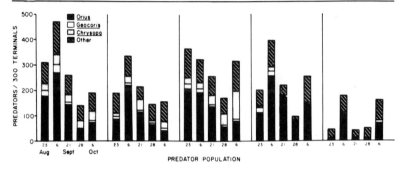

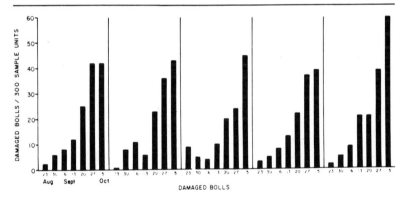

Fig. 1. Resurgence of bollworms following treatment with Azodrin®, Dos Palos, California, 1965.

Note particularly the greater numbers of bollworms and greater amount of boll damage in the Azodrin® plots as contrasted to the untreated control (check).

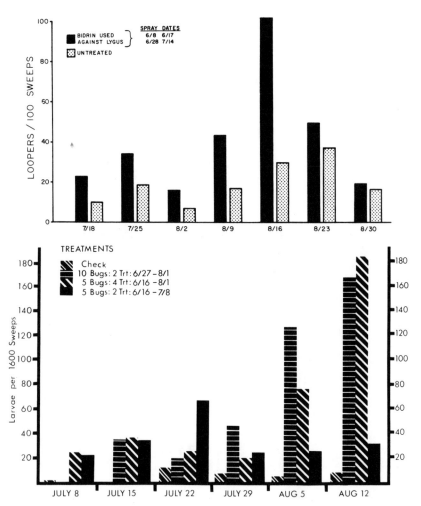

Fig. 2. Secondary outbreak of cabbage looper (top photo) following treatments of Bidrin®for *Lygus* control, Five Points, California, 1966, and of beet armyworm (bottom photo) following treatments of Toxaphene-DDT for *Lygus* control, Corcoran, California, 1969.

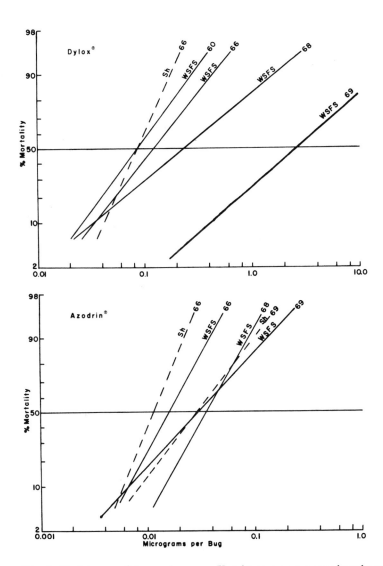

Fig. 3. Resistance of *Lygus hesperus* Knight to two organophosphorus in-secticides in California. These data show that the bug is more resistant to the materials on the west side of the San Joaquin Valley (West Side Field Station, WSFS) than at Shafter (Sh), and that resistance is increasing. (Unpublished data of T. F. Leigh and C. Jackson.)

species, (4) manipulation of these natural enemies, (5) the impact of artificial control practices on them, (6) pest phenologies as they relate to injury potentials and the timing of artificial control measures, (7) cultural and agronomic practices and their possible employment in insect population management, and (8) development of alternative or supplementive ecologically selective chemical and microbial controls.

All of these factors have been intensively investigated and substantial progress has been made in the development and utilization of certain techniques, manipulations, and materials so as to take advantage of the potentials of both biotic and abiotic natural mortality factors.

Economic Thresholds

The establishment of realistic economic thresholds for pest species is fundamental to integrated control. Artificial pest controls should only be invoked where and when their use is economically and ecologically justified. There is always a critical need for clearly defined economic thresholds to serve as the guideposts for invoking artificial controls. The general lack of established thresholds showing the need for artificial measures is a basic reason for the confusion in pest control today.

In California cotton, reliable economic thresholds have not been established for the two major pests, *Lygus hesperus* and *Heliothis zea.* This is very much at the root of the current serious pest control problems in this crop.

In actuality, with both *Lygus* and bollworm, certain specific levels of abundance have commonly been taken to indicate time for insecticide treatments, but these levels have been found to be meaningless. With other pests such as the cabbage looper, *Trichoplusia ni* (Hubner), the beet armyworm, *Spodoptera exigua* (Hubner), the salt marsh caterpillar, *Estigmene acraea,* (Drury) and spider mites (Tetranychidae), treatment criteria are even more vague. With these pests, time of season, grower whim or apprehension, salesmen's persuasion, growth stage of the crop, visible manifestation of feeding, mere presence, or similar nebulous criteria have been utilized to initiate chemical controls.

From this, it was most apparent to those involved in cotton entomology that the establishment of valid economic thresholds, particularly for the two major pests, was an imperative first step in the evolution of an integrated control program.

In assessing the over-all pest situation it became quite clear that extensive early and midseason insecticidal treatments for *Lygus* control were a direct and major cause of damaging outbreaks of the lepidopterous pests and spider mites. Consequently, early emphasis was placed on a study of *Lygus* injury to cotton, to determine whether widespread control of this pest was indeed necessary and to develop a realistic economic threshold for it.

It was quickly determined that the existing widely utilized *Lygus* treatment level (ten bugs per fifty net sweeps) was invalid. Continued research led to the promulgation in 1969 of a more flexible and meaningful treatment criterion (Division of Agricultural Sciences, University of California, 1969). This new criterion not only embraces *Lygus* density, but it also takes into consideration the cotton fruiting stage and the fruit load of the plants. In effect, it identifies the period from early June to mid July as being most critical from the standpoint of *Lygus* threat to Acala 4-42 and SJ-1 cotton varieties which are exclusively grown in the San Joaquin Valley (Fig. 4). The statement on need for *Lygus* treatment is as follows: "Therefore, it is suggested that during the critical period of June 1 to July 20, fields be inspected for *Lygus* at four to five day intervals. Where an infestation level of ten bugs per fifty sweeps is sustained over two successive sampling dates, control measures should be undertaken. After July 20, *Lygus* populations in excess of twenty per fifty sweeps may be tolerated without causing a reduction in yield or quality, provided the plants have flowered normally during the June and early July period" (*loc. cit.,* 1969).

A number of entomologist-monitored cotton fields in which this new treatment concept was used in 1969 produced good to excellent yields at greatly reduced costs. Moreover, these fields remained essentially free of damaging infestations of lepidopterous pests. The absence of the latter apparently reflected the lesser disruption of their natural enemy populations.

Grower enthusiasm is running high, and there is good prospect for future wide-scale employment of this flexible treatment threshold for *Lygus* in the San Joaquin Valley. There is a considerable feeling among members of the cotton pest control research team that as this occurs there will be increased opportunity for general implementation of the integrated control concept. In other words, the development of a flexible treatment threshold for *Lygus* appears to be a critical turning point in the integrated control of cotton pests. Not only will the new threshold lead to generally reduced pesticide use in early and mid season, but it should also result in lesser secondary pest problems, especially those involving the lepidopterous species. Meanwhile, studies of the latter have been undertaken, and considerable progress has been made along several lines.

Bollworm Studies

A striking aspect of the bollworm investigations has been our inability

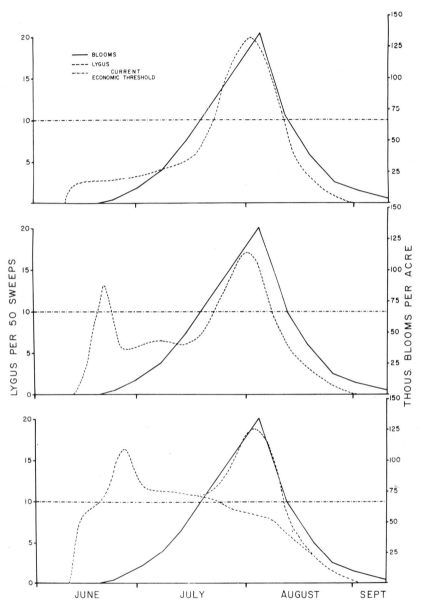

Fig. 4. Hypothetical *Lygus* infestations in relationship to the long-established economic threshold. Research indicates that only the kind of infestation depicted in the lower graph poses a serious threat to normally fruiting cotton (depicted by the blossoming pattern for the Acala, SJ-1 variety).

to correlate density of this pest with crop loss (Fig. 5). In other words, no pattern has emerged from the accumulated data which indicates that given numbers of larvae per unit crop area or number of plants will cause predictable crop losses.

In fact, in a number of experiments the untreated controls have produced yields equal to those obtained from plots in which significant bollworm reduction has been effected by insecticides and, on occasion, the untreated control plots have even out-yielded the chemically treated ones (Fig. 6). At times, of course, the reverse has also been true. But most often, in these latter cases, the increases in yields have not justified the dollar expenditures for insecticides.

Several factors appear to be involved. For one thing, the bollworm larvae in untreated cotton are normally under extreme pressure from such

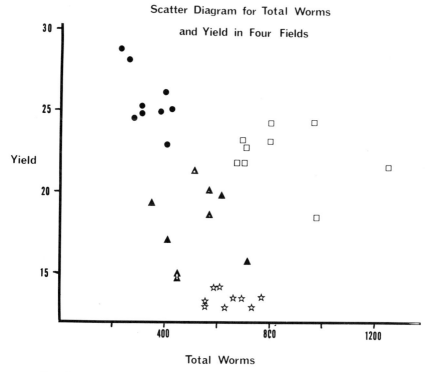

Fig. 5. Bollworm density in relation to yield in four cotton fields in Kern County, California. Note the lack of correlation between bollworm density and yield. (Unpublished data of D. Gonzales.)

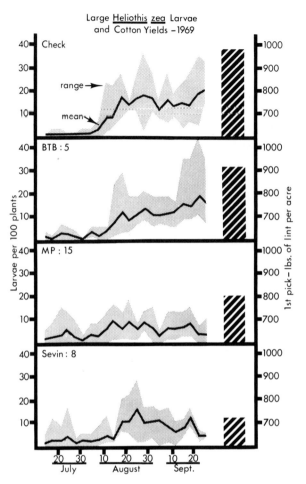

Fig. 6. Bollworm populations and first picking cotton yields in untreated plots (check) and three control programs, Rosedale area, Kern County, California, 1969. BT B:5 = *Bacillus thuringiensis* Berliner applied at an infestation level of 5 small larvae (1/2" in length) per 100 plants; MP:15 = methyl parathion applied at an infestation level of 15 small larvae per 100 plants; Sevin:8 = carbaryl (Sevin) applied at an infestation level of 8 small larvae per 100 plants. The untreated plots (check) yielded substantially more lint (horizontal bars) than did the plots treated with methyl parathion and carbaryl (Sevin), despite considerable reductions in bollworm populations under these two chemical insecticide programs. (Unpublished data of the authors)

natural enemies as *Geocoris pallens* Stal, *Nabis americoferus* Carayon, and *Chrysopa carnea* Stephens which take a very heavy toll (Fig. 7) (van den Bosch, *et al.,* 1969). Most chemical pesticides are extremely destructive to these natural enemies, and thus, much of the mortality caused by the insecticides simply replaces that caused by the predators. Furthermore, the bulk of the larvae surviving the insecticide treatments are then free to develop to maturity because of their release from the otherwise continuing predation. It is these larger larvae that are the most damaging.

Another important factor is the adverse effects of certain chemical insecticides on the cotton plants. At times, the insecticides have directly reduced cotton yields. This effect has been particularly apparent with methyl parathion, an organophosphate, but it has also been associated with a carbamate (carbaryl) (unpublished data of the authors). This result has been particularly striking where repeated applications of the insecticides have been made. The effect, which appears to be physiological, causes the plants to

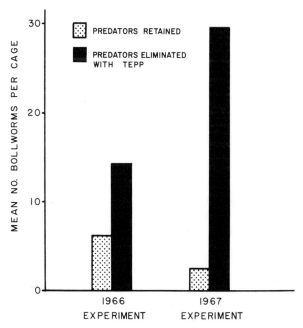

Fig. 7. Impact of naturally occurring predator populations on bollworms in caged cotton. Shafter, California, 1966 and 1967.

extend and intensify their growth phase, resulting in delayed maturity of the bolls.

Finally, the bollworm's feeding habits, its characteristic location on the plants, and the fruiting condition of the cotton frequently influence the real toll taken by this pest. In other words, bollworms often feed on surplus buds (squares), flowers, and small bolls not destined to contribute to ultimate yield anyway. These plant parts simply act as a buffer against injurious feeding by the pest.

It has been clearly established that the long-standing treatment level of 4 bollworms per 100 plants is invalid (unpublished data of the authors). A provisional level of 15 treatable larvae (first and second instars) is now employed; but, in the light of accumulating information, even this level may not represent a true economic threshold.

The weight of evidence indicates that the bollworm (*H. zea*) rarely develops to economically significant status in cotton fields not previously treated with insecticides, because in these fields its populations are repressed by predators and parasites. The dangerous situations appear to develop in those fields which have been stripped of natural enemies by chemical treatments applied principally for *Lygus* control. But even for these situations establishment of a clearly defined, real economic threshold is not yet in the offing.

Other Lepidopterous Pests

The role of natural enemies in restraining populations of other lepidopterous pests of cotton has also been clearly demonstrated. Abundant experimental evidence shows that where predators and parasites are eliminated by insecticides, such pests as the cabbage looper, the beet armyworm, and the salt marsh caterpillar frequently erupt to great abundance (Fig. 2) (Falcon *et al.*, 1968; Falcon *et al.*, 1971). Economic injury levels have never been established for these pests. Nevertheless, chemicals are frequently applied for their control. Many of these treatments are economically unjustified and therefore simply pollutive.

An important step, then, in alleviating these lepidopterous pest problems seems to lie in preserving the natural enemy populations in cotton fields. This, of course, means minimizing the disruptive effects of insecticides, either by eliminating them or by employing them at times or in ways that will limit their impact on the entomophaga. Several important advances along these lines are discussed below.

Light Trap Studies. The phenologies of the noctuid pests of cotton have been intensively studied by the use of ultraviolet (black light) light traps. The objectives of these studies have been: (1) to determine if moth collections by light traps accurately reflect the activity of the species in cotton, (2) to determine the relationship between moth flight patterns (as determined by the traps) and larval density patterns in the field, and (3) to develop appropriate mathematical models to predict the occurrence of a particular species (e.g., *H. zea*) with reference to time and location.

An analysis of the accumulated data has shown: (1) that there is a significant correlation in patterns of moth catches for traps within sites and between sites, (2) a good to excellent correlation between patterns of moth catches and larval population patterns in the fields, (3) a significant correlation between moon phase and moth counts, and (4) a mathematical relationship between lunar cycle, patterns of moth catches, and patterns of larval density (Fig. 8) (Falcon *et al.*, in preparation).

Establishment of the relationship of moth dynamics to lunar cycle has been of crucial importance to the integrated control program. Thus, it is now known that moth flight activity is greatest during the darker phases of the moon (last through first quarter), peaking at or near new moon. In any season, moth activity can be quite reliably predicted on the basis of moon phase. Knowledge of the flight patterns gives considerable insight into the timing of larval infestations which, with *H. zea*, reach their maxima about one to two weeks after peak moth flights. But most importantly, the data indicate those times when efforts should be made to avoid insecticide treatments that decimate predators and parasites. This information and the discovery that rather heavy *Lygus* populations can be tolerated during the period of peak bloom (normally, late July to early August), annually a time of heavy moth flights, have enabled us to advise growers to make every effort to avoid insecticidal treatments for *Lygus* during this critical period. This unquestionably had substantial bearing on the low incidence of lepidopterous pest problems in those fields where growers utilized the newly promulgated economic threshold level for *Lygus* in 1969.

Alternatives to Chemical Control. Intensive research has been conducted on alternatives to chemical pest control in cotton. Three artificial techniques have been stressed: (1) microbial control of the lepidopterous pests, (2) cultural control of *Lygus,* and (3) nutritional augmentation of natural enemies. These measures are important because of their lack of interference with natural enemies in the control of these pests.

Microbial control.—Microbial materials have the great virtue of selectivity. In the microbial control investigations major emphasis has been placed on experimentation with a nuclear polyhedrosis virus of *H. zea.* The virus is quite virulent, but up until 1969 it had not proved very effective in field experiments, due apparently to unfavorable pH and ultraviolet radiation in the environment. However, in 1969 rather encouraging results were obtained with virus material conditioned to withstand these adverse effects. In these tests, the residual pathogenicity of the conditioned virus materials was considerably enhanced. In one experiment virus conditioned to withstand the adverse effects of both ultraviolet radiation and unfavorable pH produced a significant reduction in bollworm larvae below the untreated control (Ignoffo *et al.,* 1972; Falcon, 1973). However, despite these promising experimental results, con-siderable developmental research is necessary before serious consideration can be

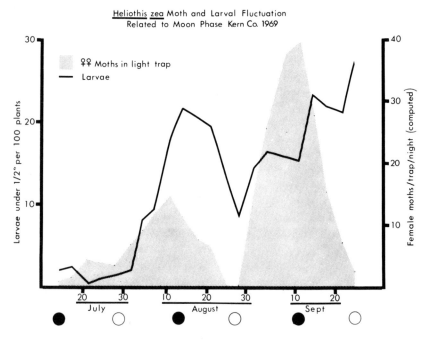

Fig. 8. Bollworm [*Heliothis zea* (Boddie)] moth flight activity and larval density pattern relative to the lunar cycle. Rosedale area, Kern County, California, 1969. Moth data represent the average numbers of ♀ *H. zea* taken in eight traps; larval density is based on the average numbers of small (mostly first and second instar) larvae sampled in untreated plots in five fields.

given to the effective utilization of this virus. Furthermore, the virus has not yet been registered for agricultural use.

On the other hand, a nuclear polyhedrosis virus of the cabbage looper has shown considerable promise and is, in fact, being used with good effect by some growers who prepare their own sprays from infected larvae collected in their fields. However, additional developmental research and federal registration of this virus will be required before it can be brought into wide-scale use and marketed commercially.

The bacterium *Bacillus thuringiensis* Berliner (a federally registered microbial material) is recommended for control of the cabbage looper, but it is expensive and as yet not widely used. In tests against bollworm, *B. thuringiensis* has shown some promise, but it has not yet effected sufficient kills to be given serious consideration for wide-scale grower use. However, more virulent preparations of the bacterium have been developed and are being tested, and these show considerable promise of wide-scale commercial use.

Cultural control of *Lygus.*—Cultural control investigations have centered on manipulation of alfalfa, *Medicago sativa* L., a favored *Lygus* host plant, either through strip harvesting of hay fields or the interplanting of alfalfa in cotton fields (Stern *et al.*, 1964; Stern *et al.*, 1969).

Alfalfa hay fields form a major reservoir for the *Lygus* which infest cotton. Frequently, the bugs move in massive numbers into cotton from the hay fields when the latter are harvested in conventional fashion (solid cut). Such harvestings are made recurrently during the summer, and so there are repeated movements of large numbers of *Lygus* into cotton during each growing season.

Research on strip harvesting of alfalfa to maintain alternate growth cycles in the fields has shown that this practice effectively prevents movements of the bugs from the alfalfa into the cotton at alfalfa harvest times. However, strip harvesting requires special care in irrigation and also other cultural and harvesting adjustments which the growers are reluctant to adopt. Consequently, very few growers have adopted the technique and it is not yet an important *Lygus* control method.

On the other hand, interplanting of alfalfa strips in the cotton fields has proved to be a quite effective *Lygus* control tactic in both experimental and commercial fields. Under this technique, narrow strips of alfalfa (e.g., 16' to 32' strips of alfalfa for every 300' to 400' of cotton) have been found sufficiently attractive to draw the bugs *out* of the cotton. The technique is particularly effective where moderate *Lygus* populations are involved. The seeding of alfalfa strips in cotton fields is a simple matter, and the strips are rather easily managed. For these reasons there is a good possibility that the technique will be widely employed in the next few years.

Nutritional augmentation of natural enemies.—As was mentioned earlier, there is abundant evidence that predaceous and parasitic arthropods take a considerable toll of the pest populations in cotton. However, for reasons not yet fully known, there is a characteristic waning in the abundance of certain key predators, particularly *Geocoris pallens,* after midseason (August). Attempts have been made to augment the predator populations during this critical period with artificial nutrients. Some success has been attained, particularly with the green lacewing, *Chrysopa carnea* Steph. With this species, adults have been attracted to the nutrient and have remained in the treated areas. There has been a correlated increase in egg and larval abundance and significant impact on bollworm populations. Injury to cotton bolls has been significantly reduced in the treated plots. An inexpensive artificial diet based on a dairy by-product will be used in large-scale field experiments (Hagen *et al.,* in press).

Supervised Control

The techniques being developed in the integrated control program will demand increasingly close monitoring of insect populations in the cotton fields. Thus, there is a critical need for qualified persons to supervise the program. In actuality, enough qualified persons are probably already involved in pest control advisement in the San Joaquin Valley, but the bulk of them are affiliated with the chemical industry and are primarily motivated to merchandise their companies' products. As matters stand, this is a fundamental impediment to expansion of integrated control. Nevertheless, about 10 per cent of the cotton acreage is now being supervised by independently practicing entomologists. Furthermore, a number of the larger growers retain persons on their staffs who function as pest control technologists.

So, there is a nucleus of persons to help implement this integrated control program for cotton in its initial phases. If there is striking success, and recent events indicate that there will be, then it can be expected that qualified persons will be attracted away from the chemical companies to fill the need for entomologists as the integrated control program expands.

INTEGRATED CONTROL AND THE ECONOMIC CRISIS
IN COTTON

As measured in dollar income, cotton is the most important crop grown in California, having an annual cash value of approximately $200 million. For many years, cotton was the major "bread winner" for numerous growers, but recently it has fallen on hard times. The reason is simple: production costs have risen while the price of cotton has fallen. In order to survive under these conditions the grower must increase his yields and cut his production costs. Efficient pest control can contribute importantly to this dual goal. Existing pest control practices are both inefficient and expensive, and because of this they are directly contributing to the economic crisis. Accumulating evidence indicates that integrated control is much more efficient and less expensive, and the hard-pressed growers are beginning to realize this. Thus, the economic crisis in cotton will almost surely add momentum to the integrated pest control program, which in turn may be an important factor in preventing economic disaster in the California cotton industry while at the same time lessening the industry's contribution to the over-all pesticide pollution problem.

LITERATURE CITED

Div. Agr. Sci., Univ. Calif. 1969. Pest and disease control for cotton. 19 pp.

Falcon, L. A. 1973. Biological factors that affect the success of microbial insecticides: Development of integrated control. Ann. N.Y. Acad. Sci. 217: 173-186.

Falcon, L. A., R. van den Bosch, C. A. Ferris, L. K. Stromberg, L. K. Etzel, R. E. Stinner, and T. F. Leigh. 1968. A comparison of season-long pest-control programs in California during 1966. J. Econ. Entomol. 61: 892-898.

Falcon, L. A., R. van den Bosch, J. Gallagher, and A. Davidson. 1971. Investigations of the pest status of *Lygus hesperus* on cotton in Central California. J. Econ. Entomol. 64(1): 56-61.

Falcon, L. A., A. Davidson, D. W. Britton, and J. Gallagher. In preparation. Influence of moonphase on adult and larval activity of *Heliothis zea* in cotton. J. Econ. Entomol.

Hagen, K. S., E. F. Sawall, Jr., and R. L. Tassan. In press. The use of food sprays to increase effectiveness of entomphagous insects. Proc. Tall Timbers Conf. Ecol. Anim. Control by Habitat Mgmt. No. 2, Tallahassee, Fla. (1970)

Ignoffo, C. M., J. R. Bradley, Jr., F. R. Gilliland, Jr., F. A. Harris, L. A. Falcon. L. V. Larson, A. L. McGarr, P. P. Sikirowski, T. F. Watson, and W. C. Yearian. 1972. Field studies of the *Heliothis* nucleopolyhedrosis virus at various sites throughout the Cotton Belt. Environ. Entomol. 1(3): 388-389.

Smith, R. F. and H. T. Reynolds. 1966. Principles, definitions and scope of integrated pest control. Proc. FAO Symp. on Integrated Pest Control 1: 11-17.

Stern, V. M., R. van den Bosch, and T. F. Leigh. 1964. Strip cutting alfalfa for lygus bug
    control. Calif. Agr. 18(4): 4-6.
Stern, V. M., A. Mueller, V. Sevacherian, and M. Way. 1969. Lygus bug control in cotton
    through alfalfa interplanting. Calif. Agr. 23(2): 8-10.
van den Bosch, R., T. F. Leigh, D. Gonzalez, and R. E. Stinner. 1969. Cage studies on
    predators of the bollworm in cotton. J. Econ. Entomol. 62: 1486-1489.

Chapter 18

THE DEVELOPING PROGRAMS OF INTEGRATED CONTROL OF PESTS OF APPLES IN WASHINGTON AND PEACHES IN CALIFORNIA

S. C. Hoyt

Tree Fruit Research Center
Washington State University, Wenatchee

and L. E. Caltagirone

Division of Biological Control
University of California, Berkeley

INTRODUCTION

On both apples in Washington and peaches in California pest control programs which rely solely on chemicals result in serious problems with tetranychid mites. In both cases some pests are present which have very low economic injury levels and which require use of pesticides to maintain economic control. The common practice of using *preventive* schedules of pesticide application has resulted in destruction of natural enemies of mites. The use of acaricides for mite control, moreover, has not been entirely satisfactory because of selection for resistance and resurgences of mite populations. (See also Chapter 1, Section III, and other chapters in Section IV.)

A. D. Pickett and his co-workers in Nova Scotia (see Chapter 13) early recognized the need for an ecological approach to pest control on apples in the mid-1940's. They found that sprays would not control all pests adequately and that other approaches to pest control should be included in the total program. They developed an integrated control program utilizing predators for the control of European red mite and oystershell scale. Their work and developments in California under the leadership of Drs. A. E. Michelbacher and R. F. Smith provided a background for the development of integrated approaches to pest control in Washington and California.

The integrated control program in Washington, conducted by the senior author, has been used commercially since 1965, while the program on peaches in California, conducted by the junior author, is of more recent origin. Both programs were developed because of the necessity for new approaches to mite control.

THE DEVELOPING PROGRAM OF INTEGRATED CONTROL OF PESTS OF APPLES IN WASHINGTON

The McDaniel spider mite, *Tetranychus mcdanieli* McGregor, the European red mite, *Panonychus ulmi* (Koch), and the apple rust mite, *Aculus schlechtendali* (Nalepa), commonly attack the foliage of apple trees in Washington. These three species were controlled by chemicals from the time of the introduction of parathion until the end of the 1958 season. Outbreaks did occur in some orchards each year, but damage was minimized by the control measures. Following the establishment of a zero tolerance for Aramite® and the selection of strains of McDaniel mite resistant to dicofol in early 1959 (Hoyt and Harries, 1961), control of this species became very difficult. Recently registered acaricides were then used and control improved temporarily, but strains resistant to each acaricide were rapidly selected (Hoyt and Kinney, 1964; Hoyt, 1966).

The numerous treatments for control of the McDaniel mite also reduced populations of the European red mite and apple rust mite and these species were rarely prevalent. However, strains of European red mite resistant to acaricides were also selected. The two tetranychids then seriously threatened apple production. Frequently, as many as five acaricide applications were required and control was often inadequate. Obviously, a more permanent and less costly means of controlling the McDaniel mite was needed.

The prevalent chemical program for controlling codling moth, *Laspeyresia pomonella* (L.), was one factor preventing use of natural enemies of mites. The ryania program used in Nova Scotia to conserve mite predators (MacPhee and Sanford, 1961; Chapter 13 herein) was ineffective for control of codling moth in Washington. During 1960 and 1961, the native predatory phytoseiid, *Metaseiulus occidentalis* (Nesbitt), was found in several orchards coincident with the declining use of dicofol. In 1961, an orchard was observed where *M. occidentalis* survived a codling moth control program in sufficient numbers to provide satisfactory control of McDaniel mites.

These developments and observations led to the research and integrated control program reported here.

Populations of Mites on Apple

With Standard Acaricide Programs. Under the prevalent, standard acaricide programs, European red mites were generally held at low levels, but with the selection of resistant strains damaging populations developed in some orchards. These populations increased during June, peaked in July, and declined in August with the increasing competition from McDaniel mites (Table 1). Apple rust mites rarely occurred in damaging numbers.

Table 1. Average numbers of three phytophagous mites and a predatory mite in an orchard sprayed with a standard spray program[†], 1966. (After Hoyt, 1969a)

Mite species	Average number of mites per leaf									
	5/10	5/23	6/6	6/20	7/6	7/18	8/2	8/17	8/29	9/12
T. medanieli	4.5	2.9	1.2	7.5	20.0	0.4	3.9	27.0	44.0	20.0
P. ulmi	0.7	0.1	4.9	5.0	43.0	1.4	1.9	0.8	0.3	1.1
A. schlechtendali	0.0	0.1	0.0	0.1	0.0	6.0	0.4	0.0	0.8	2.5
M. occidentalis	0.4	0.0	0.0	0.1	0.1	0.2	0.0	0.0	1.1	1.8

[†]Spray program (in pounds active per 100 gallons)
Petroleum oil - 1 gallon + ethion - 0.16 on 4/4
Azinphosmethyl - 0.313 + binapacryl - 0.25 on 5/18 and 7/5
Tranid® - 0.5 on 7/15

McDaniel mites frequently overwintered in high populations in trash at the base of the tree or under bark scales on the trunk. The mites ascended the tree in late March and early April, damaging watersprout foliage and leaves on spurs in the center of the tree. Where populations were very high, the blossoms on some spurs were covered with webbing and fruits did not develop. Pre-bloom sprays were generally ineffective at controlling these mites. The mites rapidly moved to the top of the tree where the highest populations later developed (Hoyt, 1969b). Acaricides applied during May and early July reduced numbers to low or moderate levels. High populations of McDaniel

Table 2. Average numbers of McDaniel spider mites per leaf in an orchard sprayed with Morestan® (0.5 lb. active ingredient per acre) for mite control. Application dates were 5/26, 6/9, 6/17, and 7/31, 1964. Rock Island, Washington.

Average numbers of mites per leaf								
5/18	6/2	6/16	6/29	7/15	7/29	8/14	8/28	9/11
20	9.5	8.5	0.6	10	36	3.3	10	44

Note: Predator activity was lacking.

mite developed rapidly in the hot weather of late July and August because regulation from natural enemies was lacking (Tables 1 and 2). Multiple applications of acaricides were made during this period in many orchards, but with the highest mite density occurring in the tops of trees control was difficult to obtain. To even use the word "control" seems improper, because the use of acaricides merely delayed population development until later in the season. Foliage was frequently damaged in the tops of the trees. If the weather remained warm and dry during early September, mite populations persisted until harvest. As a result, poorly colored fruits of lower value were produced. Overwintering mites clustered in the calyx end of the fruits and their removal was difficult. A high population of mites in the overwintering sites insured that the cycle would be repeated the following year.

With No Spray Program. Two orchards were selected to study mite populations where no sprays were applied, but where other horticultural practices were continued. Normal spray programs had been used through the 1962 season, but these were eliminated in 1963 and subsequent seasons. In both orchards, McDaniel mite populations reached high levels in late June of 1963, but *M. occidentalis* populations developed rapidly and reduced the numbers of this species to low levels by mid-July (Figs. 1 and 2), and they remained very low during the remainder of 1963, and in 1964 and 1965.

In one orchard, European red mite populations followed the same trend as the McDaniel mite (Fig. 2). In the other orchard, populations persisted until early August of 1963, apparently giving the mites the opportunity to deposit overwintering eggs. With only European red mites as a food source in early 1964, predator populations were slow to develop. Freed from competition with the McDaniel mite, the European red mite developed higher densities in 1964 than in 1963 (Fig. 1). However, during 1965 predator numbers increased early because of the presence of apple rust mites; consequently, European red mites were held at low levels.

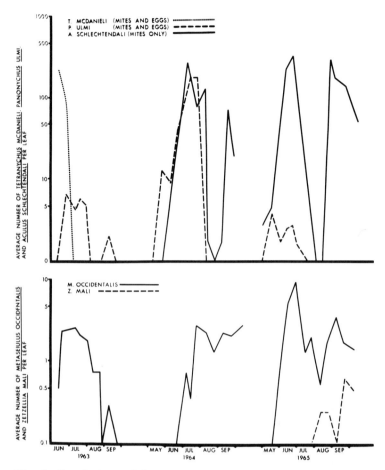

Fig. 1. Populations of five mite species in an unsprayed apple orchard. (After Hoyt, 1969*b*)

Apple rust mites were seldom seen during 1963, apparently due to the 1962 spray program. During 1964 and 1965, apple rust mites were the most numerous mite species. In each of these years, there were two peaks of rust mite populations, one occurring in mid-to-late June and the other toward the end of August (Figs. 1 and 2). Predator populations developed early and reached high levels feeding on this species.

During 1964 and 1965, the stigmaeid predator, *Zetzellia mali* (Ewing), became numerous. In 1966, this species apparently prevented apple rust mite

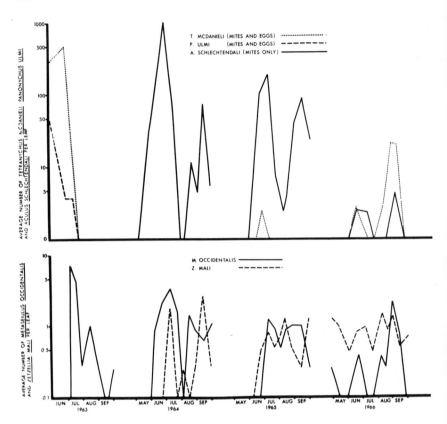

Fig. 2. Populations of five mite species in an unsprayed apple orchard. (After Hoyt, 1969*b*)

populations from developing and as a consequence *M. occidentalis* numbers remained low (Fig. 2). *Z. mali* is an inefficient predator of McDaniel mite; therefore, a moderate density of McDaniel mites developed, but was then quickly controlled by a responsive, rapidly developing population of *M. occidentalis.*

Predaceous insects, such as the coccinelid *Stethorus picipes* Casey and the mirid *Deraeocoris brevis* Uhler, and spiders were found in these two orchards, but their distribution and numbers suggested that they had only a local effect on mite populations (Hoyt, 1969*b*). While these predators played a part in the reduction of phytophagous mites, the regulation of the populations was predominantly by *M. occidentalis.*

The results of these studies suggested a very good potential for regulation of mite numbers by predators if selective materials for insect control could be found.

Distribution of Predators and Prey

The relative distribution of predators and prey is an important factor in predation. If the prey have areas of escape in space or periods of escape in time, the efficiency of the predator may be reduced. Chant (1958) stated that reliable control cannot be attained if there are areas where prey are free from attack.

The distribution of *M. occidentalis* is well synchronized with that of McDaniel mite. The two species overwinter in the same general areas and emerge at approximately the same time of year (Hoyt, 1969*b*). *M. occidentalis* inhabits all areas of the tree inhabited by the McDaniel mite, including the calyx end of the fruit. Recent studies indicate a distinct relationship between the distribution of the two species on apple foliage (Table 3). McDaniel mites inhabit both leaf surfaces in about equal numbers. During the day, 85 per cent or more of the predators are on the under-leaf surface, but during the evening hours, 50 per cent or more move to the upper leaf surfaces (Hoyt, unpublished data).

The European red mite overwinters in the egg stage, with the highest numbers occurring toward the periphery of the tree. *M. occidentalis* first inhabits the central area of the tree mainly, gradually dispersing to the periphery by June. For this reason, European red mite populations develop relatively free of *M. occidentalis* predation during May and June. Examination of the distribution of these two species in several orchards indicated no apparent relationship between the distribution of the two species on leaves (Table 3). Only when the predators become well distributed in response to one of the other two prey species will European red mite populations be regulated by this predator. This happens independently of the European red mite population density but nevertheless usually coincides with high densities of European red mite.

The apple rust mite overwinters on all aerial parts of the tree, but the highest numbers are usually on terminals and watersprouts. Populations of apple rust mites develop early, and *M. occidentalis* spreads rapidly to all areas of the tree and develops high populations feeding on this species. There is a close correlation between the distribution of the two species on foliage (Table 3).

Table 3. The incidence of *Metaseiulus occidentalis* (Nesbitt) on apple leaves infested with different species of phytophagous mites in Washington. (Summarized from Hoyt, unpublished data)

	Numbers of leaves with				
Phytophagous mite species	Phytophagous mites and *M. occidentalis*	Phytophagous mites only	*M. occidentalis* only	No mites	Chi-square significance[†]
T. mcdanieli	758	459	333	2520	***
P. ulmi	151	908	424	2097	Not significant at the 5% level
A. schlechtendali	355	635	90	260	***

[†]A significant chi-square value indicates a relationship between the distribution of predators and prey.

Feeding Behavior of Predators

M. occidentalis will feed on all stages of the McDaniel mite. However, immature predators are apparently successful in their attacks only on immature mites.

Some authors have reported phytoseiids to be inefficient predators at high prey densities (Collyer, 1958; Kuchlein, 1965; and Lord, *et al.*, 1958). Mori and Chant (1966) found a declining functional response at high prey densities, brought about by contact disturbance of feeding predators by the numerous prey. Our field results suggested that *M. occidentalis* was a highly efficient predator of appropriate prey at high prey densities, a finding supported by work of Laing and Huffaker (1969) and Flaherty and Huffaker (1970). Investigations on short-feeding periods by the predator revealed that a high percentage of eggs and larvae of the prey were killed when punctured (Hoyt, 1970). Feeding periods of 30 to 120 seconds caused a high percentage of mortality of nymphs. Feeding periods of less than five minutes resulted in less than 40 per cent mortality of adult female McDaniel mites, but substantially reduced oviposition by surviving females. Predators were observed to attack as many as 30 individual prey in 30 minutes when they were repeatedly disturbed. This suggests that disturbance of feeding at high prey densities which results in release of individual prey would increase predator efficiency in reducing prey populations.

M. occidentalis feeds on all active stages of European red mite but not on the eggs. It also feeds on active stages of apple rust mite, but has not been observed feeding on the eggs, which are extremely small and flattened dorsoventrally and may be missed by the normal searching habits of the predator.

In summary, based on its distribution and feeding habits, *M. occidentalis* is very efficient at regulating McDaniel spider mite populations at low densities or at suppressing high densities of this pest. It is not a reliable, efficient predator of European red mite, but may control this species as a side effect of its relationships with other prey mites. Since apple rust mite causes economic damage only at very high population levels, its presence in moderate numbers is desirable as a source of food for *M. occidentalis*. This predator, coupled with other factors, frequently maintains populations of apple rust mite below levels which cause economic loss.

Effects of Pesticides on Predator Numbers

A wide range of pesticides can be used in apple orchards in Washington with little direct effect on *M. occidentalis*. These include most of the organophosphorous compounds, endosulfan, Morestan,® Omite,® the dithiocarbamate fungicides, dodine and captan (Hoyt, 1969a). It is uncertain whether this low toxicity is due to natural tolerance by *M. occidentalis* or to selection of resistant strains. Huffaker and Kennett (1953) found this phytoseiid to be tolerant of parathion, and evidence in Washington suggests that a strain resistant to DDT has been selected. Some of these compounds (ethion, endosulfan, Morestan® and Omite®) affect predator numbers secondarily by destroying their food supply. However, they are normally used to reduce rapidly developing phytophagous mite populations at critical times and only misuse of these compounds would be detrimental to the integrated program. The toxicity of azinphosmethyl to predators varies with the rate of use. A dosage has been established which is effective for codling moth control yet allows sufficient *M. occidentalis* survival to maintain regulation of McDaniel mites. If oil is used prior to bloom when the predators are in protected sites, it does not interfere with the biological control. Dinocap is intermediate in toxicity to this predator, but it has the disadvantage of killing apple rust mites when used for control of apple powdery mildew. Carbaryl, binapacryl, dicofol, chlorobenzilate, and Acaralate® are highly toxic to *M. occidentalis*. Of these materials, only carbaryl is essential to the current apple production program, and its use is restricted to a single application for fruit thinning. It can be applied selectively to those trees or areas of trees which require thinning. By avoiding spraying the low, central area of the tree, the material can be used with a minimum effect on *M. occidentalis*. Other thinning materials such as dinitrocresol and naphthalene acetamide have little effect on predator numbers.

The above mentioned materials which are low or moderate in toxicity to *M. occidentalis* include pesticides effective for control of all insects and diseases affecting apples in Washington. However, there is an additive effect where more than one chemical is used in an application, and care must be taken to avoid excessive predator kill.

Insect and Disease Problems on Apple

Very few disease problems are prevalent on apples in Washington. Apple scab, *Venturia inequalis* (Cke.) Wint., is a problem of consequence in only two areas and most of the acreage in central Washington is not sprayed for scab. Apple powdery mildew, *Podosphaera leucotricha* (E. & E.) Salm., occurs in all apple areas, but the predominant Red Delicious and Winesap varieties are somewhat resistant to mildew and are rarely sprayed. Sprays are applied for Bullseye Rot, *Neotabraea malacorticis* (Cord.) Jack., but again the area treated is limited. Even where disease controls are required, a maximum of two or three applications per year are made.

Insects requiring annual treatments include codling moth, fruit-tree leaf roller (*Archips argyrospila* (Walker)), several species of cutworms, apple aphid (*Aphis pomi* De Geer), woolly apple aphid (*Eriosoma lanigerum* (Hausmann)), and San Jose scale (*Aspidiotus perniciosus* (Comstock)). Insect pests of occasional importance or limited distribution include the consperse stink bug (*Euschistus conspersus* Uhler), lygus bugs (*Lygus* spp.), western flower thrips (*Frankliniella occidentalis* (Pergande)), green fruitworm (*Lithophane* sp.), rosy apple aphid (*Dysaphis plantaginea* (Passerini)), grasshoppers and several other lepidopterous larvae. Many orchard pests prevalent in the central and eastern United States do not occur in Washington orchards.

Integrated Control Programs

Since European red mites are not controlled in the early growing season by *M. occidentalis,* it is essential to the integrated program to control this pest with sprays prior to bloom, and to hold it at low levels through June (Fig. 3). The application should also control San Jose scale, early-emerging lepidopterous larvae and aphid eggs. Use of materials which are not specific acaricides is desirable since they have less effect on apple rust mites which are essential

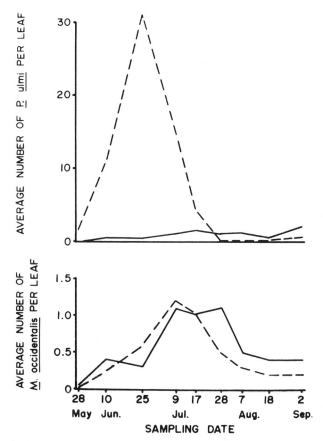

Fig. 3. Average numbers of mites per leaf on trees receiving no pre-bloom sprays (————) and on trees sprayed with 1 gallon of petroleum oil plus 0.16 pound of ethion per 100 gallons of water on April 4, 1964 (————), (After Hoyt, 1969a)

to *M. occidentalis* maintenance. Chemical thinners are applied during or just after bloom according to the selective techniques described earlier. During May and June, two or three sprays are applied for control of first-brood codling moths. Materials to control apple aphid and San Jose scale crawlers may be included with these applications. If these sprays are well-timed, codling moth should be reduced so low that further sprays are unnecessary.

If codling moth problems persist, because of improper timing or because of reinfestation from outside of the orchard, additional sprays are required. These are usually required at an adverse time in the predator:prey balance;

therefore, effective early control of codling moth is desirable. For the remainder of the season, sprays are applied only if specific problems develop. In some orchards, no additional sprays are required, but in others, control of apple aphid may be necessary. In areas where *M. occidentalis* is slow to develop because of absence of prey during the early season, it is frequently necessary to use a selective acaricide to improve the ratio of predators to prey. Unless economic damage is imminent, this corrective application should be delayed until moderate numbers of predators (an average of about 0.5 per leaf) are present. If timed properly, the surviving predators prevent a resurgence of mites for the remainder of the season (Table 4).

One problem related to need for corrective sprays for mites is the lack of adequate information on their economic levels. Precise information would be invaluable in the management of mite populations, but establishing economic levels is a slow, tedious process. For several years yet we will probably suggest corrective measures with incomplete knowledge as to whether they are warranted.

Table 4. Mite populations in an orchard using an integrated program where only *Tetranychus mcdanieli* McGregor was available as food for the predators. (After Hoyt, 1969*a*)

Mite species	1966[†] Average number of mites per leaf									
	5/17	5/31	6/13	6/27	7/11	7/18	7/25	8/8	8/16	8/22	9/6
T. mcdanieli	0.0	0.1	0.3	1.2	8.7	9.1	22.0	26.0	5.2	0.3	0.2
P. ulmi	0.0	0.0	0.0	0.1	0.0	0.1	0.2	0.2	0.2	0.1	0.0
A. schlechtendali	0.1	0.0	0.0	0.0	0.4	0.3	0.0	0.0	0.0	0.0	0.0
M. occidentalis	0.0	0.0	0.1	0.2	0.5	0.9	1.1	2.1	1.5	0.4	0.1
	1967[‡]									
	5/29	6/13	6/26	7/8	7/17	7/23	8/2	8/7	8/15	8/29	9/8
T. mcdanieli	0.1	0.1	0.6	4.3	17.0	40.0	42.0	2.0	0.1	0.1	0.0
P. ulmi	0.4	0.8	2.5	5.7	1.9	7.4	2.0	0.4	0.0	0.0	0.0
A. schlechtendali	0.0	0.0	0.1	0.0	0.0	0.0	0.0	0.0	0.0	0.0	0.0
M. occidentalis	0.0	0.0	0.0	0.1	0.1	0.2	3.8	1.2	0.7	0.1	0.0

[†]Spray program (in pounds active)
    Petroleum oil – 1-1/2 gallons + ethion – .24 per 100 gallons on 3/24 – 3/26
    Azinphosmethyl – 1.25 + dinocap – .95 per acre on 5/25 – 5/28
    Azinphosmethyl – 1.25 + endosulfan – 2 per acre on 7/27 – 7/31

[‡]Spray program (in pounds active)
    Petroleum oil – 1-1/2 gallons + ethion – .24 per 100 gallons on 3/27
    Azinphosmethyl – 1 + dinocap – 1.1 per acre on 6/7 – 6/9
    Azinphosmethyl – 1.25 + ethion – 1 + endosulfan – 2 per acre on 7/11 – 7/15
    Morestan® – .5 per acre on 8/1 – 8/3

It is not possible to give a fixed ratio of predators to prey at which selective control is warranted because this relationship is dynamic and seasonal aspects affect responses. However, if information is available on relative population changes and the distribution of predators and prey in an orchard, a realistic prediction of potential population development or decline can be made. This information can be obtained through a combination of leaf sample censuses and observations in the orchard.

Mite Populations Without Apple Rust Mites. Few McDaniel spider mites overwinter in integrated orchards. In the absence of apple rust mites *M. occidentalis* starve during April, and the population continues to decline during May and June because of a shortage of food. During hot weather in July, McDaniel mite numbers increase, followed by a delayed numerical response of the predator (Table 4). The density reached by the McDaniel mite is determined largely by weather conditions and the early-season numbers and distribution of the predator. Table 4 shows considerable differences in population development of McDaniel mite and *M. occidentalis* between 1966 and 1967. McDaniel mite numbers increased earlier and predator numbers developed more slowly in 1967 than in 1966. Temperatures were moderate in 1966, while high temperatures (in excess of $100°F$) occurred for extended periods in 1967. Force (1967) showed that *Phytoseiulus persimilis* Athias-Henriot was able to regulate populations of *Tetranychus urticae* (Koch) at 15, 20 or $25°C$ but not at $30°C$.

Relative increases in predators and prey must be carefully followed to determine if a corrective acaricide application is advisable. By July 23, 1967 (Table 4), it was obvious that a corrective spray was needed. If predator numbers are increasing rapidly (Table 4, 1966) and they are well distributed in all areas inhabited by the McDaniel mite, the latter's population then declines rapidly. Suppression to low populations usually occurs by early August and numbers remain low; thus there are very few individuals to overwinter. Predator numbers also decline and only moderate numbers over-winter. Thus, in many of these orchards, mite damage is relatively light and no economic loss is suffered if no specific acaricide applications are made. In other orchards, a single application of a selective acaricide can reduce McDaniel mite populations to a level which the predators can control.

Mite Populations With Apple Rust Mites Present. Where apple rust mite is present, its populations develop during May and June (Table 5). Populations as high as the maximum found (355 per leaf) do not cause significant foliage damage and it is doubtful if any economic loss results. *M. occidentalis* populations increase rapidly feeding on these mites (Tables 5 and 6). This predator also becomes well distributed on the tree and this presents a strong factor against increase in McDaniel mites. The latter rarely exceeds an average

Table 5.   Average numbers of mites per leaf in an orchard using an integrated program[†] where apple rust mites were present. Sunnyslope, Washington, 1969.

Mite species	Average number of mites per leaf							
	5/23	6/9	6/24	7/8	7/22	8/4	8/18	9/2
T. mcdanieli	0.1	0.2	0.2	0.7	1.0	0.1	0.0	0.0
P. ulmi	0.0	0.0	1.0	8.7	6.7	0.9	0.3	0.0
A. schlechtendali	67.0	35.0	355.0	175.0	58.0	12.0	132.0	0.2
M. occidentalis	0.1	0.2	0.3	1.0	0.4	1.6	2.3	3.3

[†]Spray program (in pounds active per acre)
    Azinphosmethyl - .75 on 5/21
    Azinphosmethyl - .75 + phosphamidon - .8 on 6/6
    Azinphosmethyl - .75 on 6/27

Table 6. Average numbers of mites per leaf where suppression of apple rust mites was necessary.[†] Columbia View, Washington, 1967.

Mite species	Average numbers of mites per leaf								
	5/29	6/20	7/6	7/14	8/3	8/10	8/21	8/28	9/19
T. mcdanieli	0.0	0.0	0.0	0.0	0.0	0.0	0.1	0.0	0.0
P. ulmi	0.0	0.1	0.0	0.1	0.2	1.0	0.2	0.8	0.1
A. schlechtendali	58.0	688.0	0.4	4.4	53.0	1.6	15.0	2.4	7.0
M. occidentalis	0.5	3.1	0.1	0.4	0.6	0.0	0.9	0.9	2.7

[†]Spray program (in pounds active per 100 gallons)
    Petroleum oil - 1 gallon + ethion - .25 on 3/28
    Azinphosmethyl - .125 on 5/25
    Azinphosmethyl - .125 + ethion - .25 on 7/1
    Endosulfan - .25 on 8/9

of 1 or 2 mites per leaf under these conditions. Though the rust mite population declines during July and early August, adequate numbers are still generally present to maintain the predator population. During late August and early September, apple rust mites increase again, with a consequent increase in *M. occidentalis.* This insures good overwintering populations of predators.

It is usually unnecessary to spray for apple rust mites since populations seldom reach damaging numbers. However, it is sometimes desirable to reduce a rapidly developing population to assure against possible economic loss (Table 6). Ethion at 0.25 pound active ingredient was used in the orchard considered here, but subsequent studies have indicated that ¼ to ½ of this amount would have given adequate reduction of apple rust mites. In any case, it is important to reduce, but not destroy, the population. For this, a low rate of application of a selective acaricide is required.

It was mentioned earlier that pre-bloom sprays are applied to reduce European red mites during May and June. If its numbers then increase during July or August, good predator distribution, based upon numerical response from feeding on one of the other two prey species brings *M. occidentalis* into sufficient contact with the European red mites to prevent their reaching damaging densities (Table 5). If, however, European red mite populations are allowed to develop during May and June, chemical control measures would be required (Fig. 3). Any application at this time for European red mite will also destroy apple rust mites and lead to predator starvation.

In many areas, winter mortality of predators is so extensive that population recovery requires a good portion of the following season. During the winter of 1968-1969, temperatures of -30°F and below were recorded in orchards of central Washington. Thus, the effects of a severe winter on overwintering survival of *M. occidentalis* populations were observed. Survival was excellent except where they were directly exposed to the cold. Mortality during late winter occurred where the predators were found above the receding snowline. The sun reflecting off of the snow caused the bark to dry rapidly and the predators to dehydrate.

Cultural Practices and the Integrated Program. Detailed studies have not been completed on the relationships of several cultural practices to the employment of biological controls. Observations indicate that orchards where sprinkler irrigation systems and sod cover crops are used have more effective integrated mite control programs than those irrigated through ditches and having only weed cover crops or clean cultivation. Overhead irrigation is practiced in several orchards, and the limited information available suggests that this method may enhance the physical control of McDaniel mite. The fertilizer program, weed control, cultivation, pruning et cetera may also effect the relationships between predators and prey but more information is necessary before we can utilize these practices towards improving integrated control.

Advantages of the Integrated Program. The integrated approach to mite control has several advantages over a totally chemical control program. Some presumed advantages are largely intuitive and very difficult to test. There is a reduction in the total amount of pesticides applied, particularly during the

latter half of the growing season. This allows an increase in the diversity of
the arthropod fauna in the orchard. Of particular importance is the increase in
general predators, such as the chrysopids, anthocorids, mirids, and syrphids.
Especially high numbers of *Chrysopa oculata* Say, *Chrysopa nigricornis*
Burmeister and the mirid *Deraeocoris brevis* have been found. In some
orchards these predators have controlled the apple aphid, eliminating any need
for a late season spray. With further study, these predators may be found to
present a predictable regulation of aphid populations. *Aphelinus mali*
(Haldeman), a very effective parasite of the woolly apple aphid, had been held
at low levels since the introduction of DDT into orchards. Substantial
increases in numbers of this parasite have occurred in integrated orchards, and
few of them have required treatments for woolly apple aphid. The reductions
in amounts of pesticides applied and in numbers of applications have
substantially reduced the cost of insect and mite control (Fig. 4). Better
control of mites, with less economic loss due to quality reduction, has been
provided by integrated control. Finally, the selection of strains of mites

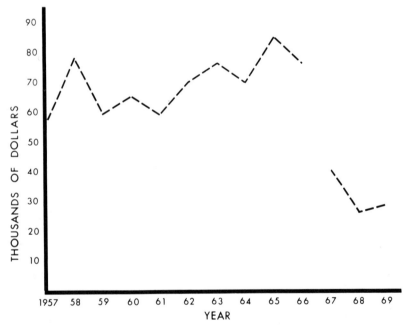

Fig. 4. Cost of materials for insect and mite control with standard spray
programs (1957-1966) and integrated control programs (1967-1969) on 1000
acres of apples.

resistant to acaricides has been slowed by the reduced selection pressure.

Disadvantages of the Integrated Program. Since the integrated method is more complex and precise, requiring deft management of insect and mite populations rather than simply near-destruction of all species present, more attention must be paid to details such as accurate timing of sprays, and correct dosages, and the faunal relationships. Much time must be spent observing population trends and distributions of predators and prey within the orchard, though this time may be offset by the release of time required for additional spray applications under the old system. While these features may be considered disadvantages by some growers, they could be overcome if grower organizations would hire individuals trained to observe and interpret faunal relationships and recommend pest management measures, as is done for some crops in California (Chapter 17) and Malaysia (Chapter 19). Mite-counting services are available to growers in both the Wenatchee and Yakima areas of Washington, but only a small percentage of growers take advantage of them. If mite counts were coupled with orchard observations by trained individuals, pest management decisions could be made on a more sound basis.

Commercial Use of Integrated Control. Integrated control was rapidly accepted by growers in Washington because results obtained under standard chemical programs were often inadequate. Integrated control required a realignment of the entire approach to mite control. Some growers rapidly understood the basic concepts behind the program, while in other cases control has been called "non-integrated" because an acaricide was applied to improve the balance between predators and prey.

The level of mite control obtained with integrated programs has varied widely. Reasons for this variation include the presence or absence of apple rust mites as food for *M. occidentalis,* growers' understanding of the program, improper applications which directly or indirectly destroy predator populations, inadequate populations of predators, and variations in susceptibility of predators to pesticides. However, in orchards using correct integrated programs, improved mite control, with a reduced number of sprays and reduced costs has been obtained (Fig. 4). Results should improve further as growers become more familiar with the techniques involved and as more information becomes available on predator: prey relationships.

THE DEVELOPING PROGRAM OF INTEGRATED CONTROL
OF PESTS OF PEACHES IN CALIFORNIA

Integrated control programs for pests of peaches in California has become a necessity, since the conventional chemical controls are at best only partially satisfactory. Although insecticides, especially organophosphates and carbamates, control most insect pests of peach satisfactorily, reducing their densities to below economic injury levels, there remain certain pests that are not controlled by these chemicals, and others that reach pest status because of the impact these chemicals have on non-target species. These pests are mainly phytophagous mites; as a group they rank as the most important agricultural pests in the state of California. In 1968, they caused an estimated loss of $54,736,112, which is about 19 per cent of the losses estimated for all agricultural insect and mite pests in California (Hawthorne, 1969). Mites are difficult to control when the program is limited to the use of chemicals because they are very prolific, and develop resistance to acaricides in a short time. It is significant that these same reasons have prompted research to develop alternative methods of control of deciduous tree fruit pests elsewhere in the United States and abroad (Asquith, 1970; Asquith and Horsburgh, 1968, 1969; Putman and Herne, 1959; Quist, 1966; also the work in Washington, *vide supra*).

The Pest Complex on Peach

There are several species of phytophagous insects and mites in California peach orchards. Some are always pests of primary importance, while others are pests only under certain circumstances. The oriental fruit moth, *Grapholitha molesta* (Busck) and the peach twig borer, *Anarsia lineatella* Zeller, are pests in most situations most of the time. Both attack the fruits in addition to the growing twigs, so it is imperative to reduce their populations to below the economic injury levels. These two moths are attacked by several natural enemies which are widespread in all peach growing areas in the State. Some of them at times destroy a high percentage of their hosts, but this is not enough. Furthermore, the effectiveness of these enemies is not increased significantly under the modified pest control program discussed below. So at the present time, the only way the grower can combat these moths is by using insecticides. Both the oriental fruit moth and the peach twig borer occur simultaneously in the same orchards, so the spray programs are designed primarily to control these pests. The chemicals most commonly used are organophosphates (e.g., azinphosmethyl, parathion), and carbamates (e.g., carbaryl) sprayed on a regular schedule during the growing season.

Other serious pests on California peaches are the tetranychid mites: the two-spotted spider mite, *Tetranychus urticae* Koch, and the European red mite, *Panonychus ulmi* (Koch). The difficulties in combatting tetranychid mites by means of acaricides are well known to farmers, entomologists, and pest control operators. It seems that mites can develop resistance to acaricides faster than the chemical industry can develop them. The efforts to combat mites make pest control in peach orchards in California a difficult, frustrating, and/or expensive operation.

The peach silver mite, *Aculus cornutus* (Banks) is an eryophiid that can reach very high population densities under certain conditions. The mites feed on the leaves, causing the foliage to turn silvery in color. Many growers consider the peach silver mite as a very important pest that has to be treated year after year.

There are other pests that at times are important: the San Jose scale, *Quadraspidiotus perniciosus* (Comstock), the consperse stink bug, *Euschistus conspersus* Uhler, the western peach tree borer, *Sanninoidea exitiosa graefi* (Hy. Edwards), the fruit tree mite, *Bryobia ribrioculus* (Scheuten), and the flatheaded borer, *Chrysobothris mali* Horn. All these can be controlled effectively by using appropriate chemicals.

Pest control on peaches in California has been limited to use of chemicals. The only instance when another approach was tried was the attempt to eradicate the oriental fruit moth by massive releases of parasites. Millions of adults of the braconid parasitoid *Macrocentrus ancylivorus* Rohwer were released from 1944 through 1946 in all the peach growing areas. This program is difficult to evaluate. When the releases were made the density of the moth was so low that the impact of the parasite on the host population was impossible to ascertain. The fact is that the moths diminished continuously until in 1947 it was difficult to find even a single specimen (Summers, 1966).

The first serious outbreak of the oriental fruit moth in California occurred in 1954. Apparently, the use of parasites was not even considered as a possibility to control the moth this second time, perhaps because the wonder insecticides, the chlorinated hydrocarbons and the organophosphates, were then available. The effort against the moth was based exclusively on use of chemicals. This is still true.

Reappraisal of Control Techniques

The way in which pests have been combatted in California peach orchards has left what van den Bosch (1966) calls a "vacuum in research upon the fundamental nature of agricultural ecosystems and the role of naturally occurring unmanipulated parasites and predators in pest suppression." The uncertainty of outcome, and the increasing costs of pest control operations as carried out today have prompted research on integrated control possibilities.

Information already gathered indicates that such a program must be centered on the phytophagous mites by modifying the practices used against other pests so as to avoid or reduce the deleterious effects of insecticides on the natural enemies of the mites.

When the amount of insecticides used against the moths is reduced by cutting down the frequency of applications and/or the rate of application, the phytophagous mites have remained at relatively low densities. We attribute this to better action of natural enemies on the mites. Our most effective predator of both the European red mite and the two-spotted spider mite is the phytoseiid mite *Metaseiulus occidentalis* (Nesbitt). The active instars feed voraciously on all stages of two-spotted spider mite, on nymphs and adults of European red mite, and also on peach silver mite. Females lay their eggs among the prey on peach leaves. Apparently, females prefer colonies of two-spotted spider mite as sites for oviposition, but they also scatter their eggs when the prey is at low density. They appear on peach foliage by the third week in April and are active until leaves drop late in the fall. *M. occidentalis* overwinters as mated females in the peduncles (fruit stems) left in the trees after the fruits are picked (Caltagirone, 1970). They hide in winter under loose bark at the base of the peduncles, and in the spongy tissue at the tips of the peduncles.

If there is a sufficiently large population of *M. occidentalis* early in the season, and if treatments to control oriental fruit moth and peach twig borer are carefully timed and kept to a minimum concentration of toxicant, it is possible for *M. occidentalis* to keep both European red mite and two-spotted spider mite at densities below economic injury levels. This has been the approach in experimenting with integrated, modified spray programs.

The tests have been conducted in a three-acre orchard of mature trees (variety Carolyn) in a peach area in Butte County. No acaricides have been used in the experimental orchard for five seasons. The spray program for 1968 and 1969 consisted of a winter treatment with dormant oil at 4gal/100 plus diazinon, 50% WP at 1 lb/100 gal, and two treatments with azinphosmethyl during the growing season each year. The treatments with azinphosmethyl were timed by following the procedure suggested by Summers (1966): (1) monitoring the oriental fruit moth flights using terpenyl acetate-brown sugar bait traps that attract adult moths, and (2) sampling twigs to detect the activity of the larvae in each generation.

The Program and Results

The results of this modified spray program for control of both the oriental fruit moth and the peach twig borer have been quite promising, and the two-spotted spider mite and European red mite have been kept under good control by predators (Fig. 5).

Smith (1969) defined integrated control as "a pest population management system that, in the context of the associated environment and the population dynamics of the pest species, utilizes all suitable techniques and methods in as compatible a manner as possible and maintains the pest population at levels below those causing economic injury." In integrated control it is implied that phytophagous insects and/or mites are manipulated in such a way as to bring them to, or maintain them in, a status of non-pests. The aim is thus not at eliminating the pest, but to adequately reduce their numbers.

The entomologist choosing integrated control as his strategy is confronted with the complex problems of determining economic injury levels for the pests, and the densities of natural enemies required to maintain the pests below those levels. The complexity of agro-ecosystems makes it difficult to know the factors that determine the population dynamics of the species involved. To develop the ideal integrated control program this knowledge is necessary. Obviously, it would not be realistic to wait until all information is in before attempting an integrated control program; the grower needs an answer to his pest problems *now*. The approach should then be to gather the most essential information, make educated guesses, and devise new control approaches accordingly. These will be in turn modified as new information is gathered.

Although the knowledge of pests on peaches in California is meager, contrasted to what is not yet known, this knowledge is sufficient to modify the pest control program towards integrated control. Some of the pertinent information follows.

Oriental Fruit Moth and Peach Twig Borer. The damage caused by the oriental fruit moth is similar to that caused by the peach twig borer. The larvae of both species mine the growing twigs, and sometimes the growing fruits, especially in spring and early summer; later they bore into the ripening fruits. The economic injury level in terms of population density for these two species has not been established. Tolerable damage to fruits may be somewhere between zero and four per cent. Some growers, and probably most if not all canners, feel that no amount of worm damage to fruits should be tolerated. Apparently, damage to twigs is of little importance in bearing trees. So, this gives some idea of how much damage we can accept. But there is another dimension to the problem. In order to implement a modified pest control program, it is necessary to predict the extent of damage that a given early season population will eventually cause, so the tactics can be modified, if necessary, without endangering either the crop or the program. The

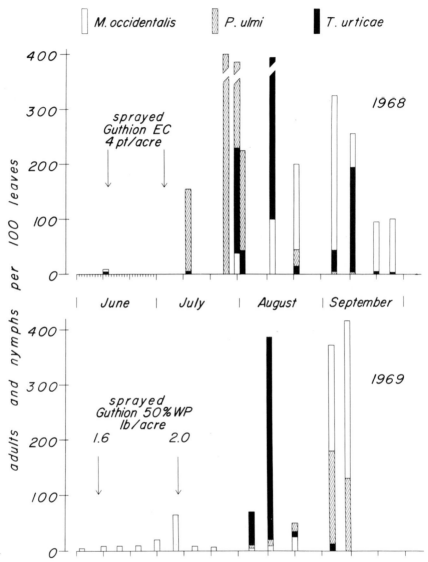

Fig. 5. Population fluctuation of the two-spotted spider mite, *Tetranychus urticae* Koch, the European red mite, *Panonychus ulmi* (Koch), and the phytoseiid predator *Metaseiulus occidentalis* (Nesbitt). Samples of 20 leaves from each of 5 trees of the variety Carolyn, Butte County, California.

question is: how small should the moth population be early in the season to be sure that their offspring will not cause economic damage? At this moment we do not know. What Summers (1966) stated for the oriental fruit moth is also valid for the peach twig borer: "No simple statistic or magic number can be proposed to describe how low the moth counts must be at any early period to guarantee that infestations will not exceed the tolerance at a later day." The only approach at this time is to reduce the numbers of moths as low as possible prior to early July without causing undesirable side effects. The life cycles of these two moths do not coincide, particularly in the overwintering instars. While the oriental fruit moth overwinters as a diapaused full grown larva in a tightly spun silken cocoon in crevices in the bark, or in the soil, the peach twig borer overwinters as a first or second instar in a small cavity (hibernaculum) bored in the bark, especially in the crotches or limbs. The overwintering peach twig borer larvae are not in diapause, but feed slowly on the bark. There is no efficient way to destroy the overwintering oriental fruit moth larvae, but the peach twig borer larvae are susceptible to some insecticides. A dormant spray of oil plus diazinone will control this pest satisfactorily, making it possible to skip the early spring treatment, thus avoiding the later spray with organophosphates which decimates the *Metaseiulus occidentalis* population at a critical time, i.e., when they are coming out of overwintering. Treatments against the oriental fruit moth are not applied until later in the season; by then, the predators have distributed themselves in the trees and started their reproductive cycle.

The European Red Mite and The Two-Spotted Spider Mite. Most peach growers have experienced the sometimes dramatic increase in numbers of these phytophagous mites following applications of chlorinated hydrocarbon, organophosphate, or carbamate insecticides where the pest control program has been based solely in the frequent, sometimes unnecessary, use of toxicants. Then to control the mites acaricides are used. This, besides increasing the cost of pest control, gives only temporary relief; in most cases the mites develop resistance to the acaricides in two or three seasons.

It seems that in this frustrating fight against mites, growers have become conditioned to regard any number of mites as dangerous, so as soon as they see them they grab the spray gun. We think this approach unjustifiable because many more mites can be tolerated than is commonly thought. Unfortunately, at this moment we do not know the economic injury levels for European red mite and two-spotted spider mite on peaches in California. But we do know that the trees on which we have conducted our integrated control studies have sustained infestations of about 5 to 12 mites (nymphs and adults) per leaf from mid-July to mid-September for five consecutive seasons without showing any harmful effects either in the size of the crop, the amount of growth, or the longevity of the leaves.

   The Predatory Mite Metaseiulus occidentalis. The predator *M. occidentalis* occurs commonly in deciduous orchards in California. It is an effective predator of two-spotted spider mite and European red mite, and also of Pacific mite on grapes (D. L. Flaherty, personal communication), and of McDaniel mites on apple in Washington (*vide supra*). Apparently, it is the key regulatory factor of these mites on peaches in California. It also preys on peach silver mite, but its role as a regulatory factor in the population dynamics of the eryophiid is not known. Organophosphates and carbamates decimate its population, but in situations where the density is high enough, particularly early in the season, the individuals that survive the treatments increase in numbers very quickly (Fig. 5). In relatively undisturbed situations the overwintering *M. occidentalis* can reach high numbers (Table 7 and Fig. 6). These individuals and their offspring can reduce the populations of the tetranychids to very low densities.

   The Peach Silver Mite. The peach silver mite, an eryophiid, plays an important role in the population dynamics of *M. occidentalis*. When tetranychids are scarce, this eryophiid provides the food needed for the large population of predators. Under these conditions, a large population of *M. occidentalis* goes into hibernation, and consequently a large number of them will be present in the orchard early in spring. If the peach silver mite is eliminated with acaricides, the population of *M. occidentalis* would starve, very few would overwinter, fewer if any at all would be present early in spring, and the tetranychids would increase to much higher numbers than if a large number of predators were present.

   The importance of peach silver mite as a pest of peaches has been

Table 7.   *Metaseiulus occidentalis* (Nesbitt) adults overwintering in peach peduncles. Mature trees of the variety Carolyn, Butte County, California, winter of 1968-1969. (After Caltagirone, 1970)

Date	Peduncles examined			Total mites found	Avg. no. mites/ peduncle	Avg. no. mites/ inhabited peduncle
	Total no.	With mites				
		No.	%			
Oct. 31	38	11	28.94	79	2.07	7.18
Dec. 5	47	15	31.91	70	1.48	4.66
Jan. 7	153	37	24.18	181	1.18	4.89

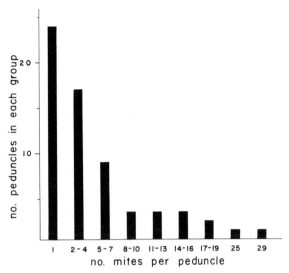

Fig. 6. Distribution of overwintering *Metaseiulus occidentalis* (Nesbitt) in peach peduncles. Mature trees of the variety Carolyn, Butte County, California (winter of 1968-1969). (After Caltagirone, 1970)

overestimated. Our observations suggest that the economic injury level for this eryophiid is several times higher than generally accepted. We have not made a quantitative analysis of the populations of silver mite in our experimental orchard, but every year the population has reached levels high enough as to cause silvering of roughly 50 per cent of the foliage. However, no premature defoliation has been noticed in these trees.

Comments. The preliminary program for integrated control of pests of peaches in California, in those orchards where *Metaseiulus occidentalis* is present, is based on the following:

(1) Control of peach twig borer, San Jose scale, and mites (partially) with a dormant spray.

(2) Elimination of the early spring spray.

(3) Control of oriental fruit moth, timing the treatments very carefully; two sprays during the season may be sufficient.

(4) Reduction or elimination of the sprays to control two-spotted spider mite and European red mite; if there are enough predators, treatments are not necessary.

(5) Limitation of the treatments against peach silver mite to the

absolute minimum; very few cases, if any at all, should require spraying.

There are many uncertain critical points in this program, the most important ones being the significance of the various phytophagous species in the orchards, i.e., their economic injury levels have not been determined. Besides, their post harvest effects on following years crops remain to be studied.

LITERATURE CITED

Asquith, D. 1970. Encouraging predators of mites in apple orchards. Penn. Fruit News 49(3).

Asquith, D., and R. L. Horsburgh. 1968. Predators of orchard mites and their role in mite control. Penn. Fruit News, March.

Asquith, D., and R. L. Horsburgh. 1969. Integrated control versus chemical control of orchard mites. Penn. Fruit News 48: 38-44.

Caltagirone, L. E. 1970. Overwintering sites for *Metaseiulus occidentalis* in peach orchards. J. Econ. Entomol. 63: 340-341.

Chant, D. A. 1958. On the ecology of typhlodromid mites in Southeastern England. Proc. X Intern. Congr. Entomol., Montreal (1956) 4: 649-658.

Collyer, E. 1958. Some insectary experiments with predacious mites to determine their effect on the development of *Metatetranychus ulmi* (Koch) populations. Entomol. Exp. Appl. 1: 138-146.

Flaherty, D. L., and C. B. Huffaker. 1970. Biological control of Pacific mites and Willamette mites in San Joaquin Valley vineyards. I. Role of *Metaseiulus occidentalis*. Hilgardia 40: 267-308.

Force, D. C. 1967. Effect of temperature on biological control of two-spotted spider mites by *Phytoseiulus persimilis*. J. Econ. Entomol. 60: 1308-1311.

Hawthorne, R. M. 1969. Estimated damage and crop loss caused by insect/mite pests— 1968. Calif. Dept. Agr. Rep. to Calif. Agr. Commissioners. Nov. 7, 1969.

Hoyt, S. C. 1966. Resistance to binapacryl and Union Carbide 19786 in the McDaniel spider mite, *Tetranychus mcdanieli*. J. Econ. Entomol. 59: 1278-1279.

Hoyt, S. C. 1969a. Integrated chemical control of insects and biological control of mites on apple in Washington. J. Econ. Entomol. 62: 74-86.

Hoyt, S. C. 1969b. Population studies of five mite species on apple in Washington. Proc. Second Intern. Congr. Acarology, Sutton Bonington, England (1967): 117-133.

Hoyt, S. C. 1970. The effect of short feeding periods by *Metaseiulus occidentalis* (Nesbitt) on fecundity and mortality of *Tetranychus mcdanieli* McGregor. Ann. Entomol. Soc. Amer. 63: 1382-1384.

Hoyt, S. C., and F. H. Harries. 1961. Laboratory and field studies on orchard-mite resistance to Kelthane. J. Econ. Entomol. 54: 12-16.

Hoyt, S. C., and J. R. Kinney. 1964. Field evaluation of acaricides for the control of the McDaniel spider mite. Wash. Agr. Exp. Sta. Circ. 439, 13 pp.

Huffaker, C. B., and C. E. Kennett. 1953. Differential tolerance to parathion of two *Typhlodromus* predatory on cyclamen mite. J. Econ. Entomol. 46: 707-708.

Kuchlein, J. H. 1965. A reconsideration of the role of predacious mites in the control of the European red spider mite in orchards. Boll. Zool. Agr. Bachicolt. Univ. Studi Milano, Ser. II, 7: 113-118.

Laing, J. E., and C. B. Huffaker. 1969. Comparative studies of predation by *Phytoseiulus persimilis* A.-H. and *Metaseiulus occidentalis* (Nesbitt) on populations of *Tetranychus urticae* Koch. Res. Pop. Ecol. 11: 105-126.

Lord, F. T., H. J. Herbert, and A. W. MacPhee. 1958. The natural control of phytophagous mites on apple trees in Nova Scotia. Proc. X Intern. Congr. Entomol., Montreal (1956) 4: 617-622.

MacPhee, A. W., and K. H. Sanford. 1961. The influence of spray programs on the fauna of apple orchards in Nova Scotia. XII. Second suppl. to VII. Effects on beneficial arthropods. Can. Entomol. 93: 671-673.

Mori, H., and D. A. Chant. 1966. The influence of prey density, relative humidity, and starvation of the predacious behavior of *Phytoseiulus persimilis* Athias-Henriot. Can. J. Zool. 44: 483-491.

Putman, W. L., and D. C. Herne. 1959. Cross effects of some pesticides on populations of phytophagous mites in Ontario peach orchards and their economic implications. Can. Entomol. 91: 567-579.

Quist, J. A. 1966. Approaches to orchard insect control. Colo. State Univ. Agr. Exp. Sta. Bull. 517S, 93 pp.

Smith, R. F. 1969. The importance of economic inury levels in the development of integrated pest control programs. Qual. Plant. Mater. Veg. 17: 81-92.

Summers, F. M. 1966. The oriental fruit moth in California. Univ. Calif. Agr. Exp. Sta. Circ. 539, 17 pp.

van den Bosch, R. 1966. The role of parasites and predators in integrated control. Proc. FAO Symp. on Integrated Pest Control 2: 143-147.

Chapter 19

DEVELOPMENT OF INTEGRATED CONTROL PROGRAMS FOR PESTS OF TROPICAL PERENNIAL CROPS IN MALAYSIA

Brian J. Wood

Chemara Research Station
Layang Layang, Johore, Malaysia

INTRODUCTION

Malaysia comprises two regions, West Malaysia (the former Straits Settlements and Federated Malay States), consisting of the Malayan Peninsula, and East Malaysia (the States of Sarawak and Sabah), consisting of the northern part of the island of Borneo. The country has a predominantly agricultural economy with vast areas devoted to monocultures of a range of crops. The climate is wet tropical (hot and humid). The temperature is without marked fluctuations or extremes (Dale, 1963) and rainfall occurs throughout the year (Dale, 1959). Despite the apparent favorability of this climate to their increase, a variety of potential insect pests remain at low numbers most of the time, particularly in perennial crops. Circumstantial evidence strongly suggests that this is because climatic conditions also favor a controlling balance between the pests and their insect natural enemies. On the other hand, pests can increase rapidly and continuously if this natural balance is disturbed, particularly by pesticides which, in some circumstances, may relatively favor the pest insect in the long run. Dramatic results occurred when broad-spectrum, long-residual contact insecticides were applied in both oil palms and cocoa, leading to pest explosions. This was a principal indicator of the need to deal with all these pest problems in a manner taking full account of ecological factors, supplementing and utilizing natural control agents to the fullest extent. The need for such an integrated approach to pest problems is now widely accepted and practical programs are being devised, as will be illustrated with examples taken from three major perennial crops, oil palms, cocoa and rubber. Conway (1971*a*) has previously discussed the problems from an ecological viewpoint.

OIL PALMS

Oil palms were first planted in West Malaysia in the early part of this century. By 1950, there were 100,000 acres and this is now approaching 500,000. They are planted at about 50 palms/acre, always in large plantations because the enterprise is dependent upon there being a processing factory in the vicinity. A few large estates have been established in Sabah in the last ten years or so, and of late the acreage there has also rapidly increased. Wood (1968) described the pests so far recorded and discussed some ecological factors involved in outbreaks.

Leaf-Eating Caterpillars

West Malaysia. Commencing in the late 1950's, certain estates suffered severe outbreaks of bagworms (Psychidae) and other leaf-eating caterpillars. These led to very severe defoliation and a good deal of spraying was done with broad-spectrum, long-residual contact insecticides such as DDT, dieldrin and endrin in an effort to contain the pests. Several species of bagworms occur, the commonest being *Cremastopsyche pendula* Joannis and *Metisa plana* Wlk. The caterpillars feed on the leaves. Isolated individuals do little or no harm but large numbers can lead to loss of virtually all phytosynthetic tissue, both by direct consumption and onset of necrosis.

Bagworms have an unusual biology, with features which determine many of the characteristics of outbreaks. The caterpillars live in cases constructed of silk, often including pieces of material taken from the host plant, and they are built up as the caterpillars grow. The cases are attached to the plant material except during movement, and may remain so fixed for some time after death or maturation. Eventually, pupation occurs in the same case, which can then be called the cocoon. This stage assumes an appearance distinctive for the species on any particular host plant. The adult male becomes a typical flying moth but the female remains within her cocoon (the "bag"), becoming little more than a sac of eggs. The eggs all hatch together and the young caterpillars disperse so that where plants are in contact, they spread outwards as they mature. Some of the new larvae appear to be carried off on air currents or by other accidental means, and can thus form nucleus infestations in new areas. There is generally a complete spread of broods, but when an outbreak occurs there is, characteristically, an even-brooded condition, the great majority of individuals being in the same stage at the same time. In the species studied on oil palms, each generation lasts about three months, and females produce from a few dozen to several thousand eggs, depending on the species. Thus, in the even-brooded condition, there can be a sudden many-fold increase (explosion) at each new generation.

The other leaf-eating caterpillars which have been recorded in large numbers have a more typical life cycle, both males and females being winged moths. However, they do not appear to travel far and an outbreak tends to be concentrated in an area and spreads out relatively slowly. The commonest group are nettle caterpillars (Cochlidiidae). Some species have a voracious appetite and can rapidly and completely defoliate palms. The most serious are *Setora nitens* Wlk. and *Darna trima* (Moore). The first named can be particularly damaging.

Entomological investigations specifically on oil palm problems were sponsored from 1962, after the outbreaks had become serious on some estates. Interpretation of earlier events was, then, dependent upon a study of estate documentation. Wood (1966) examined the history of five estates more or less heavily affected by leaf-eating caterpillar problems in the period up to 1964 and it is clear that the outbreaks first occurred or assumed serious consequences only *after* the treatments with broad-spectrum, long-residual contact insecticides were commenced. In some cases, dieldrin or endrin were applied against minor outbreaks of the pests because, having seen the results of *induced* explosions elsewhere, a grower would assume that unless he took action he would be faced with a similar problem. That this often happened "despite" the measures, served to confirm, for him, the suspicion that he had been visited by a plague which was breaking out in various parts of the country. Sometimes, outbreaks arose after chemical use against limited outbreaks of other pests, for example on one estate where spraying and heavy defoliation had affected 6,000 acres, the cause was, in part, traced back to the use of DDT against cockchafers. The time at which outbreaks occurred, and the localities, were quite different and the only feature in common was the insecticides. Several major oil palm estates were involved and once such a situation developed, outbreaks continued to spread at varying rates and, sooner or later, they would recur. Spraying was often not done very adequately, especially at first; the terrains are commonly uneven, ruling-out the use of tractor-drawn equipment, and the palms were in many cases too tall for use of shoulder-mounted devices such as mistblowers. Spray coverage was thus often patchy, and it was in these circumstances that outbreaks spread and recurred most rapidly, and where the worst defoliation took place. One estate, for example, sprayed virtually the whole of its 7,000 acres at least once and some fields five or six times from 1960-1963, experiencing severe defoliation and losing around 40 per cent of the crop in many hundreds of acres in one year, with full recovery taking several years.

In most cases, initially, both bagworms and leaf-eating caterpillars became serious, but eventually the latter ceased to be the main problem, probably due to increasing efficiency in chemical application, and although they tended to continue at higher numbers than usual, with occasional outbreaks, bagworms became the chronic pests. The interpretation of the estate histories is obscured in respect of the two bagworm species because they were both referred to as *C. pendula.* There is evidence that *C. pendula* was initially most common, while *M. plana* built up mainly as a result of spraying.

This history, and its parallel with the insecticidally-induced outbreaks in numerous other crops in the world, left little doubt that the insecticides had caused the attacks by their adverse effects on natural balance, and from the end of 1962 it was recommended that spraying of broad-spectrum, contact insecticides be stopped. Residual populations of bagworm and, more spasmodically, nettle caterpillars then existed in many of the areas where spray campaigns had been carried out, and high populations of natural enemies were associated with them. In addition to pests and their natural enemies, many insects neutral to the crop were in much larger numbers than usual. Studies were initiated on the natural enemies, both predators and parasites, and a good deal of information was acquired (Table 1) although more detailed knowledge is still needed of their life cycles, host preferences and relative abundance to assess which are the key enemies in maintaining a balanced situation.

The hope was that by stopping spraying the source of disturbance would be removed and the pests would come under complete natural control again, but despite their commonness at times, natural enemies seemed to remain at a disadvantage in the new situation so that outbreaks often recurred. The even-brooded condition which is somewhat self-perpetuating is probably the main cause of this, and itself is likely to have been brought about by insecticide use, the bagworms of a particular age being differentially spared from the effect of the sprays so that they constitute the ancestors of future generations. Further sprayings would tend similarly to re-emphasize even-broodedness. Although their biologies have not been studied in detail, the parasites evidently tend to have life cycles a good deal shorter than those of their bagworm hosts, and thus would be dependent upon a spread of broods (apparently a usual situation in low density populations) in order to find hosts

Table 1. Summary of studies on the natural enemies attacking leaf-eating caterpillars which were present in relatively dense populations on oil palms during a period after insecticide-induced outbreaks. Johore, 1963 onward.

Host Species	Stage	Enemy	Comments
Metisa plana Wlk. (Psychidae)	Larvae	*Apanteles metesae* Nixon (Braconidae)	Common. In one study, heavy natural mortality was recorded, between 60 and 100% in several collections and overall this species accounted for 60% of the mortality. High percent parasitization by this species also occurs in low density populations.
		Various Braconidae, Ichneumonidae, Chalcididae, Eupelmidae, Eulophidae, Elasmidae and Ceraphronidae	Complex of parasites and hyperparasites including 11 recorded species ranging from rare to common.
		Callimerus arcufer Chap. (Cleridae)	A common polyphagous predator. Eggs and larvae are often found inside the host cases.
	Cocoons	Various Braconidae, Ichneumonidae, Chalcididae, Eupelmidae and Eulophidae	8 recorded species, mostly in common with those which attack larvae. Cocoon parasitization less frequent than larval parasitization for this host - only 16% of some 5,000 examined was parasitized.
Cremastopsyche pendula Joannis (Psychidae)	Larvae	Eulophidae	Only one specimen of a large number examined was parasitized.
		Sycanus dichotomus Stal (Reduviidae)	In the one widespread outbreak, this predator was commonly observed feeding on the pest. In a laboratory trial, in a total longevity of 5 months, 20 specimens consumed an average of 430 field-collected bag-worms each.
	Cocoons	Various Braconidae, Ichneumonidae, Chalcididae and Eulophidae	Species mostly in common with those attacking *M. plana*. Samples over a period revealed an average of 26% parasitization.
Setora nitens Wlk. (Cochlididae)	Eggs	*Trichogrammatoidea nana* Zehntner (Trichogrammatidae)	Of 18 eggs field-collected at different times and places all had signs of extant or past parasitization. About 25-30 individuals hatch from each egg but no unparasitized egg could be found (although it should be noted that parasitized eggs are more conspicuous).
	Larvae (May emerge from cocoon)	*Spinaria spinator* Guer. (Braconidae) plus various Ichneumonidae, Braconidae, Chalcididae, Eurytomidae, Ceraphronidae, Tachinidae and Bombyliidae	12 species have been recorded. *S. spinator* is the commonest. In one field collection of 68 caterpillars from a minor outbreak the fate was: parasitized - 62%; died from unknown causes - 31%; survived to produce cocoons - 7%. Similar or even higher parasitization was found in other collections made in similar circumstances.
Various leafeating caterpillars (Cochlididae and others)	Eggs and larvae	Various parasites	Many of the *S. nitens* parasites are recorded from other species, plus other Hymenoptera and Diptera.
	Larvae	*Sycanus dichotomus* and Asopinae (predatory pentatomids) e.g., *Cantheconidea furcellata* Wolff	Prey on various caterpillars - common at times.

suitable for egg laying. For example, *Apanteles metesae* Nixon emerges from a *M. plana* larva which is approximately half grown (it lays its eggs in younger caterpillars). Presumably at times, therefore, although there may be numerous individuals present of the host species, the parasites are largely unable to locate suitable stages for egg laying.

This failure of parasite/host synchrony, once an outbreak developed, could explain why the increase and outward spread of bagworms continued for some generations after the initial upsurge, and also their resurgence some generations later after an active natural enemy population appeared to have re-established. It became clear that if good natural control of bagworms were to be restored without the risk of defoliation of a consequence serious to crop yield, it would not be possible in all areas just to stop spraying and wait for natural enemies to resume control. Instead, selective control measures were needed which would substantially reduce bagworm numbers while permitting the survival of natural enemies.

Selectivity in chemicals may depend on the way they are used, on certain intrinsic characteristics, or on a combination of these. When the decision was made to avoid non-selective insecticides against pest resurgences, our knowledge of the biological and ecological factors was insufficient to permit any attempt at obtaining selectivity by any method depending on use of chemicals in a discreet manner. The immediate hope lay in applying chemicals themselves possessing a degree of selectivity. The most likely candidates were stomach poisons, contact poisons with a degree of selectivity against Lepidoptera or those with short-lived residues. Lead arsenate, a stomach poison, seemed promising. Trials demonstrated a moderate but not dramatic kill of both bagworms and nettle caterpilllars, with reductions of from 60 to 90 per cent in doses of 3 lb/acre or more. It was used against nettle caterpillars (*Darna trima*) where there had been repeated re-outbreaks, alternating with use of the broad-spectrum contact insecticide, endrin. Although fluctuations in numbers subsequently occurred there were no further outbreaks of economic consequence and eventually the pest disappeared virtually completely on that estate, until it is now rarely seen (Fig. 1). Other similar incidents occurred with this group of pests.

In 1964, a bagworm build-up commenced in about 1,000 acres in one estate. It was obviously going to lead to serious defoliation unless contained, and in February 1965, lead arsenate was applied from the air in order to cover the whole area rapidly, following the emergence of a new generation. The results are illustrated in Fig. 2. It is apparent that the spray had no rapid effect but the pest declined to below the economic threshold (*see* Table 2, estate 2, *vide infra*).

Trials with other possible integrating chemicals continued and the use of

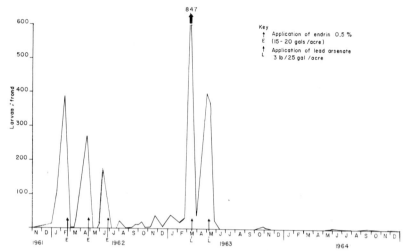

Fig. 1. Population of nettle caterpillars—mainly *Darna trima* (Moore) (Lepidoptera, Cochlidiidae)—on 1959-planted palms during a period when endrin sprays were applied and after the use of lead arsenate. Johore, 1961-1964. [All applications using shouldermounted mistblowers. Counts are averages for 96 fronds on 55 acres.]

trichlorphon (= Dylox or Dipterex) was considered, following Conway's applications of it in cocoa (*vide infra*). This chemical is a contact/stomach insecticide with a fugitive contact residue and it has proved selective in a number of situations. Trial work showed that it gave an excellent kill of young bagworms when applied from shoulder-mounted mistblowers at doses of 13 oz/acre and higher, matching the previously used 4 oz endrin in this respect. It was applied from the air at 26 oz/acre on a 2,900-acre area in October, 1965. Success was achieved and little trouble was experienced over most of the area subsequently (Fig. 3). Since then, outbreaks have been recorded in various estates, virtually always either in or adjacent to areas where broad-spectrum contact insecticides were previously used. Several aerial spray campaigns with trichlorphon have been carried out successfully (Wood, 1967) and events on four estates are summarized in Table 2. No upsets or immediate resurgences have occurred and further treatments have only been necessary occasionally. In one such case (Estate 4i in Table 2), the spray was

Table 2. Aerial spray programs using trichlorphon against the bagworm *Metisa plana* Wlk. since 1965 on four oil palm estates in Johore, where severe outbreaks had previously been associated with the use of broad-spectrum long-residual contact insecticides, in particular endrin, during the period 1960–1963. Compiled November, 1969.

Estate	Date	Acreage	Dosage Trichlorphon/acre	Average no. larvae/frond† Precount	Average no. larvae/frond† Postcount	Use of non-selective chemicals prior to 1963? ‡	Comment§
1 (11,000 acres)	i Oct 65	2,900	26 oz/2 gal	See Fig. 3	3	Yes	Some lightly infested margins were unsprayed at *i*. *ii* includes these plus 300 acres of *i* which had to be resprayed presumably due to spread back into treated area. Mostly they remained free but occasional small-scale flare-ups occurred. Rather more severe build-up in part of the area led to *iii*. Since then the estate has remained free of further signs of attack.
	ii Mar 66	600	19 oz/1-1/2 gal	26	6	Yes	
	iii Feb 68	1,000	13 oz/1 gal	25	5	Yes	
2 (7,000 acres)	i Feb 65	1,000	4 lb/3-1/2 gal (lead arsenate)	See Fig. 2	2	Yes	*ii* was alongside *i*. *iii* was in the same area, re-covering about 500 of *i*. Since this time, apart from some minor flare-ups which subsided without treatment, no new outbreaks have occurred.
	ii Feb 66	530	19 oz/1-3/4 gal	17	2	Yes	
	iii Jul 66	1,500	13 oz/1 gal	28	1	Yes	
3 (6,000 acres)	i Mar 66	740	19 oz/1-1/2 gal	11	1	Nearby	*ii* was adjacent to *i*. No further attack in *i*, but some flareup in *ii* during 1969, with ground spray being required. Another block on this estate of some 500 acres in an area of original disturbances is now experiencing a bagworm resurgence. Isolated flare-ups have occurred there from time to time since 1963.
	ii Jan 67	160	13 oz/1 gal	12	1	Nearby	
4 (7,000 acres)	i Mar 66	200	19 oz/1-1/2 gal	18	29	Nearby	*i* was sprayed before full emergence of young. *ii* included this area. No further flare-ups since these sprays. *ii* is adjacent to areas which had received massive doses of non-selective contact insecticides. The bagworm population rapidly declined after the spray and remained very small except in *iii*, adjacent to *ii*.
	ii Apr 66	1,000	19 oz/1-1/2 gal	29	<1	Nearby	
	iii Apr 69	660	13 oz/1 gal	14	<1	Nearby	

†For census system *see* Wood (1968, Ch. 5)--6 fronds/c.7 acres samples. Postcounts from 2-4 weeks after spray. The counts represent over-all average populations including lightly infested margins. Much heavier populations sometimes existed within the sprayed area.

‡Yes = previous upsurges and non-selective spraying in that area; Nearby = the area is part of a contiguous block in which upsurges occurred elsewhere.

§Except for a few isolated ground sprays, this includes reference to *all* bagworm outbreaks in that particular estate.

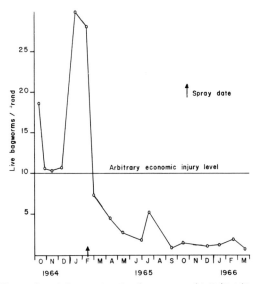

Fig. 2. Effect of aerial spraying lead arsenate (4 lb/3-1/2 gal/acre) on the population of *Metisa plana* Wlk. (Lepidoptera, Psychidae) on 8-9 year-old oil palms. Johore, 1964-1966. [Counts are averages for 732 fronds in 1000 acres.]

applied at the wrong time, when the bagworms were still in cocoon stage. This was done intentionally as a trial, and showed that active residues of the chemical would not remain long enough to deal with emerging young caterpillars. In the other case (Estate 1i), coverage was not extended to lightly infested margins and it appeared that build-up took place there in the next generation and there was a limited spread back into part of the original sprayed area.

It is not easy to carry out small scale experiments with aerial spraying but in the earlier campaigns a safety-margin was allowed in the dosage. However, it was gradually found that this could be reduced.

The situation today is that bagworms sometimes build up, still almost always in areas with a previous history of induced outbreaks, even though this may have been several years ago, and a spray campaign is required. Many estates, particularly those which have had a prior history of trouble, continue census rounds (Wood, 1968—his Chapter 5) which give an index of average numbers of bagworms according to developmental stage per frond. This forewarns them if action is required, and permits careful timing of the spray to coincide with the peak emergence of young larvae. A typical sequence of

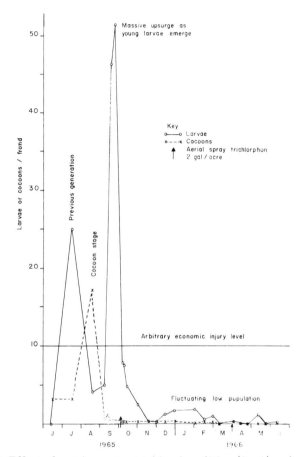

Fig. 3. Effect of aerial spraying trichlorphon (26 oz/2 gal/acre) on the population of *Metisa plana* Wlk. (Lepidoptera, Psychidae) on 7-9 year-old oil palms. Johore, 1965-1966. [Counts are averages for 480 fronds in 500 acres.]

census figures at a period when a spraying was done is shown in Fig. 4. No severe outbreak of nettle caterpillars has been recorded in recent years, and such limited outbreaks as do occur, mostly in nurseries and young plantings, are tackled with lead arsenate or trichlorphon, so far with no marked repercussions.

At the time that entomological investigations on oil palm pests were begun, an attempt was made to assess experimentally the importance of natural enemies in regulating bagworms, by employing a variation of the

CATERPILLAR CENSUS, HAY ESTATE

Field – 62
Acres – 120
No. of points/fronds checked – 16/96

| | Live Bagworm Stages | | | | | | Other Caterpillars | | | |
| | Larvae | | Cocoons | | | | | | | |
Date	Newly Hatched	Mature	*M. plana*	*C. pendula*	*M. corbetti*	Others	Total	/Frond	Predominant Species	Remarks
Jan. 8	0	3.8	0.4	0	0	0	4.2	0		Approaching maturity
Feb. 12	0	0.3	1.1	0	0	0	1.4	8	*Setora nitens*	Cocoons, male flown Expect larvae
Mar. 18	37.1	12.0	0.1	0	0	0	49.2	4	" "	High emergence – Treat
Apr. 25	0	0.1	0	0	0	0	0.1	0		Trichlorphon sprayed Mar. 23 at 19 oz/1-1/2 gal.
June 6	0.1	0	0.1	0	0	0	0.2	0		Can reduce census frequency
Sept. 17	0	0.1	0	0	0	0	0.1	0		

Fig. 4. Estate census record, illustrating the sequence of events in an infestation of *Metisa plana* Wlk. (Lepidoptera, Psychidae) during which aerial spraying with trichlorphon was carried out. Johore, 1966.

insecticidal check technique which has had success with a variety of pests, mainly those with restricted powers of dispersion such as scales and mites (DeBach, 1958). Because of the greater mobility of bagworms and, probably, their principal natural enemies, in comparison with the latter pests, single tree test treatments were considered to be insufficient. Instead, a plot of two acres was regularly given a low dosage dieldrin treatment in intentionally patchy cover. Bagworms were counted along three planting lines passing through the sprayed plot (palms are planted on a triangular system) for a distance of 50 palms each side of the center (= 500 yd). Counts were then averaged for groups of ten palms 1-10, 11-20, et cetera, adding the equivalent groups from the three transects. Thus, the relative numbers of bagworms could be compared for palms at the farthest distance from the sprayed area (groups 1-10 and 91-100), for intermediate groups of increasing closeness to the spray, and for those actually affected by it (groups 41-50, 51-60). Counts for two such trials are represented in Figs. 5a and 5b. The technique was developed as work proceeded and the possibility of fortuitous results cannot be ruled out for certain. For example, in the experiment shown in Fig. 5a, no pre-counts are available, since the need for such long transects was not recognized until pest increases began to occur. Nevertheless, the marked peak of the bagworm population around the sprayed area is clear in both cases. Fig. 5b shows the population before the commencement of spraying and afterwards, and it seems that the pre-treatment population was evenly spread and that the peak was induced by the spray. The second trial was continued for further

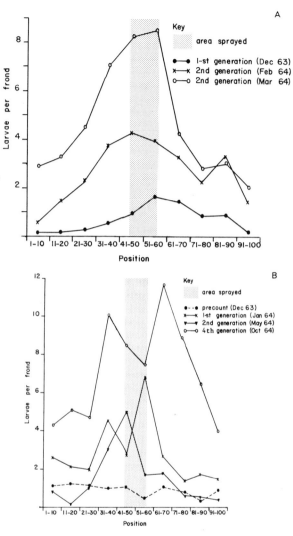

Figs. 5a and 5b. Number of live larvae of *Metisa plana* Wlk. (Lepidoptera, Psychidae) on 6 year-old oil palms in and around a two acre plot given low volume dieldrin sprays: Ulu Remis estate: Field 37 (5a, above); Field 29 (5b, below) 1963-1964. [Position is the position of palm along 3 equi-angled transects, each extending 50 palms in either direction from center of sprayed plot. Counts are averages of 2 fronds in each group of 10 palms, totalled for equivalent groups on each transect.]

generations. The peak did not become much more marked but instead the graph line as a whole simply moves upwards, presumably due to the spread of bagworms out from the center. A slightly smaller number in the sprayed area itself (Fig. 5b–Oct., 1964) is possibly due to an adverse effect from the spray against very young bagworm larvae.

East Malaysia. So far, only one major incident concerning oil palm pests has occurred in Sabah, but this one case led to extensive and severe damage (Wood and Nesbit, 1969). Two pests are involved, a bagworm (*Mahasena corbetti* Tams) and a nettle caterpillar (*Darna trima*). The former is a good deal bigger than *Metisa plana,* and individuals are consequently much more damaging. Further, each female produces from 3,000-4,000 eggs, so the potential rate of increase per generation is very high. Except in this instance, *M. corbetti,* although commonly seen on isolated oil palms in both East and West Malaysia, never increased to any extent, the majority of individuals being parasitized by tachinid flies. This situation continues in most areas and the evidence is that these natural enemies are responsible for the low numbers of the pest.

The bagworm was first noticed in large numbers in an estate in eastern Sabah during early 1965, and a devastating infestation rapidly built up, some 300 acres being completely defoliated and many more partially so. In September, a new generation emerged and there was a clear need for treatment before defoliation occurred over additional large acreages. The presumed importance of natural enemies in regulating this species, coupled with experience in West Malaysia, suggested that broad-spectrum, long-residual contact insecticides should be avoided. A hurriedly executed trial showed trichlorphon to give a reasonable kill and an aerial spray was decided upon. Through lack of forewarning, consequent unpreparedness and the remoteness of Sabah, it was impossible to execute the spraying in the most satisfactory way. There was insufficient chemical available to allow a safety margin in the dosage and application was rather late in the generation when the bagworm caterpillars were quite large and well established. About 1,000 acres were sprayed and there was a fair reduction of bagworms, although by no means complete elimination. Subsequently, further outbreaks took place in the same area and *D. trima* also increased so that during 1966 several sprayings of trichlorphon or lead arsenate were applied, both by air and from the ground.

Two striking features distinguished this situation from that which developed in West Malaysia consequent upon the widespread use of long-residual, broad-spectrum contact insecticides. First, in the early stages, virtually no natural enemies of either pest could be found. There is no evidence to suggest how these outbreaks commenced, due to a paucity of records concerning chemical application or early outbreak history, but it may be supposed that originally the situation was similar to that prevailing elsewhere, with the pests at very small numbers and their parasites and predators consequently being extremely rare (notwithstanding their numbers *relative* to the pest). Once the balance was upset the realization of the bagworm's huge increase potential at each generation apparently allowed it to reach super-abundance before the few natural enemies initially present had yet built up to the point where they were readily detectable to an investigator. The second contrast was the failure of all other insects to increase noticeably, whereas in West Malaysia a variety of "neutral" insects, as well as pests and their natural enemies, became exceptionally common for a time after the chemicals had been applied. In Sabah, only the two species of pests became common, suggesting that whatever factor or factors may have led to the initial upsurges of the pests, they were not responsible for general disturbances to natural balance. The immediate aim in the chemical applications was to prevent the recurrence of devastating damage, but Wood and Nesbit (1969) felt that the best hope for finally controlling the pest without the need of regular spray programs (perhaps having additional side-effects) was to continue with selective chemicals so that such natural enemies as did exist could have the opportunity to restore control at a low density. The best immediate hope of achieving this appeared to be by increasing the efficiency of the trichlorphon application to the point where an extremely heavy kill of the pest could be obtained, thus greatly altering the relative numbers of the abundant pests and their still scarce enemies.

By the end of 1968, new outbreaks of both *M. corbetti* and *D. trima* occurred in the originally affected area and in addition, presumably due to "seeding" from there, a new outbreak of *M. corbetti* was occurring some distance away. Both areas were sprayed in early 1969. This time, a safety margin was ensured by spraying trichlorphon at a relatively high dosage (26 oz/3 gals/acre), while timing was given a good deal of consideration. The two areas were slightly out of phase in terms of bagworm development, the originally infested area (A) being about two weeks ahead of the new area (B). The spraying was timed to coincide with the peak emergence in area (B). It was felt that it might be slightly late in area (A) but that this was a lesser risk than spraying while a high proportion of the brood was still in the cocoon stage. The expense and difficulty of bringing the spray-plane 1,000 miles from W. Malaysia ruled out the possibility of spraying on two separate occasions. Results are shown in Figs. 6a and 6b, respectively. At the time of the treatment, a very small number of cocoons had still to emerge in area (B), in order to catch the vast majority of young at the most susceptible time, and it seemed they supplied the nucleus for increase after the application; in other words, the spraying was a few days early. As it happened, the desired result of

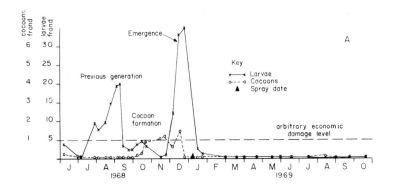

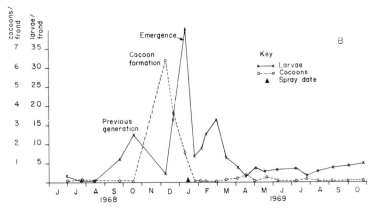

Figs. 6a and 6b. Effect of aerial spraying trichlorphon (26 oz/3 gal/acre) on populations of *Mahasena corbetti* Tams (Lepidoptera, Psychidae)–Eastern Sabah, 1968-1969.

Fig. 6a (top)–7 year-old oil palms. [Counts are averages for 588 fronds in 560 acres. Well-timed applications.]

Fig. 6b (bottom)–5-9 year-old oil palms. [Counts are averages for 858 fronds in 900 acres. Application slightly too early. Larvae emerged from few cocoons present at spraying. Ground applications continued to be made in worst affected fields.]

a virtually complete kill was obtained in area (A).

Up to this time the scarcity of natural enemies had continued, with only isolated signs of parasitization by tachinids, but in May 1969, one sample of *M. corbetti* taken from the areas included in the recent spray showed 12.6 per cent parasitization by this group.[1] Subsequently, more parasitization was

[1]Sankaran, T. Restricted circulation reports, Sabah Sub-Station, Commonwealth Institute of Biological Control.

noted. *Exorista psychidarum* Bar. was the commonest species, with *Palexorista solennis* (Wlk.) and *Eozenillia equatorialis* Townsend also occurring. Some hymenopterous parasites have also been recorded. During 1969 there was no sign of re-increase of the pest to economic levels in the area (A) where initial kill had been high, despite the passage of sufficient time for three further generations. The population gradually declined over most of area (B) with no further major resurgence. Increase did take place in some adjacent fields, perhaps because a few bagworms had spread and multiplied prior to entry of the natural enemies.

D. trima remains something of an enigma. The outbreaks appear to have been a consequence of trichlorphon application. They have been brought to a conclusion quite easily by very low dosage applications of that chemical and have resurged commonly. Frequently, heavy infestations have come to an end through the sudden spontaneous death of virtually all caterpillars. The latter is accompanied by symptoms resembling those associated with a virus in West Malaysia (G. M. Thomas, pers. comm.), although the existence of a virus has not been confirmed in caterpillars from East Malaysia.

Red Spider Mites

Occasional outbreaks of red spiders (Tetranychidae) occur in oil palm nurseries, particularly in dry weather. Two species have so far been recorded, *Oligonychus* sp. and *Tetranychus piercei* McGregor. These pests, in the heaviest attacks, cause browning and early die-back of the leaves of the young palms, and can retard growth severely. During the 1950's and early 1960's there was a tendency to use virtually any available insecticide against them. Where a broad-spectrum, long-residual contact insecticide was chosen, the infestation frequently worsened and there were outbreaks of leaf-eating caterpillars in the nursery, including nettle caterpillars and bagworms, as well as locally severe attacks by polyphagous caterpillars such as *Spodoptera litura* (F.). These attacks led to more spraying and sometimes, for a time, nurseries became heavily infested with a variety of pests.

A trial carried out in 1962 compared the effect of a specific miticide, tetradifon (Tedion), with that of some of the insecticides then in use, assessing the effect not only on the spider mite—in this case *Oligonychus* sp.—but also on its coccinellid predator, *Stethorus siphonulus* Kapur. The results (Table 3) show that tetradifon gave the best kill of mites, as well as confirming its selective action, in that it was the only chemical used which did not eliminate the predator. Since that time, tetradifon has been regularly used for red spider mite control in many estate nurseries, sometimes on a

Table 3.   Effect of various chemicals against the red spider mite, *Oligonychus* sp. (Acari: Tetranychidae) and its predator *Stethorus siphonulus* Kapur (Coleoptera: Coccinellidae) in an oil palm nursery.   Johore, October 1962.   (After Wood, 1966)

Treatment[†]		Average number of spiders per pinna[‡]		No. of *Stethorus siphonulus* [§]	
Chemical	Concentration (% a.i.)	Prespray	Postspray (5 days)	Prespray	Postspray (5 days)
Tetradifon EC	0.016	9.2	.1	40	3
Tetradifon WP	0.02	8.0	.2	32	18
Thiometon	0.25	10.4	3.9	29	0
Menazon	0.042	10.6	1.2	42	0
Malathion [¶] + DDT	0.125 0.25	12.1	1.8	29	0
Untreated check		7.7	4.9	24	19

[†]Hand-operated knapsack sprayer applying 6 fl. oz./palm (= 180 gal./acre).

[‡]9-palm plots x 6 replicates/treatment.   Counts on 5 pinnae selected at random from center palm in each replicate (30 pinnae/treatment); all motile stages counted with 8x binocular.

[§]On 4 fronds selected at random from center palm in each replicate (= 24 fronds/treatment), counted in the field.

[¶]Mixture in commercially available formulation.

prophylactic basis. Where this has been the practice, spider mites have not resurged and no induced insect problems have occurred.

The Bunch Moth

The caterpillars of the bunch moth, *Tirathaba mundella* Wlk. (Pyralidae) develop in the fruit bunches of oil palms. They bore into fruit tissue and can cause loss of kernels. In large numbers the lesions thus caused can lead to complete rotting of the bunch. Attacks are common on young palms when the first-produced fruit bunches are left to rot unharvested, presenting ideal conditions for the development of the pest. This often happens in new oil palm areas where the building of a factory has been delayed until reasonably heavy crops are being produced. Heavy attacks have also been recorded after the period when broad-spectrum, long-residual contact insecticides had been used in association with leaf-eating caterpillar attacks (*vide supra*), suggesting that natural enemies may generally be of some importance in its regulation, although little is yet known about its specific natural enemy fauna.

Wood and Ng (in prep.) described the pest, with accounts of experiments on control. The possibility of inducing outbreaks of other pests was considered a strong reason for avoiding the use of broad-spectrum, long-residual contact insecticides in oil palms, apart from the circumstantial evidence that natural enemies are important in control of bunch moth itself. Trichlorphon was used initially and was reasonably effective (about 90 per cent reduction at 8 oz/acre). A later series of trials showed that several organo-chlorine insecticides gave virtually 100 per cent kill (Table 4). Among these was endosulphan (Thiodan)®, a chemical with theoretically good characteristics for integrated control (Huttenbach, 1969, on which grounds it was chosen for use against the spasmodic bunch moth outbreaks. So far, it has proved effective against *T. mundella* and there has been no indication of induced increase of this or any other pest as a result of its use.

Rhinoceros Beetle

Oryctes rhinoceros (L.) has been a major pest of coconuts throughout Southern Asia for some years and has become a potentially devastating pest of young oil palms. Palms in replantings are especially at risk because the decaying logs of a former stand of oil palms, rubber or coconuts provide breeding grounds for the larvae, which develop in rotting organic materials. The beetles attack the heart of the palm, boring at the base of unopened leaves (spears) so that newly emerging leaves are broken off, truncated or distorted. Pathogenic organisms may enter the lesions, leading to long periods of distorted leaf production or even death of young palms.

Attempts at control by drenching the spears with insecticides were often made but these were not very effective, and in some cases this use was associated with a subsequent outbreak of leaf-eating caterpillars, constituting the beginning of a major problem over a large area (*vide supra*). Usually, control was attempted by breaking up the breeding grounds and removing developing stages (Wood, 1968, p. 131). This proved very expensive and not always highly effective. A series of experiments was commenced in 1962 investigating the possibilities of rendering the rotting tissues of a former stand unusable by the beetle, of speeding up the rotting away of logs or of destroying them completely at replanting. Planters had for long held that a heavy cover of ground vegetation was associated with reduced attack by the beetles, and this agreed with observations in coconuts in the South Pacific

Table 4. Performance of selected insecticides in various trials against the oil palm bunch moth, *Tirathaba mundella* Wlk., 1967. (Condensed from Wood and Ng, in prep.)

Trial	Treatment[†]		Average *Tirathaba* larvae/bunch[‡]		C.R.I.[§]
	Chemical	Concentration	Precount	Postcount	
A. Chemicals considered very likely to be potentially suitable for integrated control.					
(1)	DDVP	0.05	51	36	192
		0.1	52	17	89
(1)	Bidrin	0.05	76	72	258
		0.1	53	69	354
(2)	Ryanicide	0.2	29	2	23
		0.4	43	3	23
(1)	Thiodan	0.035	61	2	9
		0.07	123	0	0
B. Chemicals considered less likely to be suitable for integrated control.					
(2)	Heptachlor	0.025	37	8	71
		0.05	41	0	0
(2)	BHC	0.025	26	2	25
		0.05	29	0	0
(3)	Aldrin	0.025	13	1	60
		0.05	29	0	0
(3)	Endrin	0.025	32	9	220
		0.05	27	0	0

[†]Application 1/2 pt/bunch = c.33 gal/acre.

[‡]Single-palm plots, counts on one bunch per plot. Trial (1) - 6 replicates. Trials (2) and (3) - 4 replicates. Postcount from one to two weeks after spray.

[§]C.R.I. = $100 \frac{ad}{bc}$ where a and c = counts before and after standard spray, b and d = counts before and after experimental spray. (After Ebeling, 1947)

(Owen, 1959). Studies on this were included in the experimental program (Wood, 1969*b*).

Ground cover in young oil palm plantings may be either a "natural" cover, comprising self-perpetuating species of grasses, ferns and creepers, or a planted leguminous cover. The trials confirmed that under heavy cover (irrespective of type) there was substantially less breeding of beetles. Table 5 for example shows the difference in the numbers of grubs found in logs under a "natural" cover and on bare ground, respectively. Further, it appeared that, irrespective of breeding populations, damage to the palms was greatly reduced by such a cover. Table 6 shows the damage to palms, classified according to severity. In this instance, cover works against the adult's searching for a feeding site, since very intensive inspections of rotting logs ruled out the possibility that the effect was due to a greater number of beetles developing on the bare-ground plot. The way in which the cover works is not certain, but it seems probable that it impedes the flight or movement on the ground of the adult beetle.

This potentially devastating pest is now kept well within economic limits simply by encouraging the early establishment of a ground vegetation cover

Table 5. Effect of ground cover on number of *Oryctes rhinoceros* (L.) (Coleoptera: Dynastidae) collected from rotting oil palm stems and stumps in an oil palm replant. Johore, May 1965 - October 1966. (Condensed from Wood, 1969*b*)

Vegetation[†] cover	Mean no. *O. rhinoceros*/acre[‡]			
	All stages	Adults	Pupae	Larvae
Bare ground	2,453	95	64	2,294
Dense natural vegetation	287	8	22	257

[†]Two one-acre plots of each type.
[‡]From 96 stumps + stems/acre, totalled for 5 inspections over 18-month period.

Table 6. Effect of ground cover on damage by *Oryctes rhinoceros* (L.) (Coleoptera: Dynastidae) to 2-1/2-year-old oil palms replanted after oil palms. Johore, June 1968. (After Wood, 1969*b*)

Cover type	No. palms examined	Percentage of palms with indicated degree of damage				Percentage of palms replaced due to death since initial planting
		Severe	Medium	Light	Nil	
Dense natural cover, up to 3 ft. high	297	1.0	5.1	30.6	63.3	0
Bare ground	297	31.3	51.8	12.5	3.0	1.4

(which is in any case desirable for agronomic reasons) and leaving it undisturbed while the palms are young.

Other Pests

A number of other insect pests occur on oil palms and locally severe attacks may occur from time to time. Wherever possible, chemicals or application methods offering a degree of selectivity are used against these pests. For example, trichlorphon proves effective against leaf-eating cock-chafers (Rutelidae and Melolonthidae) while topical application of dieldrin to the trunks kills the termite, *Coptotermes curvignathus* Holmgr., a pest which may in some circumstances feed on the tissues, with a potential of killing the palms. Occasionally, as happened with grasshoppers, no selective method may be available, while very severe damage is occurring. In such a case, until further investigations reveal a selective method, the application of a broad-spectrum, long-residual contact insecticide may be demanded. Where this is done, a close watch is kept on the area subsequently and at the first sign of build-up of leaf-eating caterpillars, selective chemicals are applied.

Rats, too, are currently well controlled on oil palms by baits containing warfarin, using a technique developed after extensive ecological investigations to assess rat numbers and the extent of the losses they cause (Wood, 1969*a*), and on the effectiveness of various control measures (Wood, 1969*c*). In that this control generally has no side-effects on the remainder of the fauna, so far as can be judged, it fits in with the concept of integrated control in the broadest sense. At only one point does the rat problem require correlating with pest control as a whole, in that rats do little damage to young palms on *bare* ground. However, it very rapidly became apparent that the rats which do occur under ground vegetation constitute less of a problem than would *O. rhinoceros* on bare ground, so heavy ground cover is used and baits are applied against the rats.

COCOA IN SABAH

Apart from some early attempts which did not come to fruition, cocoa was first introduced to Sabah in the 1950's and commercial plantings were started in 1956. Ten years later, there were about 6,000 acres of this crop, mostly in the south-eastern part of the State on volcanic soil. Some of the cocoa is planted on estates which were already growing tropical perennial crops but most is in current clearings in the rich primary dipterocarp forest that is the climax vegetation of the area. The plantings are still surrounded by jungle or uncontrolled secondary forest. Commercially valuable timber is extracted, and most of the other trees are felled and allowed to rot on site, but a number of second-story trees and saplings are left to give the necessary shade for the cocoa.

Conway (1971*b*) documented the pest problems in some detail. A cocoa research station was established at Quoin Hill in 1958, covering some 220 acres. During the period 1958-1960 a problem was experienced with three bark-boring caterpillars (a hepialid ring-bark borer, *Endoclita hosei* (Tindale), and two cossid branch borers, *Zeuzera* spp.). These created considerable die-back of branches and death of trees. Attempts were commenced to kill the pests by the application of broad-spectrum long-residual, contact insecticides to the woody parts of the cocoa trees, after it was found that control by hand picking was too costly. Both dieldrin (at 15-25 oz/acre) and DDT (12-15 oz/acre) were applied as high volume sprays. From this point, events took place which showed that a range of potentially serious pests existed in the cocoa and dramatically illustrated the way in which the potentiality can be realized if natural enemy control is disturbed. The events are described in the following paragraphs and illustrated in Fig. 7.

During 1960, new pests began to appear of a type not previously common in cocoa, and the amount of spraying carried out was increased accordingly, overall applications of dieldrin, endrin, DDT, BHC, lead arsenate

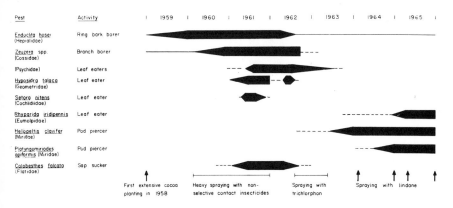

Fig. 7. Schematic representation of pest outbreaks on a heavily sprayed cocoa planting. Quoin Hill, Sabah, 1959-1965 (after Conway, 1971*a*). [Very schematic; bars do not directly indicate size of the population, only the periods during which the species was of economic importance.]

and white oil all being used at various times. Such pests as leaf-eating caterpillars, aphids, mealybugs and plant hoppers all increased in numbers during this period. The position continued to get worse, and in mid-1961 bagworms appeared and extensive defoliation took place as they built up and spread. Spraying with the above-named insecticides continued throughout 1961; many of the other pests came under control but bagworms became predominant and devastating. At this stage, the probability that the serious insect problems were at least in part a repercussion of the use of insecticides became apparent, and the spraying was stopped. Conway reasoned that, irrespective of any possible role in *causing* the outbreaks, certainly these chemicals were preventing the natural enemies from having a chance in re-establishing control.

Almost immediately a looper caterpillar was heavily parasitized by a braconid. It came under control and although within some months there was a resurgence, the population again dropped and subsequently continued at a negligible level. Other pests also declined—sometimes the decrease was clearly associated with parasitic attack, sometimes less obviously so.

The ring-bark borer continued to be a problem, as did the bagworm. The reasons why the latter might have continued even after the discontin- uance of spraying of broad-spectrum, contact insecticides may well be similar to those suggested as the cause in oil palms (*vide supra*), lying perhaps in the initial rarity of pests and hence their natural enemies, in the huge increase potential of the pest from generation to generation, and in the created even-broodedness, all mitigating against any rapid re-establishment of control by parasites. Integrated chemical control was attempted against the borer by making localized applications of dieldrin only to the bore-holes; but at about this time the pest was discovered also to be developing in a secondary forest tree, *Trema cannabina,* which was a common shade tree and had grown up in almost pure stands along roadsides and other places. These trees were then eradicated, and ring-bark borer damage dropped rapidly, so that from 1962 onwards trouble from this pest has been negligible. Conway searched for a more selective insecticide to apply against bagworms and found that the most suitable was trichlorphon. This was applied at 3/4 lb/acre and it was immediately effective in reducing the bagworms and permitting new growth to appear on the cocoa. Regular treatments had to be carried out on a monthly basis for some time and gradually the bagworm population was reduced to insignificance, after which spraying was stopped. Little problem has occurred in subsequent years from any of these pests.

Taking their lead in choice of chemicals from the Research Station, other estates in the vicinity adopted similar measures against initial borer problems, and they experienced a similar cycle of events. One estate, which decided that it would continue to use the broad-spectrum, long-residual contact materials after they were discontinued elsewhere, continued to have branch borer problems and outbreaks of bagworms and other leaf-eating caterpillars. Estates which avoided using the broad-spectrum contact insecticides from the beginning also remained free of the severest manifestations of pest attack.

Subsequently, in 1963, two mirid bugs appeared on the cocoa at Quoin Hill and other estates and slowly spread through it. These are the mosquito bug, *Helopeltis clavifer* Wlk. and the bee bug, *Platyngomiriodes apiformis* Ghauri (Conway, 1964). Even in small numbers they can do serious damage by producing lesions in the developing cocoa pods, thus permitting entry of pathogenic fungi. It became clear that some form of control would be required, but previous events warned against the too ready adoption of insecticide programs. Experience with related pests in Africa indicated that lindane could be effective and fairly selective. This chemical has a fairly fast-fading residue and dosages were kept very low (4 oz/acre) and restricted to two applications per year, prior to the fruiting season. If feasible, application was localized only to patches of cocoa where the infestations occurred.

This procedure proved effective in controlling the mirids and has not led to substantial repercussions. An eumolpid leaf-eating beetle has increased perhaps as a result of these sprays but to date it has not been very serious. Attempts at control by selective means are being made.

Minor pests occasionally trouble cocoa growers but they have usually responded to control by insecticides with a degree of selectivity, and in the last five years no major problem has arisen.

RUBBER IN WEST MALAYSIA

Rubber has been grown in Malaysia since 1877. It occupies by far the biggest acreage of any perennial crop, with 4.4 million acres in West Malaysia at present. So far, rubber plantations have been largely free of serious insect problems, although a variety of pests can be found on the tree, including leaf-eating caterpillars, beetles and grasshoppers; sucking pests such as mites, thrips, mealybugs and scales; and wood feeders, in particular termites. Occasionally, locally severe outbreaks of these pests may occur, for example various coccids when in association with ants.

In addition to the rubber itself, a leguminous ground cover comprising one or more species of creepers, sometimes with bushes, is beneficial to the crop and is sown and cultivated for some years after planting (Wycherley and

Chandapillai, 1969). This crop is subject to various pests. In particular, a leaf-eating caterpillar, *Nacoleia diemenalis* (Guen.), has frequently occurred and done severe damage, particularly after broad-spectrum, long-residual contact insecticides have been applied. Therefore, avoidance of over-all applications of these insecticides in rubber, especially where there is a leguminous ground cover, is desirable. The need for an ecological approach to pest problems in rubber has been realized from an early date. Rao (1965) described the pests of both rubber and the cover crop and outlined a general approach to their control.

Melolonthid cockchafers of various species occur and may do very severe damage. Two species, *Lachnosterna bidentata* (Burm.) and *Psilopholis vestita* Sharp are most common. The grubs of these cockchafers feed on roots. The pests have an annual life cycle. The adults emerge from the forest (hence outbreaks are restricted to areas bordering forest) during the early part of the year and lay their eggs in the soil. The developing grubs consume the roots of the rubber and cover plants, leading to severe weakening of the rubber trees, with consequent marked crop reduction and slowed maturity, and death is not an uncommon consequence. All ground vegetation is completely destroyed, often leading to soil erosion. Pupation takes place in the soil and the newly emerged adults return to the forest where they feed and mate, after which the rubber is re-entered to set off a new annual cycle.

Prior to the 1940's, digging and hand-collection was the only form of control practised. The method is only partially effective and itself may involve some root destruction and add to any erosion problem. Later, effective soil insecticides such as heptachlor and aldrin became available and careful use has been made of them. In first year plantings, about 1 litre 0.1 per cent heptachlor emulsion is poured into 6-8" deep holes around the base of each tree (Anon., 1968). This method cannot be continued when cover becomes established and requires protection since it would involve widespread distribution of insecticide, which would be highly expensive and has been known to lead to upsets in which *N. diemenalis* destroys the cover that was supposedly being protected (J. Palmer, pers. comm.).

Instead, during the period when the trees are maturing and the leguminous cover crop is at its most valuable, adult beetles can be effectively controlled by light traps (Rao, 1964). The behavior of the cockchafer lends itself to applying this form of control cheaply. There is a well-defined swarming period and trapping need only be carried out for about two months in the early part of the year. Additionally, beetles fly in substantial numbers only for a few hours after dusk. Black light is most effective and enormous numbers can be caught (Table 7).

Rao (1964) assessed the effectiveness of the light trap method by

Table 7. Number of *Lachnosterna bidentata* (Burm.) (Coleoptera: Melolonthidae) adults caught at UV-light traps in rubber estates on affected jungle margins. West Malaysia, 1960-1964. (After Rao, 1964)

Year	Dates	No. of traps	Site	Acreage	Total catch
1960	2/10-4/13	3[†]	1	100	59,245
1961	2/10-4/30	2	1	100	42,050
		2	2	20	33,220
		2	3	10	14,606
1962	2/27-3/16	2 ⎫	4	100	82,262
	3/17-3/26	3 ⎭			
1963	2/28-4/7	8	5	150	139,113
		6	6	150	110,675
1964	2/25-3/3	4 ⎫	7	20	54,263
	3/4-4/2	5 ⎭			

[†]One of the traps, taking 12,076, used a mercury vapor lamp.

comparing the number of grubs in the offspring generation against those in a nearby place where lights were not used. He found averages of 9 against 46 grubs/sq. yd., respectively. In addition to not disturbing biological balance in general, this technique allows unhindered activities of scoliid wasps and tachinids which attack and reduce populations of cockchafers to some extent (Rao, 1965).

As the rubber trees become mature and crown density and trunk girth increase, light traps cease to attract cockchafers over a wide area. By this time, the quality of ground cover is not particularly important and heptachlor granules can be broadcast on the soil surface at 1 lb/acre (Anon., 1968). No adverse side-effects have occurred. The application is primarily to kill newly matured adult beetles as they leave through the soil; the treatment is delayed until shortly before emergence is due to commence, permitting maximum survival of scoliids and tachinids which mature before the cockchafer adults.

DISCUSSION

The Need for and the Manner of the Integrated Approach

Smith (1968) listed two fundamental principles of integrated control;

the first is—consider the agro-ecosystem; and the second—utilize economic damage levels. The need for an ecological approach in oil palms and cocoa was made forcibly apparent by the serious pest outbreaks which occurred due to the unwitting disregard of the first principle, and a similar need is indicated for other tropical perennial crops in the Malaysian region, either by less dramatic or well-documented examples or by analogy.

Bartlett (1964) pointed out that a good deal can be learned by study of the exact details surrounding insect outbreaks following pesticide use. A consideration of the circumstances in which major pest outbreaks occurred in oil palms and cocoa led to the deduction that disruption to natural balance was responsible for the outbreaks. This in turn indicated the importance of natural enemies in maintaining the more normal, largely pest-free situation, and provided a basis for immediate action and the long term approach to pest problems.

The principles of integrated control are now kept foremost in the pest control programs for the three crops discussed here, both to restore natural balance in upset situations and to avoid similar incidents in tackling further pest problems. This has involved integrated control in the narrower original sense of the term (Stern et al., 1959) of supplementing natural enemy control by the use of selective chemical measures, and in the broader sense of developing control techniques which utilize all suitable measures in as compatible a manner as possible, as described by Smith (1968), an approach which Geier (1966) called "pest management." None of the crops is native to Malaysia, all three having been introduced within the last 100 years. However, the pests are primarily derived from the native fauna and have previously existed on other crops or on non-economic plants, but as they have moved onto the new crops so have their natural enemies and these and other natural regulating mechanisms have continued to be operative. Conway (1968) raised this point and forcibly argued the need for an ecological approach to pest control in the tropical environment. Rao (1965) mentioned it in connection with rubber. The mere fact of establishing a monoculture in an area previously occupied by a mixed vegetation has not automatically led to environmental disturbance whereby pest outbreaks have occurred. This has only happened when additional factors predisposed towards pest increase.

Stern and van den Bosch (1959) stated that each pest problem has unique characteristics which must be analyzed and understood before any integrated control program can be undertaken. While this is desirable and should be the ultimate objective, it is unnecessarily rigorous for tropical perennial crops where the factors restricting pest numbers tend to be strongly and continuously biotic and where biological upsets can be so devastating. Here, attempts at integrated control are essential right from the outset.

Initially, they can be based on *a priori* considerations such as provided by Bartlett (1964) on the characters of insecticides *vis-a-vis* their likely potential as selective insecticides. As Smith (1968) said, the belief that it takes too long to develop the necessary information and technology to meet current crop protection needs is not supported by the cases where integrated control is established.

Faced with an emergency situation, both in sprayed oil palms and in cocoa, due to the dramatic spread and increase of leaf-eating caterpillars and other pests, it was necessary to apply control measures immediately to prevent total devastation. A good deal of desirable background information obviously was missing but the need to avoid nonselective insecticides was soon very clear. Insecticides of potentially selective action (e.g., stomach poisons or ones presenting only short-lived residues) were tested for effectiveness and then applied. The extended success of measures, which had been presumed to have a chance of being selective in the particular circumstances, supports to some extent the initial assumptions regarding the causes of outbreaks (Conway, in press *a*).

Conway (pers. comm.) likened pest management to business management, where from time to time it is essential to make a decision despite inadequate premises. It is then necessary to make the most appropriate decision possible. The outcome of the step taken is itself a source of further information for the future. The great complexity of tropical agro-ecosystems makes it unlikely that the outcome of a potentially integrative control practice, at least where chemicals are involved, could be forecast with certainty on *a priori* grounds, however much information is available. Moreover, small plot trials will not always be likely to clearly indicate the effect of the same measures used on a large scale because with moving populations, any effect on relative numbers of pests and their enemies in the plot may be diluted within the area as a whole. A chemical which causes a disruption in a small plot will almost certainly cause one over a large area, but absence of an upset in a plot does not mean that none would occur if a whole planting or a large area were involved. Prior to attempting control of the mirid bugs on cocoa (*vide supra*), Conway considered all available indications and decided to use lindane at low dosage. The fact that no dramatic pest upsurges of the kind which had been induced by broad-spectrum, long-residual contact insecticides occurred after several treatments in different areas constitutes the only ultimately acceptable evidence that lindane application was integrative in these circumstances.

The major outbreaks of leaf-eating caterpillars and other pests which occurred in oil palms and cocoa as a consequence of insecticide use are not simply intensified versions of problems which are regularly faced in these crops. The integrated control in these cases consisted in applying chemical measures to prevent severe damage during periods when, it was hoped, full natural balance could be restored so that the pests would return completely to their more usual state of innocuousness. Had such measures not been attempted, and the use of these broad-spectrum, long-residual contact insecticides continued, effort being concentrated on improving the kill obtained through more adequate coverage, attention to timing, dosage and so on, then the inevitability of increasingly intense and regular spraying programs could well have become accepted. This actually occurred in some oil palm estates in W. Malaysia for some years, and still continues in Sumatra, a large island near to and climatically and biologically similar to W. Malaysia, where regular DDT treatments are accepted on some estates as necessary in growing oil palms. A vicious circle develops, where each treatment has the desired effect of controlling the target pest but also makes inevitable the necessity for subsequent control. The problem is often further complicated by the onset of resistance to the insecticides being used so that a continual search for new insecticides has to be undertaken. Such situations are not uncommon in agriculture where regular spraying is practised against a target pest which would be far less important if different control regimes were followed (Bartlett, 1964). The original pest may have been quite different from the eventual target pest and the critical factor in such cases is whether after the earlier situation is restored the primary pest can be controlled by integrative means. This has proved to be the case in the Malaysian situation, a notable example being the control of the wood-boring caterpillars in cocoa by removal of the alternative host plant, after earlier attempts at chemical control had been responsible for massive increases of a range of leaf-eating pests. Similarly, the minor pests in oil palms, control of which was implicated in setting off massive pest outbreaks, can all be dealt with by integrated means.

Not all pest outbreaks in perennial crops in Malaysia are induced by indiscriminate use of chemicals, and some insecticidal control is required from time to time against pests that build up for other reasons. Chemicals are chosen as far as possible on the possession of attributes tending to make them selective. After trials with various insecticides, it is not necessarily the one giving the highest kill that is recommended against a particular pest nor is control of an outbreak the end of the matter—a close watch is maintained after several applications for any signs of biological upset before the treatment is finally accepted. In oil palms, leaf-eating caterpillars, the bunch moth, leaf-eating cockchafers, spider mites and other pests sometimes build up, and treatments for them using selective chemicals have been developed which have shown no tendency to induce further, more serious troubles. In cocoa, lindane selectively controls the mirids, and here limited frequency and low dosage also contribute to selectivity. Use of non-selective insecticides against root-feeding cockchafers in rubber and against termites in that crop and oil palms leads to control, but side-effects are avoided by topical applications.

Attention is paid to the second of Smith's (1968) principles—"utilize economic damage levels" (a concept which is discussed in more detail in Stern *et al.,* 1959). The census system to monitor the incidence of the important leaf-eating caterpillars in oil palms gives a clear indication of population size at any time and place, and although values for economic thresholds are somewhat arbitrarily assumed, if followed they permit control at a time when parasites have been allowed a chance to build up but before any very serious damage has been done. The situation is less clearly determined with other pests, but a control measure is never initiated at the first sign of attack. Instead, it is delayed for some time to decide if the trend is towards increase and spread, with significant damage, or to subsidence. The use of ground cover against *O. rhinoceros* in oil palms and light trapping of cockchafers in rubber, good integrated control measures, considering the crop and its biota as a whole, are suitable methods because the survival of a few individuals is acceptable in terms of the limited damage done, when compared against the expense (and relative impracticability) of attempts at total control. Losses as high as 40 per cent of the year's crop occur after a single full defoliation by bagworms. The value of studies on economic damage levels, of course, depends on the further correlation of pest numbers with the amount of damage done and the relationship to yield. In cocoa, the problem of a very low damage threshold for the mirids was overcome by application of integrated control measures, which have to be applied against numbers which for most pests would be considered insignificant.

Illustration of Some General Principles of Biological and Integrated Control

Recently, much has been published on the theory and practice of biological and integrated control. The experiences in tropical perennial crops described here supply further evidence for some of the general conclusions reached, while conversely those general principles provide the basis for interpreting much of what has happened in these crops and give a background for developing integrated control programs. Some of this is covered in the preceding section. Additional points are discussed in the following paragraphs.

DeBach (1964) points out that the end result of an outstanding example of biological control may be "a rare species being attacked by a rare natural enemy." He calls this "invisible" biological control. No importation or augmentative manipulation of natural enemies has been made in the case of the pests involved in the various insecticidally induced outbreaks on oil palms and cocoa, but the potential for increase which was demonstrated, when compared with the general rarity of the pests, illustrates such a situation in respect to the undisturbed naturally-occurring biological control which exists in the absence of upsets. Conservation of existing natural enemy effectiveness was employed here.

Huffaker and Messenger (1964) discussed the relative importance of density-dependent and non-density-dependent factors in the regulation of pests. They stated that if the abiotic forces are more constantly stable, there is nothing to check indefinite increase except some natural form of action related to densities. The climate in Malaysia is regularly and continuously conducive to the development and increase of a wide range of potential pests. In oil palms and cocoa, the removal of the check on increase imposed by natural enemies, due to the use of insecticides, allowed some formerly inconspicuous leaf-eating insects to rise to the point where total defoliation occurred. The next level of regulation was starvation through this destruction of the food source. Doutt and DeBach (1964) also raised the question of the greater favorability of tropical latitudes towards regular and undisturbed natural balance. They mentioned (as one cause of the converse) that the more discrete annual cycles of species in temperate climates require more precise synchronization of host/natural enemy voltinism. Its absence is not normally of consequence in the tropical environment because the majority of pests have complete overlap of generations so that synchronization is not a necessary attribute for an effective natural enemy. However, when the even-brooded condition arises, as was seen to occur in bagworm outbreaks, parasites, although they may become numerous, may be unable at times to find sufficient hosts at the right ages to maintain their populations at a suitably responsive level. Such a mechanism appears to be responsible for the long continuance of the bagworm infestations, once they have begun, and for the re-enhanced outbreaks which occurred several generations (up to some years) later. Flaherty and Huffaker (1970) described the perpetuation of spider mite outbreaks following pesticide upsets on grapes in California, that required four years to be corrected.

Another consequence of climate being regularly favorable to pest increase but regularly counteracted by natural enemy action is that such outbreaks as do arise spontaneously are spasmodic rather than specifically related to certain times of the year or at particular places, being dependent more on chance factors, both biotic and abiotic. Undoubtedly, climatic variations have some effect on insect numbers even in the tropics. For example, leaf-eating caterpillars are generally a little more common on oil palms at about the turn of the year and although there is no definite evidence of the mechanism, this is when the northeast monsoon is blowing and rains are somewhat heavier, and shortly after the arrival of massive numbers of migrant insectivorous birds. These increases are not the massive changes in insect numbers which occur in strongly seasonal climatic regimes and are perhaps due to slight disturbances working through an influence on host/ enemy balance, as described by Huffaker and Messenger (1964). *Economic* outbreaks do not seem to be more common at this time than any other. Local climatic extremes may cause a temporary disruption in natural balance, which subsequently subside to bring small outbreaks thus caused to a conclusion. Rao (1965) mentioned that localized pest increase in rubber often comes to an end with a change in weather. Weather may sometimes directly affect pest status. Spider mite outbreaks in oil palm nurseries, for example, often terminate quite abruptly at the onset of heavy rain, with no apparent intermediate mechanism. Some studies indicate that such species may simply be washed from the foliage (Huffaker *et al.,* 1969).

The Establishment of Integrated Control in the Tropics

Where the ecological approach is ignored, regular expense and sometimes catastrophic pest attacks can ensue. Examples of this are documented here, and others occur. For instance, the beginnings of a leaf-roller attack were noted on one occasion in 1962 over a few acres on a tea estate in W. Malaysia. This appeared to be associated with the prior use of DDT sprays (Wood, privately circulated report). It was the first recorded attack of any consequence in that area and was at an early stage of increase. Spraying was simply stopped, and since then there has been no further problem. This contrasts with certain tea estates in Sumatra where regular dusting with DDT is carried out against this same pest, and where a whole complex of other pests occur, regularly devastating the crop.

There is cause to look at the pest control procedures in many tropical crops, for integrated techniques can often be developed without the prior need for highly sophisticated, time-consuming and expensive ecological studies (notwithstanding their ultimate desirability) to give far more effective as well as cheaper results. The work discussed here is on perennial crops which provide a stable situation. Varley (1959) and van den Bosch and Stern (1962) pointed out that there is a greater possibility for utilizing biological control agents in perennials than in annual or shorter-term crops since the planting, maturing, harvesting, cultivation and replanting of the latter are generally synchronized over large areas, and this hinders the formation of a stable natural balance. Nonetheless, the possibilities for integrated control in short-term crops are being considered, for example—rice, the major annual crop in Malaysia, seems to offer possibilities for success (Lim, 1970.) Undoubtedly, more selective chemicals are preferable to less selective ones, and in many tropical areas, including Malaysia, the planting times are scarcely governed by weather or other seasonal factors and a degree of continuity in pest and natural enemy relationship could be brought about by the simple procedure of staggering planting times.

The deliberate upset trials in oil palm tended to confirm the importance of natural balance and it would be useful to pursue these techniques further in various tropical crops as a source of information about the ecology and biology of potential pests and their natural enemies as well as to demonstrate the importance of the enemies. Rubber, so far free of major induced pest attack, might be especially interesting.

Unfortunately, attempts at deliberate upset cannot be restricted to individual trees as with scales or other more sessile pests because the insects on these crops generally have greater mobility and are usually very scarce at the equilibrium level, perhaps absent on many trees. The trial reported here was in an area where bagworms were relatively common (although, at the time, below the economic level) after heavy outbreaks. The induction of a build-up *ab initio* might require the spraying of even larger acreages with an attendant risk of setting off widespread economic outbreaks. This makes the entomologist reluctant to persuade a grower to allow such experiments and they would need to be done in isolated, abandoned or non-cropping areas unless financial support to deal with the possible consequences could be guaranteed.

The position of the insecticide manufacturer and importer also requires consideration. In regular integrated control programs the aim is to keep chemical usage to a minimum, and in the case of restoring an upset situation, large quantitites of insecticides are required in orders that in all probability will not be repeated. Therefore, there is the risk of either insufficient insecticide being available, or an importer finding himself with a large stock which he cannot sell. Possibly some form of government support to importers might be indicated here, such as is suggested for manufacturers in the developed areas by Smith (1968). Obviously, there is little likelihood of insecticides being "tailor made" when there is small prospect of substantial repeat orders. It is then necessary to search among existing insecticides for ones with a satisfactory degree of selectivity in the particular circumstance. The aim of most developing countries is to establish industries and this includes manufacturing plants for producing insecticides. The natural tendency will be to produce the chemicals with the widest market, i.e., those with the broadest spectrum of activity. Governmental action here too could co-ordinate the entomologists' views to provide the necessary guidance as to which are the most suitable chemicals to manufacture.

A problem in establishing integrated control programs is education of the growers. It is fortunate that the integrated control concept was introduced in Malaysia fairly early. In rubber, this was done before any major outbreaks had occurred but in the other crops there was a tendency to apply chemicals against minor pest attack. It is interesting that those most difficult to convince of the role played by insecticides were those who had been spraying for the longest time, for, having seen the major induced outbreaks, they were persuaded that once they got any caterpillars they had to spray in order to prevent them from being regularly devastating. Some growers may be less willing to pay directly for background knowledge and expert advice than for chemicals, but the atmosphere is good in Malaysia and various institutes exist to carry out agricultural research and give advice. There is manifestly considerable advantage in having entomologists responsible for the whole program of control in a particular crop rather than for a particular type of problem in a range of crops. Such men are far less likely to develop partisan feelings for biological, chemical or any other method of control, for the repercussions from dealing with any one problem would also be theirs to tackle. Happily, the trend in Malaysia is to assign entomologists in this way and they are ideally placed to appreciate the ecological factors at work in their crops and determine the cheapest and, over-all, most effective long-term procedures. This is by definition, integrated control.

Evidently, integrated control is not only possible, but a virtual necessity in tropical perennial crops, as has been illustrated here by numerous examples. A good degree of success has been obtained but economic outbreaks still occur from time to time. It is by no means certain that more refractive problems will not arise, and there are still wide gaps in our detailed knowledge of the complexities of the agro-ecosystems of the various crops. Further scientific effort will provide an increasingly broader basis for our ability to regulate pest numbers with maximum effectiveness at minimum cost and with minimum contribution to the problem of environmental pollution by persistent broad-spectrum pesticides.

LITERATURE CITED

Anonymous. 1968. Cockchafers. Plrs. Bull. Rubb. Res. Inst. Malaya, 97: 102-109.
Bartlett, B. R. 1964. Integration of chemical and biological control. Chap. 17. *In*

Biological Control of Insect Pests and Weeds, P. DeBach (ed.). Reinhold Publ. Co., N. Y. 844 pp.

Conway, G. R. 1964. A note on mirid bugs (Hemiptera: Miridae) and some other insect pests of cocoa in Sabah, Malaysia. Proc. Conf. Mirids and Other Pests in Cacao, 1964 (Ibadan, Nigeria). West Africa Cocoa Res. Inst. (Nigeria), Ibadan.

Conway, G. R. 1968. Crop pest control and resource conservation in tropical South East Asia. Pp. 159-163. *In* Conservation in Tropical South East Asia, L. M. and M. H. Talbot (eds.). Intern. Union Conserv. Nature and Natural Resources, Morges, Switzerland.

Conway, G. R. 1971a. Ecological aspects of pest control in Malaysia. *In* The Careless Technology—Ecology and International Development, M. R. Farvar and J. P. Milton (eds.). Natural History Press, New York.

Conway, G. R. 1971b. Pests of cocoa (*Theobroma cacao* L.) in Sabah and their control (with a list of the cocoa fauna). Kementerian Pertanian dan Perikanan Sabah, Malaysia (Dept. of Agr., Sabah), pp. 1-125.

Dale, W. L. 1959. The rainfall of Malaya, Part I. J. Trop. Georgr. 13: 23-27.

Dale, W. L. 1963. Surface temperatures in Malaya. J. Trop. Geogr. 17: 57-71.

DeBach, Paul. 1958. The role of weather and entomophagous species in the natural control of insect populations. J. Econ. Entomol. 51: 474-484.

DeBach, Paul. 1964. The scope of biological control. Chap. 1. *In* Biological Control of Insect Pests and Weeds, P. DeBach (ed.). Reinhold Publ. Co., N. Y. 844 pp.

Doutt, R. L., and Paul DeBach. 1964. Some biological control concepts and questions. Chap. 5. *In* Biological Control of Insect Pests and Weeds, P. DeBach (ed.). Reinhold Publ. Co., N. Y. 844 pp.

Ebeling, W. 1947. DDT preparations to control certain scale insects on citrus. J. Econ. Entomol. 40: 619-632.

Flaherty, D., and C. B. Huffaker. 1970. Biological control of Pacific mites and Willamette mites in the San Joaquin Valley vineyards. Hilgardia 40(10): 267-308.

Geier, P. W. 1966. Management of insect pests. Ann. Rev. Entomol. 11: 471-490.

Huffaker, C. B., and P. S. Messenger. 1964. The concept and significance of natural control. Chap. 4. *In* Biological Control of Insect Pests and Weeds, P. DeBach (ed.). Reinhold Publ. Co., N. Y. 844 pp.

Huffaker, C. B., M. van de Vrie, and J. A. McMurtry. 1969. The ecology of tetranychid mites and their natural control. Ann. Rev. Entomol. 14: 125-174.

Huttenbach, H. 1969. Selective insecticides in integrated pest control as illustrated by Thiodan® (endosulphan). Zeitschrift für Pflanzenkrankheiten und Pflanzenschutz 76: 667-677.

Lim, G. S. 1970. Some aspects of the conservation of natural enemies of rice stem borers and the feasibility of harmonizing chemical and biological control of these pests in Malaysia. MUSHI 43(11): 125-134.

Owen, R. P. 1959. Proposals for vegetative barrier experiments. South Pacific Commission, Trust Territory of the Pacific Islands, Koror, Caroline Islands. Pp. 1-3. (Multigraph)

Rao, B. S. 1964. The use of light traps to control the cockchafer *Lachnosterna bidentata* Burm. in Malayan rubber plantations. J. Rubber Res. Inst. Malaya 18: 243-252.

Rao, B. S. (with Hoh Choo Huan). 1965. Pests of *Hevea* plantations in Malaya. Kuala Lumpur—Rubber Res. Inst. Pp. 1-98.

Smith, R. F. 1968. Recent developments in integrated control. 4th Br. Insecticide and Fungicide Conf., Brighton (1967), 2: 464-471.

Stern, V. M., R. F. Smith, R. van den Bosch, and K. S. Hagen. 1959. The integrated control concept. Hilgardia 29: 81-101.

Stern, V. M., and R. van den Bosch. 1959. Field experiments on the effects of insecticides. Hilgardia 29: 103-130.

van den Bosch, R., and V. M. Stern. 1962. The integration of chemical and biological control of arthropod pests. Ann. Rev. Entomol. 7: 367-386.

Varley, G. C. 1959. The biological control of agricultural pests. J. Roy. Soc. Arts 107: 475-490.

Wood, B. J. 1966. Annual report for 1963-64. Entomology Section, Layang Layang–Chemara Res. Sta. Pp. 1-143.

Wood, B. J. 1967. Aerial spray campaigns against bagworm caterpillars (Lepidoptera: Psychidae) on oil palms in Malaysia. Proc. Oil Palm Tech. Seminar, Sabah Plrs. Assoc. Pp. 1-24. (Mimeo.)

Wood, B. J. 1968. Pests of oil palms in Malaysia and their control. Kuala Lumpur–Incorp. Soc. of Planters. Pp. 1-204.

Wood, B. J. 1969a. The extent of vertebrate attacks on the oil palm in Malaysia. Pp. 162-184. In Progress in Oil Palms, 2nd Malaysian Oil Palm Conf., 1968. Kuala Lumpur–Incorp. Soc. of Planters.

Wood, B. J. 1969b. Studies on the effect of ground vegetation on infestations of *Oryctes rhinoceros* (L.) (Col., Dynastidae) in young oil palm replantings in Malaysia. Bull. Entomol. Res. 59: 85-96.

Wood, B. J. 1969c. Population studies on the Malaysian wood rat (*Rattus tiomanicus*) in oil palms, demonstrating an effective new control method and assessing some older ones. The Planter 45: 510-526.

Wood, B. J. and D. P. Nesbit. 1969. Caterpillar outbreaks on oil palms in Eastern Sabah. The Planter 45: 285-299.

Wood, B. J., and K. Y. Ng. In prep. Biology and control of the bunch moth, *Tirathaba mundella* Wlk. (Lepidoptera: Psychidae) on oil palms. Malaysian Agric. J.

Wycherley, P. R., and M. M. Chandapillai. 1969. Effects of cover plants. J. Rubb. Res. Inst. Malaya 21: 140-157.

Chapter 20

DEVELOPMENT OF INTEGRATED CONTROL PROGRAMS FOR CROP PESTS IN ISRAEL

Isaac Harpaz and David Rosen

*Department of Entomology, Hebrew University
Rehovot, Israel*

INTRODUCTION

Up to some 15 years ago the usual approach to pest control in Israel, as elsewhere, consisted mainly of providing *ad hoc* solutions to isolated problems and generally lacked a sound ecological basis. At best, entomologists endeavored to develop effective, economically feasible control programs devised to destroy, essentially by chemicals, each isolated target pest species on the individual crop involved. The disadvantages and great dangers inherent in such an approach, characterized by a complete disregard of the entire agro-ecosystem as a coherent, interdependent entity, have been exhaustively discussed in recent literature (Smith and van den Bosch, 1967; Food and Agriculture Organization, 1966, 1968), as well as in other chapters of this book (Chapters 1, and 14-19), and need no further elaboration here.

However, in the early 1950's entomologists in Israel became aware of being slowly but steadily drawn into facing the now well-known vicious circle of old pests becoming resistant to the so-called most promising pesticides, while previously harmless species rapidly attained injurious levels following indiscriminate destruction of their restraining natural enemies by the misuse of non-selective pesticides. As a means of finding a way out of this imminent impasse, the newer concept of integrated control was embraced as a possible replacement of the long-standing attitude that was still deeply entrenched in the minds of all those concerned with pest control (Cohen, 1969). At the same time, it was also realized that a comprehensive knowledge regarding the various components of the agro-ecosystem, including all actual and potential pests as well as their natural enemies, is one of the primary prerequisites for establishing effective integrated control programs. Evaluation of the potential effects of available pesticides and the various modes of their application on the natural enemies in an agro-ecosystem, and the modification of chemical control practices in order to minimize their adverse effects on these natural enemies, were recognized as another prerequisite (Rosen, 1967b). Active employment of biological control, especially an intensive program of importation and release of exotic parasites and predators, was considered an essential component of the integrated control approach.

Efforts to develop integrated control programs in Israel have so far concentrated mainly on citrus, which is the most important crop and main export commodity of the country. Modified measures for other crops were sometimes initiated due to concern over the deleterious effects of pest control operations, as normally carried out on such crops, relative to adjacent citrus groves.

CITRUS PESTS

Fortunately, the citrus industry of Israel is organized in a unique manner which renders it most suitable for development and successful implementation of novel pest control strategies based on integrated control principles (Cohen, 1969). The Citrus Board of Israel is by law the sole organization through which all the citrus crop of the country is marketed, whether for export or local consumption. The Board is empowered to collect certain levies imposed on the proceeds of citrus fruit sales, and through its Agrotechnical Division the Board finances research and development projects covering all aspects of citrus production and marketing, among which pest control is the most prominent. With the funds appropriated to it, the Agrotechnical Division of the Board maintains a modern, well-equipped insectary for the importation and mass-rearing of natural enemies of pests. It also runs its own extension service, staffed with about half a dozen university-trained entomologists, in whom the growers have unqualified confidence. In addition, the Citrus Board is authorized to carry out pest control operations on citrus on a countrywide scale. It also acts as a major purchaser and distributor of pesticides on behalf of the country's growers for reasons both of economy and quality control (Cohen and Kamburov, 1968). The impressive achievements which this kind of organization has secured since its inception in 1954 are well reflected in the striking changes that have taken place in the over-all picture of citrus pests in Israel during these 15 years.

In 1954, for instance, the major pests of citrus were, in order of importance, as follows:

1. The Florida red scale, *Chrysomphalus aonidum* (L.), against which several thousand tons of mineral spray oils were applied annually.

2. The Mediterranean fruit fly, *Ceratitis capitata* (Wied.), which inflicted extensive losses on the citrus industry by causing fruit drop and fruit decay. Furthermore, in order to meet quarantine requirements of citrus importing countries, fruit shipments had to be treated with ethylene dibromide or kept in prolonged cold storage at a nearly-prohibitive cost. Moreover, the picking season frequently had to be shortened in order to evade the massive wave of fly infestation coming usually in late spring. The then-existing recommendations for control of the fly consisted of full-coverage sprays of persistent contact insecticides such as DDT, methoxychlor or dieldrin.

3. The third pest of economic significance was the citrus rust mite, *Phyllocoptruta oleivora* (Ashmead), which was recorded in Israel for the first time in 1944 and within ten years had spread over the entire citrus area of the country's coastal plain (Avidov and Harpaz, 1969). Treatments against this mite included sulfur dusting, or spraying of non-selective acaricides such as dicofol or wettable sulfur.

Other scale insects and spider mites were of secondary importance.

The successful introduction of *Aphytis holoxanthus* DeBach from Hong Kong by the Citrus Board in 1956 for the control of Florida red scale has brought about a complete change in the above-described picture. This project, discussed in some detail in Chapter 7, has been one of the most outstanding successes of biological control in recent years. It is illustrated in Fig. 1 by the steep decline in the use of spray oils on citrus in Israel from 1955 to 1958. Subsequent fluctuations were due to treatments for California red scale and soft scales. The economic value of this achievement to Israel's citrus and banana growers, as well as to other citrus growers in South Africa (Bedford, 1968), Florida (Muma, 1969), and Mexico (Maltby *et al.*, 1968), to whom this parasite was subsequently shipped, can hardly be overestimated. However, the parasite could not have been able to practically free the entire citrus area of Israel from losses due to Florida red scale within less than three years had it been necessary for it to operate under a regime of repeated cover sprays of chlorinated hydrocarbons used against the fruit fly and non-selective acaricidal treatments against the rust mite. Hence, an entirely new program for Mediterranean fruit fly control had to be developed, aimed at minimizing the degree of interference with

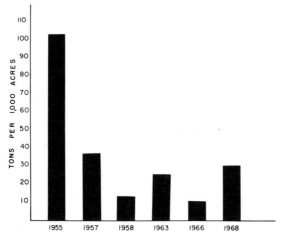

Fig. 1. Amounts of oils used on citrus in Israel in certain of the years, 1955-1968. (Data on use of oils purchased by the Citrus Board of Israel kindly supplied by Mr. Israel Cohen.)

the beneficial activity of *Aphytis holoxanthus* and other significant natural enemies among the citrus fauna. This entailed use of strip sprays of a poison bait containing protein hydrolysates as a powerful attractant for the fly females and malathion as a poison. The bait is sprayed from the air at an ultra low rate of application—i.e., down to less than one pint per acre.[1] These applications, which are made by the Citrus Board on a countrywide scale, are timed according to well-defined threshold levels based on numbers of flies caught in a dense network of plastic trap-jars covering the entire citrus-growing area of the country. The lure used in these traps is "Trimedlure," a synthetic sex attractant for *Ceratitis* male flies (Nadel and Peleg, 1965; Cohen and Cohen, 1967). Incidentally, the ultimate results of controlling the fly by this method were far superior to those obtained by full-coverage sprays of chlorinated hydrocarbons. In many instances it has secured 100 per cent control of the fly on citrus, resulting in no further need to treat exported fruits with ethylene dibromide, whether by fumigation or by dipping in aqueous emulsions of this fumigant.

The aerial application of poisoned baits was found to have no greater detrimental effect on the natural enemies of citrus scale insects than the alternative, economically less feasible, ground application (Avidov *et al.,* 1963). In preliminary tests, which still require further confirmation, Samish

[1]The exact formulation of this spray is 1000 ml per hectare, comprising 760 ml of protein hydrolysate and 240 ml of liquid malathion (Cohen, 1970).

(1973) showed that the protein hydrolysate baits did not seem to attract *Aphytis* or other economically important scale parasites, which is indicative of the great advantage of this control method for the fly as regards preservation of biological balance in the ecosystem.

The same program is currently being developed in Israel for control of Mediterranean fruit fly on deciduous and other summer fruits, where the hitherto-employed full-coverage sprays have caused severe upsurges in scale insect infestations, such as the olive scale [*Parlatoria oleae* (Colvée)], the brown soft scale (*Coccus hesperidum* L.), and the black scale [*Saissetia oleae* (Olivier)]. Since most of these crops are grown on hilly terrain, it appears that helicopters will be preferable to fixed-wing aircraft for this particular purpose. It is hoped that effective control of fruit fly infestations on summer fruits will further reduce the fly populations on citrus.

As an auxiliary to the bait spray program for Mediterranean fruit fly control, particularly for backyard gardens and non-commercial host trees, some 13 different parasites of the fly were imported from Hawaii and released in Israel. It is yet too early to evaluate the success of any of these importations, but one species, namely the chalcidid pupal parasite *Dirhinus giffardii* Silvestri, seems to have become established (Cohen and Kamburov, 1968).

The natural enemies of all significant citrus pests in Israel were studied during the last decade in a series of extensive surveys, covering the main citrus-growing regions. The identity, phenology and relative abundance of numerous parasites and predators were determined, and the biologies of the more important species were studied in the laboratory. These extensive studies were recently reviewed by Rosen (1967*b*).

Three homopterous pests were found to be kept under effective biological control on citrus by introduced natural enemies. Except for the Florida red scale (see above), these include the cottony cushion scale (*Icerya purchasi* Mask.) and Green's mealybug (*Pseudococcus citriculus* Green). The cottony cushion scale was brought under excellent control by the vedalia ladybeetle, *Rodolia cardinalis* (Muls.), introduced from Italy as early as 1912. Likewise, Green's mealybug, which became a major pest of citrus in Israel in the late 1930's, has become extremely rare following the successful introduction of a parasite, *Clausenia purpurea* Ishii, from Japan in 1940 (Bodenheimer, 1951; Rosen, 1967*b*; Rivnay, 1968).

Locally-established parasites were also found to keep three additional homopterous pests at subeconomic levels on citrus: The purple scale [*Lepidosaphes beckii* (Newm.)], formerly a serious pest, is now being held under satisfactory control by *Aphytis lepidosaphes* Compere, which has recently appeared in Israel as a result of unintentional importation or accidental ecesis (Rosen, 1965, 1967b; Rivnay, 1968; see also Chapter 7). The brown soft scale, if not upset by the misuse of pesticides, is usually kept under control by the combined action of several indigenous parasites; *Metaphycus flavus* (Howard) and *Microterys flavus* (Howard) are the dominant natural enemies at low host densities, whereas *Coccophagus* spp. are dominant in dense, ant-attended colonies of the scale (Rosen, 1967c, 1967d). The citrus black aleyrodid [*Acaudaleyrodes citri* (Priesner and Hosni)] is kept well below the threshold of economic injury on citrus by the indigenous parasites, *Eretmocerus diversiciliatus* Silv. and *Prospaltella lutea* Masi (Rosen, 1966). Natural enemies were also found to be abundant among populations of other citrus pests, and the importation of many additional species of parasites and predators has been attempted in recent years (Rosen, 1967b; Rivnay, 1968).

The citrus rust mite was found to be attacked by several local and two introduced species of phytoseiid mites. *Amblyseius swirskii* Athias-Henriot is the dominant phytoseiid predator on citrus along the coastal plain of Israel, whereas *Typhlodromus athiasae* Porath and Swirski is common in the interior valleys and Galilee; *Iphiseius degenerans* (Berlese) is also common on citrus (Porath and Swirski, 1965; Rosen, 1967b). However, the efficiency of these predators was greatly diminished by the use of sulfur or dicofol against the rust mite. At the same time, sulfur preparations are also most harmful to the *Aphytis* species and other parasites engaged in the biological control of scale insects on citrus. More selective acaricides, such as Zineb®, therefore had to be found for use in those instances where natural control of the rust mite was still inadequate after the spray programs for control of scales and the fruit fly had been stopped or greatly altered. Fortunately, however, the number of such cases is on a constant decline, and the mite has recently ceased to be a pest of major economic importance.

All commercial pesticide formulations in current use, or suggested for use against citrus pests in Israel, have been tested in the laboratory for possible effects on the survival and fecundity of parasitic Hymenoptera. At the same time, the effects of commercial acaricide formulations (for possible use against rust mites) on phytoseiid predators have been evaluated in field and laboratory tests. In fact, a permanent pattern has since been instituted in Israel whereby no pesticide, whether insecticide, acaricide, fungicide or herbicide, can be registered, or at least recommended for use on citrus, before being screened for its effects on *Aphytis* parasites, predatory mites and coccinellids (Rosen, 1967a,b; Swirski *et al.*, 1968, 1969).

Now that such a less toxic environment has been established and maintained for some time in our citrus groves, the chances seem rather promising for achieving biological control of the remaining citrus pests, notably the California red scale [*Aonidiella aurantii* (Mask.)], the chaff scale

(*Parlatoria pergandii* Comstock), the Florida wax scale (*Ceroplastes floridensis* Comstock), and the black scale [*Saissetia oleae* (Olivier)]. All these still require oil sprays for their control upon occasional build-ups of infestations.

For quite a long time, namely since the mid-1920's when the late F. S. Bodenheimer began his pioneering work on the ecology of citrus pests in Palestine, the prevalent notion had been that natural enemies were usually of no decisive importance in the general epidemiology of most citrus pests. Climatic conditions were considered to be the primary governing factors in this respect, and the prospects for biological control in subtropical, semi-arid Israel were believed to be rather dim (Bodenheimer, 1951). This view of Bodenheimer was held with particular emphasis to the California red scale.

The outstanding success in the biological control of Florida red scale served as a firm rebuttal of these assumptions. More recently, the application of the insecticidal check method (DeBach, 1946, 1955) (see also Chapters 5 and 7) using DDT to eliminate natural enemy activity on citrus in Israel has put an end to this extended controversy. This work clearly demonstrated that natural enemies, mainly parasites of the genus *Aphytis,* were indeed capable of effectively retarding the build-up of California red scale (Ben-Dov and Rosen, 1969) and Florida wax scale populations on citrus. This in turn led to renewed, concerted efforts in surveys, identification and study of the life histories of local enemies of this scale, as well as to introduction of a number of exotic parasites. One of the latter, *Aphytis melinus* DeBach, is currently displacing its congeneric competitors, *A. chrysomphali* (Mercet) and *A. coheni* DeBach, at such a rate, and with a control effect, such that we have reason to hope it will ultimately reduce the California red scale population to a level below economic significance.

In many areas in Israel, citrus groves are situated adjacent to cotton fields. Our prospective integrated control program for cotton pests is still in its infancy, so that aerial sprays of organophosphorus and chlorinated-hydrocarbon pesticides are still repeatedly applied to cotton fields. Aerial drift of these pesticides into neighboring citrus groves has often upset the favorable biological equilibrium maintained in these groves, resulting in remarkable outbreaks of California red scale in the strips adjacent to the treated cotton fields. Appropriate legislation has therefore been enacted, forbidding aerial spraying of cotton with non-selective pesticides within a distance of 200 meters (219 yards) from a citrus grove border.

In conclusion, the introduction of exotic natural enemies and the thorough revision of chemical control practices have resulted in the drastic decline of the three most injurious citrus pests in Israel. In general, an environment much more favorable to natural enemies has been created and the prospects for further successes and improvements in biological control in this overall integrated control program appear brighter than ever.

COTTON PESTS

The main pests of cotton in Israel are the following three noctuids: the Egyptian cottonworm [*Spodoptera littoralis* (Boisduval)],[2] the spiny bollworm (*Earias insulana* Boisduval) and the African bollworm [*Heliothis armigera* (Hübner)].

The integrated control program for cotton is still in its very initial stage. However, as a first step, a country-wide system of supervised control has been instituted for cotton. The timing of pesticide treatments is done according to the actual level of pest populations, and is linked to properly defined economic threshold levels for each pest concerned. This has resulted in a considerable reduction in the total number of applications per season. A study of local natural enemies of *Spodoptera littoralis*, the major pest of cotton, is well under way. They include several species of parasites (Gerling, 1969), a nuclear polyhedrosis virus (Harpaz and Ben-Shaked, 1964; Ben-Shaked and Harpaz, 1966), and a number of entomogenous fungi. In addition, a recent exploratory trip to East Africa and Madagascar yielded a number of natural enemies which at present are being investigated for the purpose of biological control.

OLIVE PESTS

In olive groves, full-coverage sprays against the olive fly [*Dacus oleae* (Gmelin)] that were used in the past resulted in severe outbreaks of soft and armored scales, often causing damage, the value of which quite frequently exceeded the benefit derived from control of the fly. Additionally, the leopard moth, *Zeuzera pyrina* (L.), which is also a severe pest of pome fruit trees, appeared to have increased the extent of its wood boring in irrigated groves following chlorinated hydrocarbon treatments against the fly. This has been suggested by Lisser (1967) to be the result of mass destruction of certain unidentified predatory ant species that normally feed on the moth eggs laid into the bark crevices. Another possible cause for the rise in leopard moth infestation is the pesticidal suppression of parasitization of its caterpillars by two local parasites, the eulophid *Elachertus nigritulus* (Zett.) and the braconid *Apanteles laevigatus* (Ratz.) (Lisser, 1967). In addition to these, Lisser mentioned the Syrian woodpecker [*Dendrocopos syriacus* (Hemprich and

---

[2]In the past, this species was erroneously referred to as *Prodenia litura* (Fabricius). The latter, however, is a distinctly different species, limited in its distribution to the Far East, Australasia, and certain Pacific Islands (see Viette, 1963).

Ehrenberg)] as an efficient predator of leopard moth caterpillars. They extensively peck them out of their galleries in the trees. Although the woodpecker is much more active in apple and pear trees than in olives, it is still considered as an appreciable mortality factor of leopard moth larvae on olive. Hence, excessive poisoning of these birds through the misuse of pesticides could have also contributed to the increase in leopard moth damage in these groves. At any rate, the situation has been remarkably improved following the replacement of full-coverage sprays against the fly by bait spot sprays, similar to the method used against the Mediterranean fruit fly described earlier (Nadel and Golan, 1966).

DECIDUOUS FRUIT TREE PESTS

In deciduous orchards, the development of an effective integrated control program has been hitherto badly impeded by the apparent need to frequently treat with chemicals against the codling moth, *Laspeyresia pomonella* (L.), which is the key pest in the spray schedule for pome fruit trees in Israel. Second to the codling moth in economic importance on these fruit trees is the olive scale, *Parlatoria oleae* (Colvée). At least six species of predators and five different parasites of this scale insect have so far been recognized in Israel (Applebaum and Rosen, 1964; Sadeh and Gerson, 1968), and the importation of additional parasites is being contemplated, following the impressive success in the biological control of this pest in California (Huffaker and Kennett, 1966; Kennett, 1967) (see also Chapter 7). However, these natural enemies cannot be expected to exert sufficient pressure on the olive scale population, or even survive, in an environment "saturated" with organophosphorus and chlorinated hydrocarbon pesticides which are often applied at a frequency of 12 applications or more per season (Plaut, 1967b). A heavy spray schedule like this, including chemicals such as carbaryl and chlorinated hydrocarbons, has inevitably aggravated the problems of red spider mite (Plaut and Feldman, 1967) and leopard moth infestations on apple. The first step in rationalizing this kind of situation has already been taken. Spray schedules are being modified so that pesticide applications are timed according to trends in the numbers of fully developed codling moth caterpillars trapped in burlap bands wrapped around the trunks of sampling trees. This has helped to considerably reduce the number of pesticide applications (Plaut, 1967a). The next move would be to replace some of the currently used pesticides by more selective ones, and finally to attempt to control the codling moth by methods other than chemicals so that natural enemies of this and other pests can be utilized.

LITERATURE CITED

Applebaum, S. W., and D. Rosen. 1964. Ecological studies on the olive scale, *Parlatoria oleae,* in Israel. J. Econ. Entomol. 57: 847-850.

Avidov, Z., and I. Harpaz. 1969. Plant pests of Israel. Israel Universities Press, Jersalem. 549 pp.

Avidov, Z., D. Rosen, and U. Gerson. 1963. A comparative study on the effects of aerial versus ground spraying of poisoned baits against the Mediterranean fruit fly on the natural enemies of scale insects in citrus groves. Entomophaga 8: 205-212.

Bedford, E. C. G. 1968. An integrated spray programme. S. Afr. Citrus J. No. 417, pp. 9-28.

Ben-Dov, Y., and D. Rosen. 1969. Efficacy of natural enemies of California red scale on citrus in Israel. J. Econ. Entomol. 62: 1057-1060.

Ben-Shaked, Y., and I. Harpaz. 1966. Protection of a susceptible insect host against a nuclear-polyhedrosis virus by ether extracts from insect larvae. J. Invertebr. Pathol. 8: 283-285.

Bodenheimer, F. S. 1951. Citrus entomology in the Middle East. Dr. W. Junk, Publ., The Hague, Netherlands. 663 pp.

Cohen, I. 1969. Biological control of citrus pests in Israel. Proc. 1st Intern. Citrus Symp., Riverside, Calif. (1968) 2: 769-772.

Cohen, I. 1970. Report on the activities of the Agrotechnical Division of the Citrus Board of Israel during 1968/69. Alon Hanote'a 24: 213-219 (in Hebrew).

Cohen, I., and J. Cohen. 1967. Centrally organized control of the Mediterranean fruit fly in citrus groves in Israel. Citrus Board of Israel, Agrotech. Div. 32 pp.

Cohen, I., and S. Kamburov. 1968. Biological Control Institute. Citrus Board of Israel, Agrotech. Div. 43 pp.

DeBach, P. 1946. An insecticidal check method for measuring the efficacy of entomophagous insects. J. Econ. Entomol. 39: 695-697.

DeBach, P. 1955. Validity of the insecticidal check method as a measure of the effectiveness of the natural enemies of armored scale insects. J. Econ. Entomol. 48: 584-588.

Food and Agriculture Organization (FAO). 1966. Proceedings of the FAO Symposium on Integrated Pest Control. 3 vols. (91+186+129 pp.), FAO, Rome.

Food and Agriculture Organization (FAO). 1968. Report of the Second Session of the FAO Panel of Experts in Integrated Pest Control. Rome, 19-24, Sept., 1968. 48 pp.

Gerling, D. 1969. Parasites of *Spodoptera littoralis* Boisd. (Lepidoptera: Noctuidae) eggs and larvae in Israel. Israel J. Entomol. 4: 73-81.

Harpaz, I., and Y. Ben-Shaked. 1964. Generation-to-generation transmission of nuclear polyhedrosis virus of *Prodenia litura* (Fab.). J. Insect Pathol. 6: 127-130.

Huffaker, C. B., and C. E. Kennett. 1966. Biological control of *Parlatoria oleae* (Colvée) through the compensatory action of two introduced parasites. Hilgardia 37: 283-335.

Kennett, C. E. 1967. Biological control of the olive scale *Parlatoria oleae* (Colvée) in a deciduous fruit orchard. Entomophaga 12: 461-474.

Lisser, A. 1967. The *Capnodis* and the leopard moth. Sifriath-Hassadeh, Tel Aviv. 156 pp. (in Hebrew).

Maltby, H. L., E. Jimenez-Jimenez, and P. DeBach. 1968. Biological control of armored scale insects in Mexico. J. Econ. Entomol. 61: 1086-1088.

Muma, M. H. 1969. Biological control of various insects and mites on Florida citrus. Proc. 1st Intern. Citrus Symp., Riverside, Calif. (1968) 2: 769-772.

Nadel, D. J., and Y. Golan. 1966. Control of the olive fly by the protein hydrolysate baiting method through aerial and ground application. FAO Plant Protect. Bull. 14: 47-53.

Nadel, D. J., and B. A. Peleg. 1965. The attraction of fed and starved males and females of the Mediterranean fruit fly, *Ceratitis capitata* Wied., to "Trimedlure." Israel J. Agr. Res. 15: 83-86.

Plaut, H. N. 1967*a*. Experiences with burlap band traps on trunks of apple trees. Entomophaga, Mémoire Hors Série No. 3, pp. 51-54.

Plaut, H. N. 1967*b*. The latest position on integrated control in apple groves and its problems in Israel. Entomophaga, Mémoire Hors Série No. 3, pp. 134-135.

Plaut, H. N., and M. Feldman. 1967. Effects of sevin and DDT on the density of field populations of *Tetranychus cinnabarinus* Boisd. Entomophaga, Mémoire Hors Série No. 3. pp. 89-93.

Porath, A., and E. Swirski. 1965. A survey of phytoseiid mites (Acarina: Phytoseiidae) on citrus, with a description of one new species. Israel J. Agr. Res. 15: 87-100.

Rivnay, E. 1968. Biological control of pests in Israel (a review 1905-1965). Israel J. Entomol. 3: 1-156.

Rosen, D. 1965. The hymenopterous parasites of citrus armored scales in Israel (Hymenoptera: Chalcidoidea). Ann. Entomol. Soc. Amer. 58: 388-396.

Rosen, D. 1966. Notes on the parasites of *Acaudaleyrodes citri* (Priesner and Hosni) (Hem.: Aleyrodidae) in Israel. Entomol. Ber. 26: 55-59.

Rosen, D. 1967*a*. Effects of commercial pesticides on the fecundity and survival of *Aphytis holoxanthus* (Hymenoptera: Aphelinidae). Israel J. Agr. Res. 17: 47-52.

Rosen, D. 1967*b*. Biological and integrated control of citrus pests in Israel. J. Econ. Entomol. 60: 1422-1427.

Rosen, D. 1967*c*. The hymenopterous parasites of soft scales on citrus in Israel. Beitr. Entomol. 17: 255-283.

Rosen, D. 1967*d*. On the relationships between ants and the parasites of coccids and aphids on citrus. Beitr. Entomol. 17: 285-290.

Sadeh, D., and U. Gerson. 1968. On the control of olive scale in apple orchards in Upper Galilee. Hassadeh, Tel Aviv 48: 819-824, 955-958 (in Hebrew).

Samish, M. 1973. The attraction of protein hydrolysate for hymenopterous parasites. Entomophaga 18: 169-174.

Smith, R. F., and R. van den Bosch. 1967. Integrated control. Pp. 295-340. *In* Pest Control: Biological, Physical and Selected Chemical Methods, W. W. Kilgore and R. L. Doutt (eds.). Academic Press, Inc., N. Y. 477 pp.

Swirski, E., S. Amitai, S. Greenberg, and N. Dorzia. 1968. Field trials on the toxicity of some carbamates and endosulfan to predaceous mites (Acarina: Phytoseiidae). Israel J. Agr. Res. 18: 41-44.

Swirski, E., N. Dorzia, S. Amitai, and S. Greenberg. 1969. Trials on the control of the citrus rust mite (*Phyllocoptruta oleivora* Ashm.) with four pesticides, and on their toxicity to predaceous mites (Acarina: Phytoseiidae). Israel J. Entomol. 4: 145-155.

Viette, P. 1963. Le complexe de "*Prodenia litura* (Fabricius)" dans la région malgache (Lep. Noctuidae). Bull. Mens. Soc. Linn. Lyon 32: 145-148.

AUTHOR INDEX

SUBJECT INDEX

477